COMPTES RENDUS

OF

OBSERVATION AND REASONING

COMPTES RENDUS

OF

OBSERVATION AND REASONING

BY

J. Y. BUCHANAN, M.A., F.R.S.

Commandeur de l'Ordre de Saint Charles de Monaco

Vice-Président du Comité de Perfectionnement de l'Institut Océanographique
(Fondation Albert Ier Prince de Monaco)

"Prove all things.   Hold fast that which is good."
I Thess. v. 21

Cambridge:

at the University Press

1917

CAMBRIDGE
UNIVERSITY PRESS

University Printing House, Cambridge CB2 8BS, United Kingdom

Published in the United States of America by Cambridge University Press, New York

Cambridge University Press is part of the University of Cambridge.

It furthers the University's mission by disseminating knowledge in the pursuit of education, learning and research at the highest international levels of excellence.

www.cambridge.org
Information on this title: www.cambridge.org/9781107653580

© Cambridge University Press 1917

First published 1917
First paperback edition 2014

A catalogue record for this publication is available from the British Library

ISBN 978-1-107-65358-0 Paperback

PREFACE

AS the title of this volume indicates, the book consists of "accounts rendered" of work done at different times, in different places and on different subjects.

In republishing papers, many of which are of almost ancient date, it was thought advisable to accompany them by explanatory notes and comments. As it was inconvenient to introduce these in the text, they have been embodied in the Contents, which form in fact a summary of the work.

The Contributions to Newspapers concerning matters of public interest at the time have been reprinted because they are of public interest still.

Of the scientific communications the most important are those concerning the Natural History of Steam and Ice, and these have been reprinted in their original form, although this has involved some reduplication of matter. This has been accepted on account of the importance of the experiments and of the consequences which follow from them and of the apparent unwillingness of the scientific public to make use of them.

In the *Chemical and Physical Notes*, which formed part of the *Antarctic Manual* of 1901, will be found a résumé of my experiments in the domain of Inorganic Natural History which, from my own experience, I judged would be found useful by the Chemists and Physicists of the Expeditions for which the *Manual* was prepared.

It was conveyed to me through an old friend and former colleague that this contribution to the *Antarctic Manual* had done much to retard the Standardisation of Research. I took it as a compliment. To standardise research is to limit its freedom and to impede discovery. Originality and independence are the characteristics of genuine research, and it is stultified by the acceptance of standards and by the recognition of authority.

J. Y. BUCHANAN.

26 *May* 1916.

CONTENTS

PAGE

No. 1. RECENT ANTARCTIC EXPLORATION. (From the *Quarterly
Review*, October 1906.) 1

The Belgian Expedition in the "Belgica" was the first to
winter within the Antarctic circle 1

The Newnes Expedition, under Captain Borchgrevink,
was the first to winter on land within the Antarctic circle.
He explored the Ice-barrier and discovered the creek in it
from which, later, Amundsen started on his journey, when he
reached the South Pole. When these two expeditions were
returning, the British Expedition in the "Discovery" under
Captain Scott and the German Expedition in the "Gauss"
set out 2

A year later the Scottish Expedition, under Dr Bruce,
sailed in the "Scotia" 3

A Swedish Expedition, under Nordenskjöld, and a French
Expedition, under Charcot, started for Graham's Land later 5

The "Challenger's" search for Wilkes' Termination Land
in 1874 6

The "Gauss" frozen in at the beginning of winter . . 7

Scott's Expedition follows the route to the South taken
by Sir James Ross in 1841–43. Remarkable difference between the appearance of Mount Melbourne and Mount Erebus
in 1902 and in 1841 9

Ross' description of the Ice-Barrier and the Parry Mountains 10

Trustworthiness of Ross' observations: in appraising such
observations, their date is of small importance; the determining factor is the competence and the experience of the
observer 11

Demonstration that the Parry Mountains exist where
Ross reported them to be 12

Confirmed by Armitage 13

Evidence to show that Ross' "Appearance of Land"
reported in February 1842 was the land seen by Scott in
January 1902, and named by him King Edward VII Land . 15

viii *Contents*

PAGE

The "Discovery" in winter quarters; mild weather . 16
Remarkable disappearance of ice in winter . . . 17
Scott's farthest South 82° 17′ S. reached 28 December 1902 18
Armitage's remarkable winter journey 20
Probability that the Barrier ice-sheet is a self-contained
névé or *firn* situated at, and, to a great extent, below the level
of the sea 21
Extensive melting of the ice of the Ferrar Glacier in
January 1902 22
Evidence of the conservative influence of an ice-sheet on
the land surface beneath it 23
The "Discovery" sails for England 24

No. 2. CHEMICAL AND PHYSICAL NOTES. (Contributed to the
Antarctic Manual, 1901.) 25

In preparing these notes it was assumed that the Chemist
and Physicist of the Expedition would be a man of considerable laboratory experience and would be fully instructed and
practised in all routine observations before the start of the
Expedition. These notes have reference rather to observations by the way, a class of work which is both useful and
fascinating. In putting them together, the chemical and
physical phenomena likely to be met with in Nature have
been regarded and treated as facts in Natural History of
which Chemistry and Physics are branches. I have en-
deavoured to imagine myself as filling the post and to recall
from my experience the kind of phenomena which I should
expect to encounter, and to frame working directions for
myself with a view to their treatment and elucidation. In
doing so I have avoided all hypothetical matter and confined
my directions to matters of observation and experiment.

Low oceanic temperatures; the extensive area of very
cold water which occupies the bottom of the sea all along
the East Coast of South America from the Straits of Magellan
to the Equator is an oceanic feature which calls for thorough
study 26
Importance of the determination of the density of this
very cold water 27
Freezing Temperature of average Sea-water . . . 28
Relation between freezing-point and percentage of Chlorine 29
Sea-Ice as met with in Polar Seas 30
Weyprecht's description of the freezing of the sea in which
the "Tegetthoff" was imprisoned 31

PAGE

Demonstration that the Ice produced by freezing Sea-water and similar solutions is pure ice 33

Analogy between Snow and Sea-ice. The freezing-point of a saline solution is analogous to a Dew-point: Cryohydrates 35

De Coppet's experiments on the freezing-points of saturated solutions 37

Experiments in the Engadine in Winter on Cryohydric points 40

Cryohydrates of Salts forming Isomorphous mixtures . 43

Land-ice and the mechanics of Glaciers 45

Importance of the study of the Grain of the Glacier. . 47

Sea-ice contains much Brine and flows easily ; Land-ice contains little Brine and flows with difficulty 49

The whiteness of the surface of a Glacier is a secondary feature, due to the disintegration of the Grains by the Solar Radiation 50

Study of boiling mixtures of steam and salts ; analogous to freezing mixtures of ice and salts 51

The boiling-point of a substance is the temperature at which it, as a vapour, condenses on itself as a liquid, and as a liquid, evaporates into itself as a vapour 55

Comparison of Barometer and Hypsometer . . . 56

To determine the boiling-point of a liquid, it should be boiled by its own steam 61

Data relating to boiling mixtures of NaCl, Table VII . 63

The thermal law, called Blagden's, is applicable to the boiling-points as well as to the freezing-points of solutions . 66

The temperature at which steam condenses on a perfectly dry, solid substance is unknown, but when it is moistened in the slightest degree even with dew, the temperature of condensation is that of water boiling under existing conditions . 67

Proof that steam produced by a boiling solution must quit it at the temperature of the boiling solution. Meteorological observations and instruments 68

The rate of cooling, or its reciprocal, the term of cooling, of a thermometer, is as important a constant as the position of the ice-point 69

The work on this subject by Newton, Lambert and Leslie is classical.

Term of cooling is an index of thermal *nimbleness* . . 70

The logarithmic law 75

Causes of deviation from it 76

Effect of silvering the bulb 79

Difference between a calm in-doors and a calm out-of-doors 80

PAGE

Thermometer as an Anemometer 81

Thermometer as Calorimeter 82

Estimation of the thermal mass of the bulb of a thermometer by mensuration 83

Relation between the volumes of mercury and glass in the
bulb 89

Particulars of air thermometer 92

Application of calorimetry to hydraulics.
Temperature of insolation 94

When the thermal mass of the bulb of a thermometer
and its term of cooling are known, it can be used as a solar
calorimeter 95

Example of the method of determining the calorific power
of the sun's rays, which strike the bulb of a thermometer and
are absorbed by it, and of the difference between different
thermometers in this respect 96

Convenience of compound names to express compound
units, such as gramme-degree for calorie 98

Value of solar constant obtained with the steam-calorimeter
in Egypt, and variation of the same during the total eclipse
of the sun, 17 May 1882 101

The necessity of the knowledge of calorimetric factors in
connection with the use of the Barometer.

Advantage of Fahrenheit's thermometric scale. The coefficient of thermal expansion of mercury is 1/10,000 per
degree Fahrenheit 102

The meteorological records of all inhabited places on the
earth can be kept in numbers having the same sign, with
very few exceptions 103

The temperature of the air immediately above the sea
surface in tropical regions is almost as constant as that of
the water 104

Effect of change of Force of Gravity on the Pressure of
the Atmosphere and on the Height of the Barometer . . 105

Table XVIII gives the true atmospheric pressure and the
boiling-point of water corresponding to a Barometric Height
of 735·5 millimetres at 0° C. and at the sea-level in different
latitudes 110

Change in the force of gravity affects the pressure of the
atmosphere but not the height of the Barometer . . . 112

Rule for detecting the deviation from the normal of the
local force of gravity by simultaneous observations of the
standard height of the Barometer and the boiling-point of
water 114

Contents

Suggestion of the use of the Hypsometer at sea for detecting changes of depth in the ocean 115
Advantages and disadvantages of the Aneroid barometer
Examples of Rapid Variations of Atmospheric Temperature, especially during Föhn 116
The curve of the temperature of the air drawn by a recording thermometer is serrated during the day and smooth during the night 117
Characteristics of Föhn 118
Föhn is independent of absolute height; it depends on difference of height 119
Well-marked Föhn at sea-level in Scotland . . . 120
Effect of Föhn in Polar regions 122
Puffs of air of temperature higher than that of the human body are characteristic of well-developed Föhn . . . 123
Necessity of acquaintance with the Term of Cooling of the thermometer when used in determining the temperature of the air during Föhn 124
Table XXII. Temperature observations at equal altitudes over the Morteratsch Glacier and on the mountain west of it 125
Remarkably high temperature of the air immediately contiguous with the ice of the glacier during Föhn . . . 126
Observations of the temperature of the air when ascending Piz Languard during Föhn 128

Note. The Föhn described in the paper is fine-weather Föhn, with cloudless sky and strong sun. The Föhn occurring with overcast sky and rain is not treated. It is this Föhn particularly that wastes glaciers.

No. 3. ON ICE AND BRINES. (From the *Proceedings of the Royal Society of Edinburgh*, 1887, Vol. XIV, p. 129.) . . . 130

This investigation was undertaken in order to determine experimentally whether the salt which occurs in the ice formed by freezing sea-water and other saline solutions is a part of the solid ice or belongs to the liquid brine which adheres to the ice. The fundamental principle on which the experiments were based is that, if a saline solution, such as sea-water, is partially frozen and the temperature of the mixture is observed, then, if the ice so formed is pure ice, it may be removed and be replaced by pure ice of independent origin, such as snow, having the same temperature, and if heat is then supplied, the snow will melt in the solution at the same temperature as that at which the ice which was produced in the solution was found to melt. This was found to be the case. As a corollary, it follows that the temperature at which pure ice melts depends on the medium in which it melts. The subject is dealt with under two heads ; namely : (*a*) the temperature at which sea-water and some other saline solutions freeze, and the chemical constitution of the solid and the liquid into which they are split by freezing ; and (*b*) the temperature at which pure ice melts in sea-water and in a number of saline solutions of different degrees of concentration 133

An important condition of success in these experiments is to freeze the solutions gently. The temperature of the freezing bath was usually about 2° C. below the freezing temperature of the solution.

Table I. Freezing sea-water. Analyses of Fractions. Constancy of Ratio SO_3 : Cl in the original sea-water, in the ice formed, and in the residual Brine 135

Table II. Calculation of Ice formed on the basis of the salinity of the original water and of the residual Brine . 136

Table III. Agreement of weight of ice formed calculated from the salinity of the solution and from the thermal exchange during freezing 137

Impossibility of solidifying sea-water by any cold occurring in Nature.

Melting of pure ice in sea-water and in other saline solutions 138

Demonstration that the mere density of the solution has no direct connection with the lowering of its freezing-point . 142

Experiments with concentrated solutions of hydrochloric acid and sulphuric acid 143

Demonstration that the experiments prove that the ice formed by freezing sea-water and similar saline solutions is pure ice and that the salt belongs to the brine . . . 144

Application of this discovery to the phenomena occurring in the freezing of fresh water and other solutions of very high dilution 147

No. 4. ON STEAM AND BRINES. (From the *Transactions of the Royal Society of Edinburgh*, 1899, Vol. XXXIX, Part III, No. 18.) 151

The object of the research was to study, generally, the natural history of boiling mixtures of steam and salts, analogous to the freezing mixtures of ice and salts ; and, particularly, to settle the question of the temperature at which steam leaves a boiling saline solution. The experiments were continued so as to give the concentration and the

PAGE

boiling temperature of solutions of different strengths below
that of saturation 151

If a solution can condense steam from the outside, it
necessarily must be able to condense its own steam, because
the two substances are identical. Therefore, steam introduced from the outside must continue to raise the temperature
of the solution even after it has passed above the temperature
of the boiling water from which it has been obtained and
until the temperature of the solution has been raised to that
at which the solution itself liberates steam of atmospheric
pressure, and, by consequence, allows steam of atmospheric
pressure introduced from the outside to pass freely through
it. It follows further that, to determine the boiling point of a
saline solution, it suffices to pass a current of steam of atmospheric pressure through it until it raises the temperature of
the solution to a maximum, and that is the boiling-point of
the resultant solution.

The following law holds :—The temperature at which steam
condenses depends on the nature of the medium in which it
condenses ; and the temperature at which it is generated
depends on the medium in which it is generated ; and the
temperature at which steam condenses in a given medium is
that at which it is generated in the same medium ; the pressure
in all cases remaining uniform.

The above method of determining the boiling-points of
saline solutions gives, easily, results of the greatest precision.
When the solution is boiled over a flame the results are
irregular and untrustworthy 152

Preliminary remarks on the experiments to be made, and
examples of the solubilities of some salts at the boiling
temperatures of their saturated solutions 154

Apparatus and method of experimenting , . . . 155

The result to be obtained from each experiment is the
difference between the temperature of pure saturated steam
and the maximum temperature produced by the condensation
of such steam in the brine, or the mixture of salt and brine.
Therefore, the temperature of pure saturated steam must be
observed at the same time as that of the boiling brine. It
follows that for the conduct of these experiments observation
of the barometer is unnecessary 158

Importance of the relative dimensions of the entry and
exit of the steam-tube : experiment which proves the efficiency
of the steam-tubes used 159

It is essential that the thermometer and the inside of the

PAGE

steam-tube be chemically clean : Soap is used for this purpose, and the best way of using it is described 160

The thermometers used and their verification . . . 161

The general order of the experiment 162

The use of steam for drying the inside of complicated glass apparatus.

As the result of an experiment, we obtain the temperature of ebullition of the saturated solution and its concentration,— also the boiling temperature and the concentration of a series of more dilute solutions 163

Determination of the thermal capacity of the apparatus with steam 164

Thermal specification of the apparatus 165

Localities where experiments were made 166

The altitudes above sea-level of these localities are given : they range from 87 metres at Edinburgh to 2733 metres in the Engadine 167

The salts used in the research 168

Detailed description of the experiments, the results of which are recorded in Tables II and III 169

Description of Table IV which contains the greater part of the results of the research 171

The importance of the physical meaning of the expression $W(t-T)$.

In the equation $W(t-T)=$constant, we have Blagden's law of the lowering of the freezing-point of saline solutions extended to the raising of their boiling-points . . . 173

Table V gives the results of experiments made with dilute solutions.

Exclusion of chloride of rubidium from the Table on account of impurity of the sample. As regards the metallic component it was, as it professed to be, "spectroscopically pure," but it contained sulphate as well as chloride of rubidium. This affords a good example of the necessity of testing chemicals which are used in work that is to be published, no matter how great the reputation of the furnisher of the preparation may be.

Table VI is derived from Table V and gives the values of $W(t-T)$ for even temperatures.

The results given in Table IV in terms of temperature, are repeated in Table VII in terms of the equivalent barometric pressure 176

Table VIII gives the vapour tension of water in kilograms per square centimetre at temperatures from 90° C. to 120° C.

Contents

Observations on the behaviour of mixtures, and particularly, of isomorphous mixtures of salts 177

The device of the elastic tank of uniform depth. It provides a standard of comparison between the chemical effect of a dissolved salt and the mechanical effect of increased pressure upon the boiling temperature of water. It furnishes the extra load required to press on the surface of the pure water in the tank so as to prevent ebullition at the temperature at which a given weight of salt dissolved in the water would produce the same effect. It furnishes the mechanical equivalent of the weight of salt, in so far as the raising of the boiling temperature of water and the resistance to steam pressure are concerned 178

The specific properties of the dissolved salt are most pronounced in its concentrated solution 179

The mechanical experiment with the elastic tank represents the chemical case where the increase of steam-tension neutralised by the salt is proportional to the quantity of the salt. A case of the use of the elastic tank is described fully . . 180

Demonstration of the identity of Blagden's Law with the thermal law of mixture 182

No. 5. THE SIZE OF THE ICE-GRAINS IN GLACIERS. (From *Nature*, August 22, 1901, Vol. LXIV, pp. 399–400) . . . 226

This investigation was undertaken in order to ascertain by actual observation the maximum size of the ice-grain in a particular region of a particular glacier, and the gradation of size of the other grains from this maximum downwards to the minimum occurring in the block of ice dissected. In all published descriptions it was admitted that the shape of the ice-grain is irregular, but in dealing with the size of the grains at any part of a glacier, the descriptions were frequently incompatible with the geometrical principle that, if the shape of the grains is irregular, there must be grains of practically all sizes from the maximum downwards, in order that the ice may fill the space of a solid block.

The blocks dissected were taken from the Glacier des Bossons in the Chamonix Valley and from the Aletsch Glacier at the Mergelin Lake 227

Table I gives the weight in grammes of single ice-grains . 228

Table II enables a view to be formed of the average size of the grains in a block 229

The action of solar radiation on glacier-ice is twofold : it produces disarticulation and lamination 230

PAGE

Glacier-ice is protected from sun-weathering by ice and by water 231

It is the dependence of the melting-point of ice on the nature of the medium in which it melts that enables glaciers to move under the action of gravity 232

No. 6. ICE AND ITS NATURAL HISTORY. (From the *Proceedings of the Royal Institution of Great Britain*, 1909, Vol. XIX, p. 243) 233

Recapitulation of principles and laws established in earlier work and applicable to the matter handled in this lecture . 233

The freezing-point of a substance is defined to be : the temperature at which it, as a liquid, passes into itself as a solid ; and its melting-point to be the temperature at which it, as a solid, passes into itself as a liquid 238

When ice is melting in a mixture of ice and water, the temperature of the water must be a little higher than that of the ice, else there would be no inducement for heat to pass from the water to the ice; similarly, when ice is being formed in a mixture of ice and water, the temperature of the water must be a little lower than that of the ice produced . . 239

The temperature at which ice begins to take form in water which is cooled when in contact only with itself, or with a solid other than ice, has not been determined, and is in fact uncertain. The moment the smallest particle of ice is present, the water has the opportunity of passing, as a liquid, into itself as a solid ; but not till then 240

It is probable that in nature ice never melts and water never freezes exactly at o° C. The law which regulates the melting of ice in nature may be stated as follows :—If the pressure is constant, it varies with the nature of the medium ; and if the nature of the medium is constant, it varies with the pressure.

The influence of salt in inducing the melting of ice at temperatures between o° C. and its Cryohydric point furnishes a quantitative explanation of observed anomalies in its physical constants.

The belief that ice, at temperatures near o° C., does not contract but expands on being cooled, has been maintained by experienced observers, such as Hugi, Petzold and Pettersson. Their observations were exact but their interpretation of them was faulty. When due weight is given to the influence of the medium, the anomaly disappears, and it is found that ice does not behave in the capricious way supposed, but conforms to the usual custom, by expanding when warmed and contracting when cooled 241

Quotations from the paper 'Ice and Brines' in illustration
of this 242

These quotations from my paper of 1887 are given at
length, because, though published so many years ago, it
appears to have been but little read. I know of no treatise
on natural ice in which the dependence of the melting temperature of ice on the nature of the medium in which it melts is
even mentioned, still less taken into account. For these
reasons, I have made the development of this important subject the principal feature of this Lecture. 244

After stating the constants involved, the cryoscopic principle established in the paper on 'Ice and Brines' is applied
to the discussion of the apparent variations of volume of a
block of ice, the volume of which at $0°$ C. is 1000 cubic centimetres. It contains $1·5105$ gram NaCl, diffused through it,
and it is provisionally assumed to be in the inert state, in
which it is deprived of the power to induce the melting of
ice at temperatures between $0°$ C. and its cryohydric point
$(-21°·72$ C.). If the temperature of the block containing the
inert NaCl be reduced to $-23°$ C., its volume will be reduced
to $996·320$ c.c., and as the temperature is below the cryohydric
temperature, the salt is by nature inert. At such temperatures
ice and common salt are indifferent to each other. Returning
to the initial state, let the temperature of the block be reduced
to $-21°$ C., the ice remaining inert. The volume of the ice
will then be $996·64$ c.c. Let the NaCl now recover its activity,
it will melt $5·628$ c.c. ice producing $5·160$ c.c. water under a contraction of $0·469$ c.c. so that the apparent volume of the ice at
$-21°$ C. is $996·171$ c.c. By the aid of Table I and the other
constants, we can calculate the composition of a block of ice
of any weight or volume which contains $1·5105$ gram NaCl.
The results of such a calculation are given in Table II for a
block of ice having the volume 1000 c.c. at 0 C. The results
are given graphically in Curve B, Fig. 1. This curve has the
remarkable feature of two acute angles, giving the appearance
of the letter Z at the cryohydric temperature. Above this
temperature the ice expands with heat, but at a gradually
diminishing rate, until at $-7°·0$ C. the increase of volume
due to simple expansion of the ice is exactly balanced by the
contraction due to induced melting. At this temperature the
coefficient of apparent expansion changes sign and becomes
negative. Therefore, the coefficient of apparent thermal expansion of this ice changes sign three times when it is warmed
from a temperature below the cryohydric point of solution of

PAGE

chloride of sodium to that at which liquefaction is complete ; and the actual facts of the apparent expansion have been accounted for step by step by the cryoscopic reaction of NaCl on ice between $-21°\cdot72$ C. and $0°$ C. 249

For blocks of ice which contain per 100 parts by weight of ice less than $29\cdot97$ and more than $1\cdot7164$ parts of NaCl, the coefficient of apparent expansion is negative at all temperatures above $-21°\cdot72$ C. For solutions of higher dilution Table III gives the critical temperature at which the coefficient of apparent dilatation changes sign and becomes negative. This Table shows that in the case of ordinary potable waters this critical temperature is sufficiently removed from the temperature of congelation to be easily determined, and it furnishes the best method of arriving at the concentration of fresh water, or solutions of very high dilution. Thus, when the dissolved matter is equivalent to no more than one gram of chlorine, present as chloride of sodium, in one hundred thousand grams of water, the critical temperature is $-0°\cdot725$ C. The critical temperature of the apparent expansion of ice affords a means of detecting impurity equivalent to quantities of chlorine as small as 1 gram in *ten tons*, and even 1 gram in one hundred tons of water 251

Cryoscopic equivalence between pressure and salinity. This subject has been worked out in much the same way as the similar phenomenon of the elevation of the boiling-point of water by pressure and by dissolved salt, for which we used the device of the elastic tank of uniform depth. In both cases the same Law is found to exist. For a given elevation of the boiling-point in the one case and depression of freezing-point in the other, there is the same approximate proportionality between the area of the surface of the water exposed to the pressure and the volume of the water which holds the equivalent amount of salt 252

Influence of impurity on the apparent latent heat of ice . 255
Size of glacier grains 258
Sun-weathering of granular ice produces white surface of glacier 262
Snow, névé and glacier 263
The grain of lake-ice 265
Characteristics of an advancing glacier 267
Grooving of ice by rock 268
External work of a glacier 270
The real region of mechanical erosion and attrition is the sea-shore 272

Advantage of the study of tropical lands.

The rounding of pebbles, generally attributed to the mechanical action of streams, is due to the chemical action of subaerial weathering. When an irregular piece of rock is exposed to subaerial weathering, or any other external chemical decomposing agent, the resultant form of the undecomposed kernel of the fragment is necessarily an ellipsoid .　.　. 273

Weathering in Tenerife　.　.　.　.　.　.　. 274

Similar case observed by the Prince of Monaco in the Cape Verde Islands　.　.　.　.　.　.　.　. 275

The " Crumble " formation.

This is the name which I invented for my own use in recording the occurrence of a common surface feature of tropical and equatorial regions, when on board the " Challenger." I found in all such countries that the rocks were decomposed to a depth of many metres, the residual material often remaining in situ, with such a fresh appearance that it was difficult to imagine that it could be anything but unaltered rock. It was only necessary, however, to touch it with a stick or even with the fingers, for it to " crumble " into fragments of all sizes down to sand and clay. In almost every place within the tropics visited for the first time the rocks were logged as consisting of "the crumble formation." Outside of the tropics this formation occurs only in a rudimentary form. The crumble formation owes its existence to a high atmospheric temperature and to humidity. It is the typical formation produced by the subaerial weathering of rock in situ. Very slight mechanical disturbance is sufficient to cause degradation under gravity, with the production, first of taluses ; then as the taluses undergo further gravitational degradation, they flatten out, and the final product of the crumble formation and of subaerial weathering is the pampa, prairie, steppe or desert, which are the native names for the same formation in homologous climatic regions of the earth.

Demonstration that the generally accepted doctrine that rock-fragments, which are found so frequently covering the tops of mountains, are split off from the parent rock by the energy liberated by water freezing in its interstices, is untenable　. 278

No. 7. Beobachtungen über die Einwirkung der Strahlung auf das Gletschereis. (Extract of paper read before the physical section of the Schweizerische Naturforschende Gesellschaft at its meeting at Basel on 6 September, 1910, and printed in its *Verhandlungen*, I, p. 330) .　.　.　. 28c

PAGE

As almost all my work on ice, outside of the laboratory, was done in Switzerland, and as the Swiss Naturforschende Gesellschaft honoured me many years ago by making me one of its honorary members, I considered it my duty to communicate to it an account of the more important results of my work on the Swiss glaciers. I had also in my mind that in case of there being any doubt about the work, the Swiss were geographically in a better position to verify it than any other people. As the meeting was being held in the principal city of German Switzerland, I made my communication in the German language.

No. 8. In and around the Morteratsch Glacier : A Study in the Natural History of Ice. (From the *Scottish Geographical Magazine*, 1912, Vol. XXVIII, p. 169.) .   . 283

My first acquaintance with glaciers was made in the summer of 1867, when I visited Grindelwald. Both of the glaciers of that valley were then in the state which they had maintained for more than a century 284

Recent change in the Lower Grindelwald Glacier and in the Rhone Glacier, shown by ancient pictures and modern photographs. These pictures exhibit the wastage which has taken place in less than 50 years. M. Vallot estimates it to amount to one-eighth of the total amount of sinking since the last ice-age 286

Shrinkage of the Morteratsch Glacier in the last few years, exhibited by photographs taken during the period of 1906–1911. The greatest amount of subsidence took place in the very hot summer of 1911, amounting to between 4 and 5 metres, the average being 2·5 to 3 metres.

At the end of the glacier the rate of apparent annual retreat of the ice was less than 2 metres, while, on the western flank, it was nearly 9 metres 288

The principal source of the impurity in the intergranular water is rock-débris.   291

Usefulness of the artificial grottos which are met with in all frequented glaciers for the study of glacier-ice in its primary state 292

Intergranular melting of ice and re-freezing of the same . 293

The intergranular moisture or liquid is an important item in the economy of the glacier. It is the mother-liquor of the grain and produces the medium in which the activity of the crystallisation and dissolution of its ice develops itself. Effect of solar radiation on glacier-ice. Its principal effect is the production of the white surface-layer 294

Contents

xxi

PAGE

Its whiteness is due to the same cause as that of a field of
snow, namely discontinuity in the discrete particles or masses
of ice of which they are composed. The snow-field is produced synthetically by the atmosphere ; the white surface of
the glacier is produced analytically by the sun's rays. The
internal melting of the ice of the glacier by radiant heat
proceeds by the melting of the outer surface of the individual
grain 295

If a glacier were exposed only to heat of convection, it
would melt with a smooth surface and its colour would be
blue. As the sun's rays analyse a piece of glacier-ice,
separating its constituent grains, so they continue the
analytical process on the isolated grain or crystal and
separate it into its constituent lamellæ which are arranged
normally to the principal axis of the crystal. The Morteratsch
Grotto and the experimental use made of it 296

Important observations made in February 1894 . . 297

Delineation of the grains on the walls of the grotto
extends as far as direct skylight penetrates.

Illustration of the state in summer.

Granular feature in winter 298

Hoar-frost produces etched figures of the granulation of
the ice, especially on the roof of the grotto 299

Disintegration behind the surface of the walls.

Usefulness of photography inside the grotto. By pro-
longing exposure views may be obtained of features which
the eye cannot detect. The stereoscopic camera is especially
useful in the study of the internal disarticulation of the ice,
and of other features 300

Nature of the apparent stratification of the ice, which is
sometimes visible on glaciers 301

Importance of studying the grotto in spring or early
summer 302

Improbability that a ribbon-structure extends through the
mass of a glacier.

The case of a glacier which has never been exposed to
light 303

Water a protective medium. It is opaque to the rays
which disintegrate glacier-ice 304

The Mergelin Lake and its icebergs. They are analogous
to those of the Antarctic Ocean 305

The granulation of glacier-ice is best studied at the
Mergelin Lake and in the month of July, when the sun is
powerful 306

PAGE

Hugi's Fundamental Experiment. A piece of fresh blue glacier-ice, exposed to the rays of a powerful sun, in a very short time crumbles together into a heap of individual ice-grains 307

Hugi's greatest achievement was his *Winterreise ins Eismeer*, in January, 1832, a date which will be for ever memorable in the history of the study of glaciers. Although only a schoolmaster, it was carried out entirely at his own expense. The organisation of it was complete, covering observations in the valley with those which he and his party were to make on the mountain. It lasted fourteen days, and nearly every one of these days is the date of a fundamental discovery in the natural history of land-ice 308

Recapitulation and Conclusion 309

No. 9. ON THE USE OF THE GLOBE IN THE STUDY OF CRYSTALLOGRAPHY. (From the *Philosophical Magazine*, 1895, S. 5, Vol. XL, pp. 153–172.) 313

In this paper the usefulness of the Globe in the study of solid Geometry generally is explained. The term crystallography used in the Title refers to a particular case where polyhedra are limited by certain symmetries.

The blank globe, whether black or white, with the divided circles belonging to it, is a calculating machine, adapted to the solution of all the problems to which the analytical methods of spherical trigonometry are usually applied . . 313

The globes which I used were those of 22 centimetres diameter, which the firm of E. Bertaux of Paris supply. For measuring purposes a system of divided circles of the same radius as the sphere, called the Metrosphere, is supplied. The real usefulness of the globe is not to be learned by theory or precept, but by actual experience in the solution of problems whether the study be astronomy or navigation or geography or crystallography 314

A detailed example is given of the use of the globe as applied to a polyhedron of any number of plane faces, arranged in any way so as to completely enclose the space . 315

When the operation described has been performed, we have on the globe a number of points which form a complete catalogue of the faces of the polyhedron or crystal. Similarly the arcs connecting each pair of poles furnish a catalogue of the inclination of every single face to every other . . . 316

By another graphic process we obtain a catalogue of all the plane angles occurring on the faces of the polyhedron . 317

PAGE

The representation of the faces of a crystal by great circle planes, and that of the edges by diameters, all of which necessarily meet in the centre, facilitates the choice of a suitable system of crystallographic axes 318

A collateral advantage which the student gains by using the globe in this way is the excellent mental discipline which it affords. A very short experience developes enormously the sense of direction. *There is no operation in the geometry of polyhedra or crystals which cannot, with the greatest ease, be performed by the use of the globe* 319

Postscript containing examples of the use of the globe in dealing graphically with the relations of the faces, edges, etc., of a crystal 321

It is impossible to illustrate examples without the aid of the globe itself and the reader must have one before him in order to make for himself the constructions described. One metrosphere can be used with any number of globes, so that separate details can be worked out on separate globes and combined on others, thus avoiding the risk of mistakes due to overcrowding 325

No. 10. ON A SOLAR CALORIMETER USED IN EGYPT AT THE TOTAL SOLAR ECLIPSE IN 1882. (From the *Proceedings of the Cambridge Philosophical Society*, 1901, Vol. XI, Pt I, pp. 37-74.) 337

While engaged in discussing questions connected with the physics of the ocean, I found the want of definite knowledge of the amount of solar heat which really reaches the surface of the land or sea in a form which can be collected, measured and utilised. There was no lack of actinometrical observations, but I found it impossible from them to obtain the data that I sought. The aim of most observers has been to arrive, by more or less direct means, at what is known as the solar constant, that is, the quantity of heat which is received in unit time by unit surface, when exposed perpendicularly to the sun's rays outside of the limits of the earth's atmosphere. For my purpose the amount of radiation arriving at the outside of the earth's atmosphere was of no importance. What I wanted to know and to measure was the amount of solar radiation which strikes the earth at the sea-level and is there revealed as heat. It is the energy of this radiation which maintains the terrestrial economy. Having the opportunity of accompanying the expedition to Egypt for observing the total eclipse of the sun on May 17th, 1882, I determined to make observations for myself on the amount of heat which could actually be collected from the solar radiation in these favourable circumstances 337

I determined to use a calorimeter which should depend for its indications on change of state and not on change of temperature, and I designed a steam-calorimeter by which the sun's rays should be collected by a conical reflector of definite capacity and be thrown on an axial tubular boiler in which they should be transformed into the latent heat of steam ; and the amount of steam so produced per unit of time from water at the temperature of ebullition should be a measure of the rate of heat-receipt from the sun, in the conditions of the experiment.

Locality.

The astronomers chose a spot on the banks of the Nile close to the town of Sohag, and in Latitude 26° 37′ N. for the observation of the eclipse, and experience showed that it had been well chosen. The eclipse was total at 8.34 a.m. on the 17th May 1882, civil reckoning. The maximum duration of totality that was expected was 70 seconds, and, in fact, it lasted longer than 65 seconds. The expedition arrived on the 8th May and I began work with the calorimeter on the 11th and devoted the whole of my time to this work until the 19th May, when the camp was struck. From the outset the calorimeter worked most satisfactorily and the only alteration which had to be made was to replace the original metal dome, as steam space of the boiler, by a glass tube. From the 11th to the 15th May I was occupied in studying the instrument and learning how to use it, in the only way by which this is possible, namely, by setting it to do the work expected of it, noticing deficiencies of arrangement and mistakes made in handling, and rectifying them as they showed themselves. In the course of this educational work valuable preliminary results were obtained, and on the 16th, 17th and 18th trustworthy experiments were carried out with the apparatus in best working order and under very favourable natural conditions 338

Table I gives the sun's altitude and azimuth at noon and at every half-hour on each side of noon until sunset for the date of the eclipse 339

Illustrations of the calorimeter, Figs. 1 and 2 . . . 340

Construction of the calorimeter 341

Principal section of the calorimeter, Fig. 3 . . . 342

PAGE

In use, the calorimeter is pointed axially to the sun. Method of adjustment to compensate the rotation of the earth 343

General description of an experiment.
When everything was at the temperature of the air and the instrument was pointed to the sun at 2 p.m., the water in the axial boiler boiled in 40 seconds, and it continued to boil so long as the instrument truly followed the sun and as the sun was not obscured. The boiling proceeded with perfect regularity, even when the sun was at its hottest, as on the forenoon of 18th May ; and with the glass dome as steam-space everything could be followed minutely 344

The instrument is not intended for "snapshotting." It is essential that the distillation be kept running continuously, and the water produced in successive intervals of time be weighed or measured. If the meteorological conditions are such that the boiling is interrupted, then it is of no use attempting to make any observations. There was no trouble from this cause at Sohag. The sole object of my experiment was to ascertain the greatest amount of heat which can be obtained per unit of time from the sun's radiation at or near the sea-level. Owing to the great latent heat of steam, the immediate feed of the boiler is automatically maintained at the boiling temperature, so that, when in continuous running, the whole heat collected from the sun and thrown on the boiler is used in transforming water into steam of the same temperature 345

Illustration of the calorimeter on equatorial mounting. Fig. 4. 347

Description of the construction of the reflector . . . 348

Table giving the numerical data relating to the reflector used 353

Geometrical construction of reflector, when the position of a point on one of the mirrors and the position and length of the focal line are given. Fig. 6 354

Meteorological observations at Sohag : Table II . . 358

Description of the eclipse 361
It was witnessed from the banks of the Nile by the whole population of Sohag. The most striking and the least expected feature of the eclipse was the appearance, with totality, of a brilliant comet, between two and three sun's diameters from the darkened disc and with a slightly curved tail equal in length to the sun's diameter. Only the astronomers were prevented from witnessing this unique phenomenon, by the nature of their observations, which had to do with one branch

or another of solar photography, and with only a minute in which to make them. The principal stars, as well as the comet, were shining brightly during totality : nevertheless, I was able to read the scale of an ordinary thermometer during totality. Two solar protuberances were visible to the naked eye, but to me they appeared to be notches in the dark disc of the moon through which red light was visible. This observation has an important bearing on the interpretation of the phenomenon generally called " Baily's beads." Of all the natural phenomena which I have had the opportunity of witnessing, there is none which produces so powerful an impression as a total eclipse of the sun.

Conditions during the forenoon of 18th May, when the maximum value of the solar radiation was observed with the calorimeter 362

Discussion of the observations 363

Rates of distillation observed on 16th, 17th and 18th May: Table III 364

Graphic representation of the same 369

The weather on each of the three days was very fine, but it was best during the forenoon of the 18th, when the maximum rate of distillation observed was 1·501 cubic centi-metre per minute 370

To transform 1·5 gram of water at 100° C. into steam of the same temperature 803 gram-degrees of heat are required ; and this is the greatest amount of heat which has been collected by the calorimeter in one minute. The actual collecting area of the reflector is 903·5 square centimetres. Therefore, the heat so collected is equivalent to 8888 gram-degrees per square metre : and 8888 gram-degrees C. suffice for the generation of 16·6 grams steam at 100° C. Therefore, it has been shown by our experiments that, by the use of ordinary mechanical appliances it is possible, under favourable geographical and meteorological conditions, to collect, on a square metre of surface, at or near the sea-level, exposed perpendicularly to the sun's rays, the energy of generation of 16·6 grams of steam per minute, or 8888 gram-degrees of heat, which are equivalent to 3777 kilogram-metres of work : and, as this work is done in one minute, the agent is working at the rate of at least 0·84 horse power. Therefore, the object for which I designed the calorimeter and used it in Egypt was achieved.

Taking the area of a great circle on the earth's surface to be $129·9 \times 10^{12}$ square metres, the useful energy received by

PAGE

the whole earth is at the rate of at least 109 × 10^{12} horse power. Taking the radius of the earth's orbit to be 212 times the radius of the sun, the radiation of one square metre of the sun's surface is spread over 45,000 square metres of the earth's surface; therefore, the sun must radiate energy at the rate of at least 37,000 horse power per square metre of its surface 372

Comparison of the heat of the sun's rays with that produced by the combustion of liquid iron in oxygen . . . 373

Observations made during the solar eclipse.

Diagram of exposed surface of the sun at successive epochs after totality: Fig. 9 376

Table IV gives the rates of distillation observed at these epochs 377

Conclusion, containing remarks on the solar constant.

The experimental value found with the steam calorimeter at Sohag on 18th May 1882 was 0·89 gram-degree C. per square centimetre per minute, without making any correction whatsoever. It is certain that, using experimental means not inferior in efficiency to the steam calorimeter, heat can be obtained on the banks of the Nile from the sun at this rate.

The only physical constant used is the latent heat of steam, and this has been determined with the greatest exactness. Therefore, the principle on which the instrument is founded cannot be improved. The construction of the instrument was by the late Mr John Milne of the Milton House Works in Edinburgh, and no name in Great Britain could be a better guarantee of exact workmanship: but, independently of the reputation of the constructor, the instrument did in fact do the work with perfect efficiency. In order to save time in the construction, every part of the instrument which could be so made was made out of brass tube. Instrument-making is now so advanced that, with good will, there should be no difficulty in getting a steam calorimeter which, in competent hands, would give as good results as I got.

No. **11**. SOLAR RADIATION. (From *Nature*, September 5, 1901, Vol. LXIV, p. 456; with postscript in 1911.) 383

Only an infinitesimal proportion of the total radiation reaches the earth. The remainder, in so far as we know, is wasted by dissipation into space 383

The amount of radiant heat which we can count on as being supplied to the whole earth in unit of time is the

PAGE

constant which is of greatest importance in terrestrial
physics 384

When we have ascertained the supply of heat to the earth
we have to inquire what becomes of it.

Fundamental principle :-- The heat which the whole earth
receives from the sun in the course of a year also leaves it in
the course of a year.

The differential behaviour of the atmosphere to heat-rays
striking and leaving the earth's surface is an example of
Kirchoff's law that a body absorbs by preference the rays
which it itself emits 385

Wind and all mechanical atmospheric effects are due to
differences of density. These are produced not only by
thermal expansion of the air, but also by mixture with it of a
lighter gas ; such a gas is water-vapour 386

Description of methods of measuring the thermal effect of
the sun's rays 387

Simultaneous observations made at the summit of Mont
Blanc and at Chamonix 390

Observations on the Peak of Tenerife 392

Observations with the steam calorimeter in Egypt by the
writer and by Mr Michie Smith at a height of 7000 feet
above the sea at Kodaikanal in India. The highest rate of
distillation observed by Mr M. Smith with the same instrument which I used in Egypt was 1·754 cubic centimetres per
minute, whereas the highest realised in Egypt was 1·501 cubic
centimetres. Therefore, at a height of 7000 feet above the
sea one-seventh more heat can be collected, measured and
used than was found possible on the banks of the Nile. But
it is the amount which arrives at the sea-level, or near to it,
that is alone of interest in the study of terrestrial physics   393

In a work entitled *Strahlung und Temperatur der Sonne*
(1899) J. Scheiner sums up the discussion of this subject by
giving 4 as the most probable value of the solar constant.
This statement is disputed and it is shown that values of the
solar constant of the order of 4 must be exaggerated.

No comment on these remarks was offered, but the later
estimates of the solar constant began to fall. Attention is
called to them in a Postscript written in February 1911 . . 395

Féry, who quotes 2·40 as the value of the solar constant
generally accepted in 1909, arrives by his own experiments at
the value 1·70, and he attributes the difference between these
two values to excessive correction for the absorption by the
atmosphere 396

Conclusion :—Considering only my own observations, the final result of the experiments made with my steam calorimeter is that the calorific value of a sheaf of the sun's rays, having a section of one square metre, is at least 9698 gram-degrees Centigrade per minute at the terrestrial sea-level ; and it is certain that heat can be obtained from them there at this rate and in a useful form by mechanical appliances of simple construction.

No. 12. THE TOTAL SOLAR ECLIPSE OF AUGUST 30, 1905. (From *Nature*, December 21, 1905, Vol. LXXIII, p. 173.)   399

As this eclipse was to be visible in an easily accessible part of Europe, where the probability of finding fine weather was great and the calculated duration of totality was nearly four minutes, I determined to see it. A considerable display of protuberances was expected and I wished to form my own idea of their size by personally checking their persistence or non-persistence through the phase of totality as seen from a station situated as nearly as possible on the line of mid-totality. The display of protuberances at the moment of second contact was very brilliant ; when the time of mid-totality arrived not a trace of them was visible to the naked eye. Therefore these very brilliant protuberances had an

apparent height less than 45 seconds of arc 400
This eclipse is compared with that of May 17, 1882, which was observed at Sohag on the Nile   401

No. 13. ECLIPSE PREDICTIONS. (From *Nature*, October 19, 1905, Vol. LXXII, p. 603.) 402

The predictions respecting the solar eclipse of August 30, 1905, as issued by the British *Nautical Almanac* and by the French *Connaissance des Temps* are compared, and their want of agreement is illustrated by a Table. The most striking discrepancy between the two predictions is shown by the width of the band of totality in Spain and the adjacent Mediterranean, which is given as from ten to eleven nautical

miles greater by the French than by the British prediction . 403

The astronomical data on which these predictions were founded differed only in the accepted values of the apparent diameter of the moon, but the grounds on which the British and French authorities differed on this point have not been made public, so far as I know.

PAGE

No. **14.** THE SOLAR ECLIPSE OF APRIL 17, 1912. (From *Nature*, May 9, 1912, Vol. LXXXIX, p. 241.). 404

This remarkable eclipse had a peculiar interest of its own, and the observers of it had the advantage that the central line passed through Paris and many other important places in Northern Europe, but up to the last moment it was uncertain whether the eclipse would be total, annular or partial. The value accepted for the apparent diameter of the moon was therefore of paramount importance. I observed it in front of the school house at Eaubonne, a northern suburb of Paris, and used an ordinary binocular when the naked eye was not enough, and a hand-screen made of three coloured glasses which reduced the density of the sun's light without altering its colour. This was a very efficient instrument and I am sorry that I do not know the source from which it was obtained. I bought it from a hawker in the streets of Barcelona, on the eve of the eclipse of August 30, 1905, and of course, like all hawker's goods, it was anonymous . . 405

I used it in front of the eye-pieces of the binocular and with it the diminution of the luminous crescent could be easily followed, and the view furnished was very sharp. As the area of the luminous crescent diminished rapidly before the advance of the dark lunar disc the colour of its light suddenly changed to a deep red. It would have been impossible to perceive the red colour, intense though it was, had it not been for the perfection of the hawker's reducing glass 407

After the light of the solar crescent had become red the lower cusp became indented by black blades or teeth ; then the upper cusp showed a similar phenomenon and almost in a moment the teeth spread irregularly over the whole crescent, crossing and intersecting each other like a crystallisation. Very quickly the dark disc of the moon advanced and pushed the beautiful network over the eastern edge of the sun, and apparently at the same moment the network reappeared, coming over the western edge of the sun, attached to the black limb of the moon and at the same time held by the limb of the sun. In a few moments the uncovered crescent of the sun had increased so much that the delicate lacework could no longer bear the tension, it parted and disappeared instantly, while at the same moment the dark limb of the moon recovered its perfect smoothness of outline. This was the form which "Baily's beads" took in this memorable eclipse. The pattern observed by Baily himself in 1835 was different. I am not able to offer any satisfactory explanation of what Baily's phenomenon is, but it certainly is not due to the interruption of the solar rays by the mountains of the moon 409

No. **15**. THE PUBLICATION OF SCIENTIFIC PAPERS. (From *Nature*, August 10, 1893, Vol. XLVIII, p. 340.) 410

The complaint is frequently heard from abroad that important papers by British scientific men are almost inaccessible to the foreigner, because it has been the fashion to communicate them to local societies and to rest content with such publication as is secured by their being printed in the Society's *Proceedings* or *Transactions*, and the circulation of these is confined to exchanges with other scientific societies. They are not dealt in by the bookselling trade 410

A plan is sketched in the paper whereby what is at present inefficiently and extravagantly done by a multitude of amateur publishers scattered over the country, could at much less cost be efficiently done by a central publishing office as a matter of business 411

One great advantage of this would be that it would get rid of the censorship on the part of the Councils of the Societies, which is the disgrace of British scientific life. It is now (1916) twenty-three years since this paper was published and the publication of scientific papers in Great Britain remains as it was.

No. **16**. THE ROYAL SOCIETY. (From *Nature*, January 28, 1904, Vol. LXIX, p. 293.) 413

This paper is a summary of what I said at a special meeting of the Fellows of the Royal Society, when the constitution and functions of the sectional Committees were under consideration.

The main function of the sectional Committees is to *refer* papers received by the Society from Fellows . . . 413

In so far as the public is concerned the effect of the *reference* is to make doubtful the declared authorship of any paper taken at random in the publications of the Royal Society, inasmuch as it may have been altered to an unknown extent by a person unnamed in the title.

The practice of the Royal Society in dealing with papers by its Fellows is compared with that of the French Academy of Sciences, and to the advantage of the latter . . . 414

The following personal experience furnishes a good example of the practice of the French Academy in dealing with papers.

In the summer of 1867, while working in the laboratory of Wurtz in the École de Médecine in Paris, I made some investigations on the products of the reaction of perchloride of phosphorus on salts of isethionic acid. I collected the results in a short paper and, with Wurtz's approval, I proposed to offer it to the Academy. At that date Wurtz himself was not yet "of the Institute," but there was a standing custom that papers by his *élèves* were presented by Balard, the veteran discoverer of bromine. Accordingly I took my paper with me and made a formal call on M. Balard, who received me with the greatest kindness and courtesy in his study, wearing as had been the fashion in his younger days, a black frock-coat and a white neck-cloth taken twice round his neck. When I had expressed my desire that he would do me the honour to present my paper to the Academy, he replied at once that he would have the greatest pleasure in doing so. I handed him the paper, he presented it the following Monday and it was published in the *Comptes Rendus* of the next week[1].

No. **17.** Nomenclature and Notation in Calorimetry. (From *Nature*, May 12, 1898, Vol. LVIII, p. 30.) 416

The motive for this paper was furnished by the inconvenience caused by the use of different units of heat and of different names for the same unit of heat, by writers on calorimetric subjects 416

As a general principle, compound units should be expressed by compound names.

In the early literature of the equivalence of heat and work, the unit of work was chosen on this principle, and in it we meet with only such names as foot-pound, kilogramme-metre, and the like, which explain themselves.

The recommendation made in the paper is that the unit of heat should be chosen on the same principle and should be called gram-degree (Celsius), pound-degree (Fahrenheit), or the like 417

No fancy names should be used for such important units. An oceanographical example is given of the use of a special

[1] 'Sur quelques dérivés de l'acide iséthionique. Note de M. J. Y. Buchanan; présentée par M. Balard.' *Comptes Rendus*, 2 September 1867, Vol. LXV, p. 417.

PAGE

heat-unit, composed on this principle, in which the unit of
length replaces the unit of weight 418

    In terms of the usual British units it is the fathom-degree
(Fahrenheit), and in terms of the French units it is the metre-
degree (Celsius) ; and, by arithmetical necessity, the fathom-
degree (Fahrenheit) is equal to the metre-degree (Celsius) . 419

No. **18.** THERMOMETRIC SCALES FOR METEOROLOGICAL USE.
(From *Nature*, August 17, 1899, Vol. LX, p. 364.) . . . 420

    It was known that at the Meteorological Congress, about
to be held in Berlin, an attempt would be made to force
English-speaking meteorologists to renounce Fahrenheit's
scale in favour of that of Celsius, and the object of the paper
was to direct attention beforehand to some of the advantages
in securing accuracy and in relieving labour which Fahrenheit's scale offers over that of Celsius, when used for meteorological purposes 420

    In the most populous regions of Europe and North America
the temperature in winter presents frequent oscillations from
one side to the other of the melting-point of ice. If the
observer is compelled to use a thermometer which he must
read upwards when the temperature is on one side of that
point, and downwards when it is on the other side of it, and if
he may be called on to perform this fatiguing functional inversion perhaps several times in the same day, it is certain
that he will suffer from exhaustion and that the observations
will be affected by error 421

    These representations had no effect, though they could
not be disputed, and I asked one of the German members
how it was that, when they had already a thermometric
scale which excluded such avoidable errors, and one which
had been devised by a distinguished countryman who was
not only a German but a Prussian, they had consented to
exchange it for a foreign scale so conspicuously inferior to it.
The answer was that it was *befohlen*, and I was advised to
get it *befohlen* in England. I was surprised by the confession
and could only reply that, in my country, we do not do these
things in that way.

No. **19.** THE METRICAL SYSTEM. (From the *Times*, February,
1903.) 424

    In view of the renewal of the agitation for the universal
adoption of the French metrical system, it appears to be
opportune to furnish the public with facts regarding it, derived

PAGE

from the fountain-head. Laplace's account of it has been taken as such, because he was one of the most important members of the committee appointed by the French Government for the purpose. It is taken from his *Exposition du Système du Monde*, where it is introduced collaterally with the discussion of the length of the second's pendulum as a measure of the force of gravity at the earth's surface. The committee consisted of five of the most distinguished French mathematicians of the day. It is perhaps to be regretted that their findings were not subjected to revision by a committee of ordinary business men 424

The fundamental excellence of the system is the simple relation between the unit of weight and the unit of length, and it cannot be doubted that its far-reaching advantages would have been disengaged by such a committee from its fundamental defect, namely, the inconvenience of the unit of length selected 425

It is pointed out that, at the date of the sitting of the commission, there was already, and there had existed for ages, a common and convenient unit of length, the subdivision of which was decimal, and it was in universal use by all seafaring nations. It was the nautical mile, which is subdivided into ten cables and each cable into one hundred fathoms, whence the nautical mile was one thousand fathoms . . 426

Previous to the year 1870 it would have been easy to secure the rectification of the unit of length, because, at that date, it would have been only France that would have had to alter her system, instead of the whole of the rest of the world.

When Germany accepted the metrical system without criticism, it became immediately more difficult. Nevertheless the population which is affected by a change is still so large that it is worth while that, before surrendering their old system, they should insist on getting in exchange one less affected by easily rectifiable defects than the French metrical system . 428

No. **20**. THE POWER OF GREAT BRITAIN. (From the *Scotsman*, March 26, 1897.) 430

This letter was suggested by an article, having the same title, which appeared in the *Hamburger Nachrichten*, the German paper which was always considered to represent Prince Bismarck's personal views. Although it occupied a whole column of the *Scotsman*, the subject was not alluded to in the editorial part of the paper, and copies of the letter which were sent to two important London morning papers were declined. Therefore, the message of the letter may be taken literally as the echo of a voice from the wilderness. How different would our position be to-day (1916) if heed had been given to the warning then.

The date of the letter was not much later than that of the famous telegram from the Emperor of Germany to Mr Kruger, the President of the Transvaal, after the Jameson raid ; and this, of itself, was sufficient warning of the hostile sentiments held by Germany towards Great Britain, but the warning was unheeded.

An example is given of the extensive and precise information which Germany possessed with regard to our supplies of men and material for the Navy and the Army, and the conclusion is drawn by the German paper that England, without conscription or compulsory service, could not hope to withstand attack from any first class European power . . . 431

Exception is taken to the statement of the Hamburg paper that it is now too late for England to adopt the method of universal service, and it is pointed out that this is inexact, for a great deal can be done even in one year.

The experience of the present war shows what a formidable army can be furnished by the voluntary devotion of a large section of the young men of Great Britain. Had the British Government been equally loyal and insisted on the laggards supporting the volunteers, we should now (1916) be very little worse off than if we had had universal service for the last twenty years. Perhaps the British people will ask itself if a form of government under which such injustice, or rather iniquity, is possible shall be permitted to persist.

The civil advantages of universal military service to the people are shown to be at least as great as the military advantages to the nation. Moreover, the stronger the defensive force of a nation, the less likely is the country to be attacked 433

The recommendation on p. 433 to read the experience of the French in 1870-71, in order to learn what actual invasion of one's country and home means, sounds grotesque.

It is a remarkable fact that at the present day no German writer or speaker ever refers to the behaviour of the German troops under Wilhelm I during the invasion and occupation of France. We have it from Bismarck himself that he tried to persuade that grand old King and soldier to treat the civil population by what is now known as the methods of frightfulness and under the same specious plea of shortening the war, but he tried in vain. The King accepted his advice in political

matters, but brooked no interference with the way in which
he directed his army to wage war.

After the termination of the war of 1870–71, the military
renown which the Germans had won was almost equalled by
their reputation for humanity towards the inhabitants of the
invaded country. This reputation spread all over Europe
and in Great Britain and even to some extent in France it
prevented adequate preparation being made to meet the
attack which his successor, for purposes of conquest and
plunder, was manifestly preparing. The people had heard
their fathers say "If we are to be invaded, let us be invaded
by Germany." They did not then know the significance of
the difference between the numerals I and II.

No. 21. AND THE HOUSE OF COMMONS? (From the *Scotsman*,
October 5, 1907.) 435

This letter was written when I had finished my autumn
work on the Morteratsch glacier, and saw in the English
newspapers that the Prime Minister had fixed October 5,
1907, for the meeting in Edinburgh where he was to open
the promised campaign against the House of Lords as an
independent legislative body under the British Constitution.
It occurred to me that, for the day, Edinburgh would
be practically a concentration camp of the followers of
Mr Asquith, and that it might perhaps be useful to give them
an opportunity of reading, in the morning, views on the
constitutional Houses of Legislature different from those
which they were likely to hear in the evening. It gave me
great satisfaction when I learned that the letter had appeared
in the *Scotsman* of that morning.

The behaviour of autocrats of different types differs in
their methods of diverting public attention from the abuses
of personal rule at home. Monarchical autocrats usually
involve their country in a foreign war. Parliamentary autocrats, when a dissolution is impending, start some burning
question affecting the inhabitants of a remote foreign country,
so that the people is diverted from attending to its own
interests until the general election is over. In the present
case this geographical limitation was set aside and the people
were incited to a form of civil war in which it was hoped there
would be no fighting. The House of Lords was used for the
same purpose as Ireland had been used by generations of
unscrupulous British politicians.

PAGE

The need of an independent Upper Chamber increases as
confidence in the House of Commons diminishes .   .   . 436

The vote in Parliament of a Member of the House of
Commons represents only the will of a small proportion of a
selected minority of the people of the place and cannot be
taken as the will of the people of the constituency.   What is
true of one Member and his constituency is true of the House
of Commons and the country.   The inhabitants, male and
female, who do not possess the franchise have no representation except in the House of Lords 437

Sterilisation of the usefulness of capable business men
when they enter the House of Commons.   The uncertainty
of the duration of the House of Commons interferes with its
usefulness.   Exemption from election is an element of efficiency
in the constitution of the House of Lords   438

Suggestions are given of possible reforms in the constitution
of the House of Commons which would tend to promote the
independence of its Members.   With regard to the House of
Lords, later events have shown that I misinterpreted the value
which it put on its own independence   439

No. 22. LORD MILNER AND IMPERIAL SCHOLARSHIPS.   (From
the *Morning Post*, October 15, 1909.) 440

This letter concerns the benefit which young men from the
Overseas Dominions gain by going from their home schools
to study at an English University.   It is pointed out that the
action of the English University on the colonial or foreign
student must necessarily be accompanied by the reaction of
the scholars on the University, inasmuch as the school training
of the Rhodes Scholars has been different from that of the
English schoolboys who form the bulk of the students, and
whose previous education has been stereotyped *ad hoc.*

It is as great an advantage to a schoolmaster to have to
continue the education of foreign boys as it is for boys to have
their education continued by foreign masters.   Hence the
belief is expressed that Mr Rhodes, in devising his bequests,
expected them to have an educative effect on his University
as well as on the young men to whom he furnished the means
of attending it.

It is pointed out that the true compliment to Rhodes'
scheme is to furnish the teachers of Oxford and Cambridge,
at the beginning of their career, with outside experience of
what is to be the business of their lives ; and it is proposed
to enable the Junior Fellow of his College to spend the first

year of his Fellowship at a foreign University and to study there some subject which brings him into personal contact with the students and teachers, as is found in a University laboratory or *atelier*. A scheme for providing reasonable augmentation of the Fellow's stipend during the first year spent abroad is outlined which would provide the probable sums required during a period of twenty years. The advantage to the College would be that after twenty years the management would be in the hands of men familiar with foreign methods and kept in touch with their variations ; and, if all the Colleges adopted the system, the University would in the same lapse of time become as up-to-date as the Colleges.

Although changes can be made at once by an Act of Parliament, no real reform of a popular institution such as a University can be effected in less time than the period of a generation.

No. **23**. HISTORY IN HANDY VOLUMES. (From the *Morning Post*, October 2, 1912.) 446

This letter was the closing unit of a series of contributions by readers of the *Morning Post* on the subject of their favourite reading.

Its purpose was to recommend the practice, which I have followed during my life, of always going to the fountain-head for information about any fact of whatever kind which has passed into history, whether it be civil or military, industrial or scientific.

The History of his own time by Frederick the Great and the *Gedanken und Erinnerungen* of Bismarck give us the facts at first hand about the creation of the Kingdom of Prussia out of the Mark Brandenburg and that of the German Empire out of the Kingdom of Prussia.

Similarly, the *Illustrated London News* of 1854–55 give us the facts at first hand of the Crimean War, a Military Expedition which was full of instruction but of little else.

A comparison is made between the service of news relating to the Crimean War in 1854 and that relating to the present European War ; and the performance of the *New York Times* is cited as an example.

SUMMARY OF CONTENTS 449

LIST OF PLATES

PLATE

Portrait of the Author . . . . . . *Frontispiece*

To face page

II.   Fig. 3. A gallery in the grotto of the Morteratsch Glacier in January, 1907 260

III.   Fig. 4. Solar etching in the Morteratsch Grotto in September, 1907 261

IV.   Fig. 5. Etching of ice by hoar-frost : Morteratsch Grotto, January, 1907 262

V.   Fig. 6 *a*. San Antonio (Cape Verde Islands). View of stream bed after three rainless years. (From a photograph by H.S.H. Prince Albert I of Monaco)   *Back to* 274

VI.   Fig. 6 *b*. The same view as Fig. 6 *a* on the same day after a violent rain storm. (From a photograph by H.S.H. Prince Albert I of Monaco) . . *Back to* 275

VII. Fig. 7. Chemical and gravitational degradation of rock on Chilian coast 276

VIII. Two views of the Lower Grindelwald Glacier taken from the same position 285

        Fig. 1. Engraved in 1777. By recollection, the glacier was in this state in 1867.

        Fig. 2. Photographed in September, 1909. The degree of coincidence of the views may be judged by the church which appears in both, and necessarily occupies the same site.

IX.   Two views of the Rhone Glacier 286

        Fig. 3. From a print dating before 1856. According to an engraving of 1777, it presented the same appearance at that date. According to recollection, at least one-half of the lower glacier existed in 1883.

        Fig. 4. From a photograph taken in September, 1909.

PLATE   *To face page*

X.   Views of the entrance of the Morteratsch Grotto, both taken
     from the same position and showing the slope of Munt
     Pers behind   288

         Fig. 5. Photographed 17th September, 1907.

         Fig. 6. Photographed 5th September, 1911. The
     wastage of the ice, due to superficial melting, is shown
     by the increased exposure of the slope of Munt Pers.
     The lateral wastage is indicated by the perspective of the
     figures, and by measurement it amounted to 35·6 metres
     in the four years.

XI. The Morteratsch Grotto in Winter   298

         Fig. 7. Taken on 16th January, 1907. Shows the
     great abundance of hoar-frost on the roof and its
     absence on the walls.

XII. The delineation of the grain of the glacier   299

         Fig. 8. Shows the delineation of the grain in winter,
     by the condensation of moisture in the form of hoar-frost
     on the roof of the grotto.

         Fig. 9. Shows the revelation of the grain in summer
     on a pillar in the interior of the grotto, which received
     subdued daylight from the entrance.

XIII. Discontinuities which cause the appearance of white ice in
     the walls of the grotto in winter   302

         Fig. 10. Shows a very remarkable picture of these
     discontinuities in the north wall, about two metres from
     the entrance.

         Fig. 11. Overlaps Fig. 10 a little and shows the
     comparative absence of white ice one or two metres
     further in.

XIV. Fig. 12. The Mergelin See   305

         The waters are retained by the ice of the Aletsch
     Glacier, from which small icebergs are frequently
     detached.

         Fig. 13. On the Morteratsch Glacier.

No. 1. [*From the Quarterly Review, October* 1906.]

RECENT ANTARCTIC EXPLORATION

1. *The Voyage of the "Discovery."* By Captain Robert F. Scott, R.N. Two vols. London: Smith, Elder, 1905.
2. *Zum Kontinent des Eisigen Südens.* By Erich von Drygalski. Berlin: Reimer, 1904.
3. *The Scottish National Antarctic Expedition.* By William S. Bruce. Papers in the "Scottish Geographical Magazine," 1905 and 1906.
4. *Two Years in the Antarctic.* By Albert B. Armitage, Lieut. R.N.R. London: Arnold, 1905.
5. *The Siege of the South Pole.* By Hugh Robert Mill, D.Sc. London: Alston Rivers, 1905.

WHEN, in October 1901, the subject of the South Pole was last discussed in this Review, two great expeditions had recently left European shores for the prosecution of antarctic research; and two others had just returned from those regions richly laden with new experience and interesting scientific results. Of the home-coming expeditions, the first had been fitted out in Belgium, and was commanded by Captain de Gerlache. The ship was appropriately called the "Belgica." Captain de Gerlache's companions were of many nations, but all of them were ardent explorers. The funds available for the expedition amounted to no more than £12,000, yet the results need not fear comparison with those of expeditions costing many times this amount. To the "Belgica" belongs the honour of being the first ship to winter within the antarctic circle; and she did so under circumstances of peculiar danger, being frozen in on the open sea far from all shelter of land. In this position the ship remained for over a year, when, with great difficulty, she forced her way out and returned home.

The other expedition was that of Mr Borchgrevink, a Norwegian, who had already visited the antarctic seas on a whaling expedition. He induced Sir George Newnes to fit out a scientific expedition on board a single ship, which landed him and his companions on Cape Adare in the northeast part of Victoria Land. For the first time a winter was spent on the antarctic continent; and the conditions obtaining on the land were ascertained, the meteorological record being of particular interest. This expedition was planned and carried out in a business-like way. A Norwegian whaler was purchased, fitted, and loaded with huts and everything required by a party landing on a desert and inhospitable coast and proposing to spend the winter there. Notwithstanding the exposed character of the coast where a landing was effected, and the frequent storms which impeded the work, everything had been put on shore, the party installed in their new dwelling, and the ship had started on her return journey in the space of a fortnight. At almost the same date of the next year she returned, took the party on board again, and steamed south, visiting the sheltered inlet of Wood Bay on the way to McMurdo Bay. She coasted the great barrier from Cape Crozier, and not only reached the highest southern latitude which had been reached until then by a ship, but she was able to moor alongside the barrier at a place where it had the height of a wharf or quay, and to land her party for a day's excursion on the ice where they reached the farthest south so far attained, viz. lat. 78° 50′ S.

A month before Captain de Gerlache and Mr Borchgrevink returned to Europe, the other two expeditions, to which reference was made above, had set out. One of these was on board the British ship "Discovery," Commander R. F. Scott, R.N.; the other on the German ship "Gauss," under the direction of Professor Erich von Drygalski. Both these ships were fitted with everything that ample funds could provide. They left Europe in August 1901. It had been arranged that the "Discovery" should explore the district to which the Ross Sea gives access, lying south of New Zealand ; while the "Gauss" should endeavour to proceed southwards in the neighbourhood of the 90th meridian of east longitude, where the "Challenger" had crossed the antarctic circle in 1874, and where Wilkes, in 1840, had seen the "appearance of land," to which he gave the name of "Termination Land."

But these four expeditions, of which two started and two returned in the autumn of 1901, do not exhaust the list. For some years previously Mr W. S. Bruce, who had considerable arctic and antarctic experience, had been endeavouring to fit out a Scottish expedition ; but lack of funds stood in the way. In the end this difficulty was removed, mainly by the liberality of Messrs James and Andrew Coats of Paisley. A Norwegian whaler, the "Hekla," was bought, and, under the generous direction of the late Mr G. L. Watson, she was made practically a new ship, which was named the "Scotia." The "Scotia" left Scotland on Nov. 2, 1902, and arrived at Port Stanley, Falkland Islands, after a smart passage of fifty-nine days. She left Port Stanley on Jan. 26, 1903, and spent two months making hydrographical and oceanographical explorations in that part of the antarctic ocean which lies south-west of the Falkland Islands, and bears the name of the Weddell Sea. The rapid approach of winter forced Mr Bruce to seek winter-quarters ; and he found them in a bay in one of the South Orkney Islands. Here the ship was frozen in for eight months ; and it is a remarkable fact that this happened in so low a latitude as 60°. During the winter a complete series of meteorological observations was taken ; hydrographical and geological surveys of the island were made ; and large collections of the land and marine fauna brought together. Mr Bruce made good use of the experience he had gathered in the expeditions of the Prince of Monaco, and prepared skeletons of nearly all the animals collected, by the Prince's method of sinking them in pots to the bottom of the sea, and leaving them there until the minute crustaceans had cleaned the bones of everything edible. Owing to this division of labour, the collection of skeletons brought home by the Scottish expedition is one of its most remarkable features.

On Nov. 25, 1903, the " Scotia" was set free by the breaking up of the ice. Unfortunately it was necessary for Mr Bruce to get into telegraphic communication with Scotland to obtain the credit necessary to refit and supply his ship for a second season. To do this he had to go to Buenos Ayres, which occasioned a serious loss of valuable time. But there was some compensation. Through the friendly co-operation of Mr Davis, the head of the astronomical and meteorological department of the Argentine Republic, the Government of that country was induced to interest itself in the expedition, and, besides contributing most generously to the material necessities of the " Scotia," it appointed three observers to return to the South Orkneys with Mr Bruce and continue the meteorological work which had been carried on under the immediate direction of the well-known Scottish meteorologist, Mr Mossman. When the " Scotia" returned to the South Orkneys, Mr Mossman agreed to remain there for another winter in order to organise the meteorological service of these interesting regions. This was the beginning of what promises to be the most important network of meteorological stations in the southern hemisphere.

But much valuable time had been lost ; and the exploratory part of the work of the " Scotia" began no earlier in 1904 than it had in 1903. Nevertheless the season proved to be so open in the Weddell Sea that a large amount of useful hydrographical and oceanographical work was done ; and the party were enabled to discover and delineate a portion of the antarctic continent, which was appropriately named Coats' Land. Here the " Scotia" was beset in the ice for a week. Fortunately she freed herself, and Mr Bruce started for the north, making a very important series of observations along a meridian to Gough Island, an outlying member of the Tristan d'Acunha group. He surveyed the island itself and then went home by the Cape of Good Hope. As it will be some time before the results of this expedition are in the hands of the public, further discussion of them would be premature.

Two other expeditions visited the antarctic regions south of South America at or about the same time as that of the "Scotia." One was from Sweden, under the direction of Dr Otto Nordenskjöld, and the other from France, under Dr Jean Charcot. Both these expeditions made important additions to our knowledge of the natural history as well as of the physical and meteorological conditions of these regions; but it is impossible here to do more than mention them. The rest of our article must be devoted to the doings of the "Gauss" and the "Discovery."

The German expedition was projected as a private venture, but early in 1899 it was taken over by the Government. The ship was built in the naval yard at Kiel, and was named after the great mathematician Gauss. The expedition was under the command of Professor Erich von Drygalski. The "Gauss" left Kiel on Aug. 11, 1901, coaled at St Vincent, and arrived at the Cape on Nov. 23. On Dec. 7 she left the Cape, and, after calling at the Crozet Islands, and spending some time at Kerguelen, she left that island on Jan. 31, 1902. The course was first laid for Heard Island, which lies about 250 miles south-east of Kerguelen. The settlement of sealers which was there when the "Challenger" visited the island in 1874 had disappeared; and the sea-elephants which they displaced had reoccupied the beach.

The first object after leaving Heard Island was to search for Wilkes' Termination Land. In February 1874 the "Challenger" had looked for it without success. It may be of use to quote what is said about it in the report of that expedition (*Narrative*, i. 405, 407):

"On the 25th [February], at 3 a.m., the wind having moderated to force 5, and the weather being fairly clear, sail was made towards Termination Land. As the vessel proceeded towards the pack, the berg was passed which had been fouled early on the previous day, the score on its surface made by the jibboom remaining well-defined, notwithstanding the heavy fall of snow....After getting clear of the pack at 11 a.m., the ship sailed along its edge until noon, being from 10 a.m. until that time within about fifteen miles of the supposed position of Wilkes' Termination Land ; but neither from the deck nor mast-head could any indication of it be seen. The limit of vision as logged was twelve miles ; and, had there been land sufficiently lofty for Wilkes to have seen it at a distance of sixty miles (which was the distance that he supposed himself off it), either the clouds capping it or the land itself must have been seen. If Wilkes' distance was over-estimated, then that of the 'Challenger' would be increased, and it may still be found ; but, as the expression in Wilkes' journal is, 'appearance of land was seen to the south-west, and its trending seemed to be to the northward,' and not that land was actually sighted and a bearing obtained, it is probable that Termination Land does not exist. Still it is curious that pack-ice and a large number of bergs should have been found in nearly the same position as by Wilkes in 1840 ; and this would seem to indicate that land cannot be very far distant."

This expectation was realised by the "Gauss." Having failed to find Termination Land in its reputed position, she steered west and then south through pack-ice carrying soundings of 1500 to 2000 fathoms. On the morning of Feb. 19 shelter from the wind and snow was sought under the lee of a large iceberg; and here a sounding was taken when a depth of only 130 fathoms was found. The remainder of the day was spent in the endeavour to drive the ship southward through the pack-ice and against the wind. Towards evening a swell from the southward was met which gave hopes of open water ; and these were fulfilled. The ice rapidly opened out, and Feb. 20 was spent cruising in a sea free from ice. The depth of the water had increased to 350 fathoms. The south-easterly wind, which blew with great violence, prevented much way being made. On the morning of the 21st the weather had improved and land was sighted. The photographs show a perfectly open sea with the land uniformly covered with ice. No bare land of any kind was visible. Everywhere the inland ice ended in a cliff which rose some 150 feet above the sea.

The reader will find it difficult to understand why Drygalski, when he had discovered new land with open sea in front of it, did not devote himself to exploring it in preference to any other work. It was legitimate geographical work, which would have afforded an opportunity for himself and his companions to refresh and recruit. During the fine weather of the early part of the day the "Gauss" could have steamed well up to windward, and might have found shelter under the lee of the land. Even if the land had proved unapproachable, the oceanographical and biological survey of the sea would have afforded profitable employment for several days under shelter. In order however to obtain magnetical observations, the ship ran some four or five miles to the north-west; then dredging was done, while the ship drifted farther out towards the pack-ice, and an easterly wind arose and rapidly freshened. The ship ran before it—she probably could not have made head against it—into the pack-ice through an opening between two edges. "I confess," says Drygalski, "that in passing between those edges I experienced serious misgivings." Still, a north-westerly course was in the direction of the open water, and he could naturally expect that with luck he would work through. In the night he tried to put back to the open water off the newly-discovered land, but the ship could make no way against the storm. It mattered not how her head might lie, she drifted with the ice. This went on hour after hour. About four o'clock in the morning the motion both of the ship and of the surrounding ice diminished, and in a short time everything stood still. The open water was not more than a mile distant from the ship; and it was naturally hoped that she would get free. But the "Gauss" had gone into winter-quarters, and she remained fast for a whole year, with open water almost always in sight from the mast-head.

This long period of confinement was spent in fairly comfortable circumstances. The "Gauss" was ice-bound some fifty miles from the edge of the continental inland ice; but the travelling over that distance appears to have been remarkably easy. Frequent excursions were made to that part of the continent where the Gauss-berg protruded. No expeditions further inland were made. This was not due to any difficulty in travelling over the ice, but chiefly to the fact that the permanence of the ship's winter-quarters was open to doubt. So far as can be gathered from the narrative, a succession of north-westerly gales might at any time have broken up the ice ; and then the position of a party no farther away than the Gauss-berg would have been precarious. For-tunately the east winds held, and the " Gauss " never moved ; indeed there was every possibility that she might have to pass another winter there.

When she did get free, in the middle of March 1903, there was heavy ice on all sides. An attempt was made to penetrate southwards on the route of the " Challenger," but the ice was too close ; and in the early days of April the ship bore up for the north, reaching home on Nov. 24, 1903. The work of a busy winter in an antarctic station will be looked forward to with the keenest interest. Drygalski does not anticipate, but in the concluding chapter of his narrative he indicates that the magnetic work was particularly fruitful in results, especially in connection with the displays of aurora, which were very frequent. A short summary of the meteoro-logical observations could, one would think, have been given without indiscretion ; and it would have interested even the least instructed reader.

The British national expedition, under Commander Robert F. Scott, R.N., sailed from Cowes on August 6, 1901, in the " Discovery," a vessel built especially for the expedition. The ship arrived at the Cape on October 3, and remained there until the 14th, when she left for New Zealand. During the passage she went as far south as lat. 62° 50′ S. in long. 139° E., only about 200 miles north of Adélie Land, discovered by Dumont d'Urville. The final departure for the south was from Port Chalmers, on Dec. 24, 1901. On Jan. 3 the ant-arctic circle was crossed and the ice-pack entered. This belt of pack-ice proved to be about 200 miles in width ; and the " Discovery " did not get through it until the 8th. On the same evening land was sighted. The weather was perfect ; and by the light of the midnight sun the blue outline of the high mountain-peaks of Victoria Land was seen far away to the south and west. The members of the expedition were astonished to find that, even at the great distance of more than a hundred geographical miles, they could clearly distinguish the peaks of the Admiralty range, discovered by Ross some sixty years before.

The ship's course was now directed to Robertson Bay, which is formed by the long gravelly spit which stretches northwards from Cape Adare. It was on this spit that the expedition sent forth by Sir George Newnes, and commanded by Mr Borchgrevink, spent their winter in 1896. On leaving Cape Adare and coasting southwards, the "Discovery" was destined to experience the might of the tidal currents of these regions, and the risk of encountering them amongst heavy pack-ice. Not having the advantage of steam, Ross was unable to explore this coast closely on account of the extensive pack-ice; but he mapped all the features of the high land. The "Southern Cross," with the aid of steam, was able to follow the coast pretty closely in 1897. The "Discovery" was still more fortunate in 1902, being able to approach some interesting places which were denied to Mr Borchgrevink. With the prevailing easterly and south-easterly winds and the westerly currents, this coast is constantly a lee-shore against which the pack-ice is apt to be pressed very close.

Contrary to expectation, the "Discovery" was unable to penetrate into Wood Bay, which had been reported by the "Southern Cross" to be capable of affording snug winter-quarters, with a considerable extent of land free from ice and snow at the base of Mount Melbourne. This fine mountain rears an almost perfect volcanic cone to a height of 9000 feet; and, standing alone with no competing height to lessen its grandeur, it constitutes the most magnificent landmark on the coast. It is shown in two beautiful photographs to be covered with snow to the summit, with, however, some bare patches of rock. South of this point the character of the Victorian coast changed; and very little snow was observed on the high mountains behind it. In a beautiful sketch by Mr Davis, master of the "Terror," preserved in the Hydrographic Office, the diminution of snow on the mountains south of Cape Washington is apparent. It commences, however, with Mount Melbourne itself, which is shown as bare of snow for at least two or three thousand feet from the summit. This suggests the possibility that the volcano may have been active shortly before the date of Ross' visit; and it would tally with the fact that Mount Erebus was in considerable eruption in 1841, though quiescent in 1902. Another remarkable feature of Mr Davis' sketches is that the smoke from Mount Erebus is depicted as travelling from east to west, while during the whole of the sojourn of the "Discovery" it was observed to travel in the opposite direction. Ross estimated that at each explosion the ejected matter was thrown to a height of 2000 feet above the summit; it may therefore have reached a region where the wind was from the east.

South of Cape Washington, miniature ice-barriers were met with, due to enormous glaciers, one over fifteen miles across, which thrust their snouts many miles out to sea. In Granite Harbour a safe anchorage was found, but it was too much shut off from the south to be selected at once as winter-quarters for the expedition. From Granite Harbour the "Discovery" reached over towards the great volcano discovered by Ross and named after his ship the "Erebus." In doing this, Captain Scott examined McMurdo Bay, afterwards known as McMurdo Sound, and formed the idea that he might winter there.

The "Discovery" then proceeded along the north coast of what was afterwards called Ross Island, towards the ice barrier *par excellence*. This feature must be reckoned as one of the wonders of the world. No excuse is therefore necessary for quoting the simple narrative of its discovery by Sir James Clark Ross in 1841. It is all the more important to do so, because there is some conflict between the evidence of Captain Scott and that of Sir James Ross, which is not to be decided off-hand on the mere basis of date.

On Jan. 28, 1841, Ross writes (*Voyage*, i. 218 ff.):

"As we approached the land under all studding sails, we perceived a low white line extending from its eastern extreme point as far as the eye could discern to the eastward. It presented an extraordinary appearance, gradually increasing in height as we got nearer to it, and proving at length to be a perpendicular cliff of ice between 150 and 200 feet above the level of the sea, perfectly flat and level at the top, and without any fissures or promontories on its even seaward face. What was beyond it we could not imagine; for, being much higher than our mast-head, we could not see anything except the summit of a lofty range of mountains extending to the southward as far as the 79th degree of latitude. These mountains, being the southernmost land hitherto discovered, I felt great satisfaction in naming after Captain Sir William Edward Parry.... Whether ' Parry Mountains' again take an easterly trending and form the base to which this extraordinary mass of ice is attached, must be left for future navigators to determine. If there be land to the southward, it must be very remote, or of much less elevation than any other part of the coast we have seen, or it would have appeared above the barrier.... The day was remarkably fine; and, favoured by a fresh north-westerly breeze, we made good progress to the E.S.E. close along the lofty perpendicular cliffs of the icy barrier.

"Jan. 29. Having sailed along this curious wall of ice in perfectly clear water a distance of upwards of one hundred miles, by noon we found it still stretching to an indefinite extent in an E.S.E. direction. We were at this time in lat. 77° 47' S., long. 176° 43' E....I went on board the 'Terror' for a short time this afternoon (29th Jan.) to consult with Commander Crozier and compare our chronometers and barometers. ...After an absence now of nearly three months from Van Diemen's Land, the chronometers of the two ships were found to differ only 4" of time, equal to a mile of longitude, or, in this latitude, less than a quarter of a mile of distance."

These quotations from Ross' *Voyage* show how careful he was about his observations. The position in which he lays down this part of the Barrier may therefore be accepted with absolute confidence as determined by one of the most experienced, accurate, and cautious officers, controlled and confirmed by the captain and staff of the consort ship. In questions of this kind, and as between the years 1841 and 1902, date counts for nothing in weighing evidence; the determining factor is the competence and experience of the observer. The " Discovery " had no second ship to act as control; and none of her officers had experience of polar navigation which could be compared with that of those serving on the " Erebus " and " Terror." When, therefore, other things being equal, there is any conflict between the

evidence of the "Discovery" and that of the "Erebus" and
"Terror" regarding the determination of geographical posi-
tions, we have no hesitation in abiding by those fixed by
Sir James Ross and his consort.

Captain Scott writes (vol. i. 171):

" Already there was a strong case against the Parry Mountains ; and
later we knew with absolute certainty that they did not exist ; it is difficult to understand what can have led such a cautious and trustworthy
observer as Ross to make such an error.   I am inclined to think that, in
exaggerating the height of the barrier in this region, he was led to suppose
that anything seen over it at a distance must necessarily be of very great
altitude ; but, whatever the cause, the fact shows again how deceptive
appearances may be and how easily errors may arise.   In fact, as I have
said before, one cannot always afford to trust the evidence of one's
own eyes."

As the height of Ross' mast-head would be at least
140 feet above the water, there can be no suspicion of over-
estimation when he gives the height of the Barrier as from
150 to 200 feet.   The observation of the "Discovery" that
the edge of the Barrier on the west lies further south than it
did in Ross' time confirms that made on board the "Southern
Cross"; and the estimates of these two expeditions agree in
making its height from 60 to 70 feet.   It is obvious that, if
twenty miles of the ice have disappeared, the first part to go
would be the cliff which Ross surveyed in 1841 ; and the
belief which Captain Scott expresses, that Sir James Ross
over-estimated its height, cannot be founded on direct observation.   The only legitimate conclusion which can be drawn
from these facts is that the Barrier ice-sheet in this region
at the present day has such a thickness, or is otherwise so
circumstanced, that it exposes above water a cliff having a
height of not more than 60 or 70 feet ; whereas, in the year
1841, when it extended some twenty miles farther out to sea,
it exposed a cliff of from 150 to 200 feet in height.

The distance from Cape Crozier to the nearest point on
the 79th parallel is 90 geographical miles.   Captain Scott
often remarks on the visibility in these regions of very distant
mountain-peaks, e.g. on his first view of the Admiralty range
at a distance of over 100 miles, and again when he saw at one

moment Mounts Melbourne and Monteagle with Coulman Island to the north and Mount Erebus to the south, "that is, an included range of vision of 240 geographical miles." If, then, there are mountains of the requisite height in the required direction, we must conclude that these were the mountains which Ross saw and named the Parry Mountains.

On his map accompanying *The Voyage of the "Discovery,"* Captain Scott lays down a range of very lofty mountains between the parallels of 78° and 79° S., and he specifies the following peaks with their heights, viz. Mount Lister, 15,384 feet; Mount Hooker, 13,696 feet; Mount Rooker, 12,839 feet; Mount Huggins, 13,801 feet; Mount Harmsworth, 9644 feet; Mount Speyer, 8913 feet; Mounts Dawson and Lambton, 8675 feet; and, further to the west and nearer to the Barrier, Mount Discovery, 9887 feet; Black Island, 3456 feet; Brown Island, 2750 feet; and White Island, 2375 feet.

White Island, Black Island, and Brown Island would be distant from Ross' position between 40 and 45 miles; Mount Discovery would be distant about 50 miles; Mount Huggins about 65 miles; and Mount Harmsworth about 95 miles; and all would bear from south-west to south-south-west. There is nothing in Captain Scott's map to show that they would not be visible on a clear day, so soon as the shoulder of Mount Terror was open to the south-west. Therefore we abide by our conviction that Ross was not mistaken when he reported having seen lofty mountains to the southward, reaching nearly to the 79th parallel; and we are convinced that the above-named peaks are some of those which he saw and named collectively the Parry Mountains. Lieut. Armitage, in his book, confirms this view. When navigating the ship along the Barrier he saw over its edge these mountains from the crow's nest; and he says they "were evidently the Parry Mountains of Sir James Ross."

On Jan. 23, 1902, the "Discovery" started on her cruise along the Barrier. On Jan. 29 Captain Scott gives his noon position as lat. 78° 18′ S., long. 162° 6′ W.; and he remarks that this position is an interesting one, being to the southward and eastward of the extreme position reached by Sir James Ross in 1842, whence he reported a strong appearance of land to the south-east. But this remark of Captain Scott's is inexact.

On Feb. 23, 1842, while approaching the Barrier from the north-west, Ross reports having passed a berg with a large rock on it, apparently about six feet in diameter, followed later by some bergs and pieces of heavy ice with numerous stones and patches of soil, which raised his expectation of sighting land to a high pitch. Ross noticed also that the appearance and character of the Barrier in this locality differed from that presented by the Barrier nearer its western end. Having arrived within a mile and a half of the Barrier, he hove to, in order to allow the "Terror," which had dropped behind, to come up, when an interchange of signals between the two ships took place. The latitude of the "Erebus" was 78° 8′ S., that of the "Terror" 78° 11′ S., the mean of which, 78° 9′ 30″ S., was adopted ; and this placed the face of the Barrier in lat. 78° 11′ S., in the long. of 161° 27′ W. The "Discovery's" noon position on Jan. 29, 1902 (78° 18′ S., 162° 6′ W.), lies west and not east of this position, the difference in longitude being 39′; and it is still further to the west of the extreme position reached by Ross. In this position Ross found the height of the highest part of the Barrier to be 107 feet, and observed that from this point it gradually declined for about ten miles to the eastward, where it could not have been more than 80 feet. Ross then made sail along the Barrier to the eastward until he came to the lower part of it above-mentioned, being about ten miles east of his previous position, and therefore about twenty-three miles east and north of the noon position of the "Discovery" on Jan. 29, 1902. On his arrival at this point, Ross says (vol. ii. 202) :

"We perceived from our mast-heads that it [the land] gradually rose to the southward, presenting the appearance of mountains of great height perfectly covered with snow, but with a varied and undulating outline, which the barrier itself could not have assumed. Still there is so much uncertainty attending the appearance of land, when seen at any considerable distance, that although I, in common with nearly all my companions, feel assured that the presence of land there amounts almost to a certainty, yet I am unwilling to hazard the possibility of being mistaken on a point of so much interest, or the chance of some future navigator under more favourable circumstances proving that ours were only visionary mountains.

"The appearance of hummocky ridges and different shades, such as would be produced by an irregular white surface, and its mountainous elevation, were our chief grounds for believing it to be land, for not the smallest patch of cliff or rock could be seen protruding on any part of the space of about thirty degrees which it occupied. I have therefore marked it on the chart only as an 'appearance of land.'"

As on Feb. 23, 1842, in the "Erebus" and "Terror," so on Jan. 29, 1902, in the "Discovery," all the appearances of the Barrier suggested the proximity of land. From his noon position on Jan. 29 Captain Scott steamed along the face of the Barrier, and he says (i. 178):

"Our course lay well to the northward of east; and the change came at 8 p.m., when suddenly the ice-cliff turned to the east, and, becoming more and more irregular, continued in that direction for about five miles, when it again turned sharply to the north. Into the deep bay thus formed we ran, and as we approached the ice which lay ahead and to the eastward of us, we saw that it differed in character from anything we had yet seen. The ice-foot descended to varying heights of ten or twenty feet above the water, and behind it the snow surface rose in long undulating slopes to rounded ridges whose height we could only estimate. If any doubt remained in our minds that this was snow-covered land, a sounding of 100 fathoms quickly dispelled it. But what a land! On the swelling mounds of snow above us there was not one break, not a feature to give definition to the hazy outline. Instinctively one felt that such a scene as this was most perfectly devised to produce optical illusions in the explorer, and to cause those errors into which we had found even experienced persons to be led."

A careful consideration of the positions of the ships as above discussed shows that on the evening of Jan. 29, 1902, the "Discovery" must have arrived at a position close to that attained by the "Erebus" and "Terror" on the evening of Feb. 23, 1842; and the report which each explorer furnished of what he saw can leave no doubt that they were both looking at portions, and probably identical portions, of the same landscape. To Captain Scott, therefore, belongs the honour of confirming Sir James Ross' discovery of land in this part of the south-polar regions, and of vindicating the trustworthiness and the caution of that great navigator.

The "Discovery" proceeded along the face of the Barrier, or rather the ice-edge which now represented it. On the evening of Jan. 30 small patches of bare rock were detected appearing through the icy covering of the distant high land. After this the ship wandered among ice and fog, but on Feb. 1 she got back to the position where the rock patches had been seen; and, the weather clearing up, a good view was obtained, not only of the coastal range, but also of what was probably the summit of a distant and lofty range of mountains. The "Discovery" could now return westward with the satisfaction of having not only confirmed the existence of land on the eastern side of the Barrier ice-sheet, but of having to a certain extent delimited it.

On Feb. 3 the "Discovery" entered the same creek as, or one in the immediate neighbourhood of, that in which the "Southern Cross" moored in 1897. While lying alongside the ice-wharf for twenty-four hours, the ship and wharf rose and fell together, so that the ice-sheet was afloat. As the depth of the water was 315 fathoms, it could not well be otherwise. Captain Scott makes the important observation that the surface current set into the Barrier and under the ice for a certain time, then turned and set out again to sea. It would be very interesting to know how far "inland" this flux and reflux penetrates. The surface of the ice is smooth and undulating; an extensive view of it was obtained from the captive balloon.

On Feb. 8, 1902, the "Discovery" was brought into the bay which was to be her winter-quarters; but the weather persistently declined to freeze her in. As a matter of fact the open season was only beginning. It was not until March 24 that the ice between the stern of the ship and the shore was strong enough to bear the weight of a man; and then the bow of the ship was in open water. Almost up to the date of the disappearance of the sun (April 20), open water frequently appeared outside that point. Indeed, the behaviour of the ice in the neighbourhood of Ross Island was at all seasons very capricious. One of the first expeditions undertaken after the return of the sun was to Cape Crozier, to deposit a record for the relief ship. On Oct. 12, corresponding to our April, the party arrived at the cliffs above that cape, from which they had an extensive view over the ice-bound sea. From the 13th to the 18th the party were confined to their tents by a blizzard, during which they were almost buried by the drifting snow. When they were able to quit their tents, they found that the Ross Sea, which before the storm had been frozen over as far as the eye could see, was now a sheet of open water. Not a scrap of ice remained in sight, excepting the small shelf immediately under the Barrier, which formed the breeding-place of the Emperor Penguin. It is very difficult to account for the phenomenon, unless the whole pack was moved bodily seaward. The movements of the ice and of the water in this district deserve close study.

As a centre of exploring expeditions the winter-quarters proved very advantageous. Besides many short expeditions to the nearer islands and channels, which furnished much useful information, the principal sledge-journeys in the first season were that to the farthest south led by Captain Scott, and that to the high plateau of inland ice to the westward led by Mr Armitage. In the second season the principal expedition was that of Captain Scott to the farthest west on the lofty plateau of the inland ice. All these journeys are remarkable achievements; and they show Captain Scott at his best, as a man of indomitable pluck and energy, who not only did the hardest work himself, but was able to get others to follow suit, and to do so willingly and cheerfully.

On his journey to the farthest south, Captain Scott started on Nov. 2, 1902, from the winter-quarters of the "Discovery" in lat. 77° 52′ S. Immediately to the southward lay White Island, Black Island, and the Minna Bluff, a long ridge stretching eastwards from Mount Discovery. These necessitated a detour over the ice to the eastward. About eight or nine miles off the extremity of Minna Bluff a station, called depôt A, was made. This was not only of great importance to Captain Scott's party on their return journey,

when, besides other misfortunes, they were nearly at the end
of their provisions, but it was the means of revealing the fact
that the ice in this district moves northward at the rate of
between 500 and 600 yards per annum. Further observations,
however, will be required before it can be accepted that the
Barrier ice-sheet has a general motion at this rate.

From depôt A the route continued southwards until lat.
79° 40′ S. was reached; the mountainous coast-line on the
west having been kept at a distance of about seventy miles.
From here the course was altered to south-west in order to
close the coast and, if possible, to land. A second depôt (B)
was established in lat. 80° 25′ S.; and here attempts to land
were made, but they were defeated by crevasses and other ice
disturbances. From depôt B the course was continued southwards in discouraging circumstances, against which few would
have been able to make head.

On December 28, the camp was pitched in lat. 82° 11′ S.;
and, although the actual "farthest south" 82° 17′ S. was
reached on the 30th, the weather both on the 29th and the
30th was thick so that no distant view could be obtained.
On the 28th Captain Scott writes (ii. 76):

"It is a glorious evening, and fortune could not have provided us
with a more perfect view of our surroundings. We are looking up a
broad deep inlet or strait which stretches away to the south-west for
thirty or forty miles before it reaches its boundary of cliff and snow-slope.
Beyond, rising fold on fold, are the great *névé* fields that clothe the
distant range; against the pale blue sky the outline of the mountain
ridge rises and falls over numerous peaks till, with a sharp turn upward,
it culminates in the lofty summit of Mt Markham....The eastern foothills
of the high range form the southern limit of the strait; they are fringed
with high cliffs and steep snow-slopes....Between the high range and the
barrier there must lie immense undulating snow plateaux covering the
lesser foothills, which seem rather to increase in height to the left until
they fall sharply to the barrier level almost due south of us. To the
eastward of this again, we get our view to the farthest south; and we
have been studying it again and again to gather fresh information with
the changing bearings of the sun. Mount Longstaff we calculate as
10,000 feet. It is formed by the meeting of two long and comparatively
regular slopes; that to the east stretches out into the barrier and ends in
a long snow-cape which bears about S. 14° E.; that to the west is lost

behind the nearer foothills; but now fresh features have developed about these slopes.

"Over the western ridge can be seen two new peaks which must lie considerably to the south of the mountain, and, more interesting still, beyond the eastern cape we catch a glimpse of an extended coast-line; the land is thrown up by mirage, and appears in small white patches against a pale sky. We know well this appearance of a snow-covered country. It is the normal view in these regions of a very distant lofty land, and it indicates with certainty that a mountainous country continues beyond Mount Longstaff for nearly fifty miles. The direction of the extreme land thrown up in this manner is S. 17° E.; and hence we can now say with certainty that the coast-line, after passing Mount Longstaff, continues in this direction for at least a degree of latitude."

"Instinctively" the reader feels (as Captain Scott felt on a previous occasion) "that such a scene is most perfectly devised to produce optical illusions"; and he will reflect that, however certain the explorer may be, it might have been prudent, in dealing with these great distances, to confine his report to "appearances of land."

On Jan. 13, 1903, depôt B was fortunately found. The provisions there picked up made the conditions in respect of food favourable; but the strength of the party was diminishing, and the health of one of their number—Lieutenant Shackleton—caused serious misgiving. The dogs had long ceased to be of any use for dragging, and had had to be sacrificed. A straight course was now made to depôt A, which was reached on Jan. 28, and the party joined the ship on Feb. 3. The journey had occupied 93 days; and during it 960 statute miles were covered. The credit which is due to Captain Scott and his companions in this journey can only be appreciated by those who read the account of it, and know something of what arctic travelling is.

During the whole of his journey Captain Scott travelled on the ice-sheet, which terminates northward in what is known as the Great Barrier. This is land-ice, not sea-ice. Through four and a half degrees of latitude it maintained the same level; and Captain Scott concludes that it is afloat. If this be so, then the Ross Sea stretches at least to latitude 83° S.; and, as there were no signs to the southward of a change in the character of the scenery, it is impossible to guess how much farther it may stretch. There was no appearance of this *mer de glace* being delimited by a coast on the east. In the latitude of the Barrier, which may be taken as 78° S., the ice lies between longitude 160° E. and 160° W. It extends therefore over 40° of longitude, or about 480 nautical miles. In latitude 83° the same distance in longitude would equal 290 nautical miles, a distance which would preclude the possibility of seeing land, even if of great height. The possibility of the existence of a deep inland sea, such as that discovered by Nansen in the north, with a depth of perhaps 2000 fathoms, is not excluded ; but the source, if a source be required, of the ice that forms the sheet which ends in the Great Barrier, becomes more and more puzzling the further south it is shifted. It may lie on the other side of the Pole ; for instance, on the southern declivities of Coats' Land. If the observed dislocation of depôt A is to be taken as any indication of the movement of the ice-sheet as a whole, the supply of ice must be enormous ; and, bearing in view the scarcity of precipitation in those high polar latitudes, it is almost impossible to imagine where the supply is to be found. The subject is full of difficulties, but all of them are fascinating ; and before long the solution of the problem will attract not one or two but many, who will have to thank Captain Scott and his brave companions, Dr Wilson and Lieutenant Shackleton, for having shown the way.

Not less remarkable than his journey to the farthest south was Captain Scott's expedition in the spring of 1903 to the high continental plateau behind the lofty mountains which bounded the view from the winter-quarters towards the west. In the preceding season an important expedition had been carried out in the same direction by his second in command, Mr Armitage, who performed a mountaineering feat which would daunt most Alpine guides. He took his expedition, dragging everything in the way of provisions and shelter for fifty-two days, on sledges up glaciers and over ridges never before trodden by man, to a height of 9000 feet, at a temperature generally about that which freezes quicksilver. Captain

Scott was able to improve on the road; but he had abundance of other difficulties to overcome, and he overcame them most successfully. While he did not attain any greater height than Mr Armitage, he pushed on beyond his turning-point, and travelled over the continental plateau at a height of 8000 to 9000 feet for a distance of 12° of longitude, or 150 nautical miles, preserving an average latitude of about 77° 45′ S. It will be long before this achievement is surpassed.

It has not been possible to notice the other expeditions made by the crew of the "Discovery"; but every member was busy, and contributed his best to the great fund of new knowledge which is the result of the two years' sojourn in south-polar regions. At the end of Captain Scott's book summaries of results of observations are given by himself and several members of the staff which throw much new light on these interesting regions, and at the same time raise many questions which it is not easy to answer.

The ice of the great tabular bergs was known to be vesicular, belonging rather to the *névé* than to the glacier type. The ice of the Great Barrier appears to be of this character. This sheet consists, for at least a considerable thickness below its surface, of snow, more or less consolidated and passing into *névé*. Excavation showed many thin sheets or crusts of solid ice intercalated with the snow; and in this respect it resembles the winter snow of the High Alps. To what extent consolidation takes place in the deeper layers of the ice is uncertain. Captain Scott's impression was that the mass must throughout contain large quantities of air, an impression supported by the examination of some ice taken from the bottom of an overturned berg. Theoretically this appears to us to be likely. According to Buchanan's theory the motion of a glacier under the influence of gravity is intimately connected with the melting and regelation, or generally the metamorphosis, of the grains of which it is composed, in a medium containing varying though minute quantities of dissolved matter. The variation of the dilution of the medium is accompanied by variation of its freezing temperature. When ice is removed from it by freezing, the grain immersed in it increases, and the freezing-point of the medium falls. When ice is removed from the grain by melting, the medium is diluted and its freezing-point rises. The effect produced molecularly by variation of dilution is similar to that produced mechanically by increase and diminution of pressure. The maximum size of the grain at any point in a glacier is roughly a function of its distance from the source, and is a measure of the amount of metamorphism which the ice has experienced (*Antarctic Manual*, pp. 93, 94). If the Barrier ice-sheet is a self-contained *névé* or *firn*, situated at, and to a great extent below, the level of the sea—and we think that Captain Scott's observations clearly point to its being so—it can have no motion in its mass under gravity, and cannot therefore develop the adult grain of the glacier. Its granular structure must remain rudimentary like that of the Alpine *névé*. Being already at the lowest possible level, glaciers cannot flow from it ; and its surplus material is dispersed as icebergs, which are thus generated directly by the *névé*.

It is remarkable that, in spite of the very low temperature continuously experienced by the expedition, in certain parts, as on the Ferrar glacier, there was at the beginning of January, locally at least, extensive melting of the ice. The Ferrar glacier appeared to be stationary. There must, however, have been at some time considerable motion of ice from the inner highland to the level of the sea, or to that of the Barrier sheet, in order to furnish the abundance of moraine matter which is found on Ross Island, covers White Island nearly to the summit, and is distributed all over the lower ice-flats in the vicinity of the "Discovery's" winter-quarters. When a glacier is moving, it develops of itself the heat which initiates and promotes the metamorphosis of the ice. No information as to the size of the grain of these glaciers is given ; and indeed it is not easy to obtain it in latitudes where the power of the sun, even at midday, is insufficient to disarticulate completely the grains of a mass of ice. The ice of the inland plateau seems to be an accumulation of snow more or less consolidated ; the annual increment, if any, is probably small. The continuous and violent westerly wind appears to keep it always on the move, and Captain Scott formed the view that the snow which is evaporated probably equals, and may exceed, that which is precipitated, so that the ice-covering is not increasing, and may indeed be diminishing. All the ice and snow on the antarctic continent represents water removed from the southern ocean by the westerly winds. As in other land masses exposed to moist winds, the first high land that the air meets deprives it of the greater part of its moisture. On passing farther inland it contains so little that it has almost none to deposit. As regards the supply of material, the Barrier ice-sheet seems to be more favourably situated than the inland plateau. According to Captain Scott's experience, it is frequently swept by warm snow-laden southerly winds, which must be the return northwards of the upper westerly winds, shown by the smoke from Mount Erebus to be a very constant feature at high levels.

It is interesting to notice that Mr Ferrar, the geologist of the expedition, has arrived quite independently at the conviction that a covering of ice, so far from being destructive, is an eminently conservative agent as regards the land surface beneath. This conviction was arrived at by the writer many years ago; but it was due to the contemplation of the far-reaching destruction produced by the warm moist air of equatorial regions on unprotected rock surfaces.

From all the observations there appears to be no doubt that there has been a great diminution of the icy covering of the land in a period the length of which there are no adequate data to determine; and it is argued that the climate of Victoria Land must have changed very much in the interval. Those who have had the opportunity of witnessing during their lifetime the enormous removal of ice from the surface of Switzerland without the occurrence, according to meteorological data, of any appreciable change in the climate, will not attach too much importance to this conclusion.

Captain Scott returned from his great journey to the farthest west on Christmas eve, 1903. During his absence much work had been done in preparing the ship for sea and in attacking the ice with saws. On Jan. 5, 1904, to the surprise of every one, two relief ships appeared ; but they were unable, owing to the fast ice, to approach within less than ten miles of the "Discovery." It was not until Feb. 17 that everything was ready and the three ships left McMurdo Sound. The journey home was effected without further incident.

To the selfish reader of Captain Scott's charming and instructive book the relief of the "Discovery" comes as a disappointment. Having followed him in his first, or apprentice's, journey to the farthest south, and seen how he every day gathered more and more experience of the work ; having then followed him in his next journey to the farthest west, and observed the remarkable development of his power of covering ground against difficulties, it is impossible not to regret that he was unable to deliver the master-stroke by following up his own pioneer work and going still farther south, perhaps to the Pole itself.

We think that it would have been legitimate for Captain Scott to take the view that his expedition had a quasi-warlike character. He was engaged, as Dr H. R. Mill puts it, in the siege of the South Pole. The attack of the fortress had to be delivered on land, and claimed the presence of the chief. The service of support and relief had to be conducted by sea, and fell naturally under the command of a subordinate. Such a scheme of division of labour would have offered many advantages. The costly ship would have continued to be active in the service of the expedition, which would then have become self-supporting ; and every department of it would have come under the immediate personal control of Captain Scott. The reappearance of the "Discovery" in Australian waters in April 1902 would have relieved the promoters of the expedition of the obligation to find a second ship, and would have been welcomed by the friends of all the members of the expedition, about whom some anxiety had begun to be expressed. Indeed there would have been no necessity to evacuate Ross Land at all ; for the members of the expedition could have been relieved and replaced, and the occupation continued, until the fortress had fallen. The public is never backward in supporting an enterprise when it has begun to show sure prospects of success.

CHEMICAL AND PHYSICAL NOTES

IT is unnecessary to frame instructions to the chemist and physicist with regard to the performance of routine operations. He must be fully instructed and practised in these matters before he leaves. Still less is it necessary to frame instructions for carrying out delicate operations such as the determination of the amount of carbonic acid, and of the permanent gases present in the sea-water, because these operations can only be carried out, in the way that would justify the expenditure of time, by an expert who had previously provided himself with all the apparatus and appliances which are necessary.

But there is another class of observations which it is very desirable to make. They are not routine observations; they may rather be termed observations by the way. Elaborate provision of instruments is not necessary. What is most important is to have a definite idea of the kind of observations that are wanted, and of the way to make them.

The following notes have reference to this class of work, which is perhaps the most fascinating, and may be the most fruitful that can be engaged in. They are for a large part reprinted from two papers, one of which[1] was originally drawn up with a view to an Antarctic Expedition, and was read at the Dover Meeting of the British Association in 1899; the other[2] is a paper which was read before the Royal

[1] 'On the Physical and Chemical Work of an Antarctic Expedition,' by J. Y. Buchanan, F.R.S. *Geographical Journal*, November, 1899.

[2] 'On Rapid Variations of Atmospheric Temperature, especially during *Föhn*, and the methods of observing them,' by J. Y. Buchanan, F.R.S. *Proc. R. S.*, vol. lvi. p. 108.

Society in May 1894, and consists of the daily record of "observations by the way," such as it is most important that we should have from the Antarctic Land and Ice. The paper is reproduced almost in its entirety, because its principal usefulness consists in small matters and minute details which would be eliminated in condensation.

Low Oceanic Temperatures: Ice at Sea and on Land, etc.[1]

One of the striking features of the ocean discovered by the "Challenger" expedition, was the extensive area of very cold water which occupies the bottom of the sea from the east coast of South America to the ridge which runs north and south in the meridian of the island of Ascension. Here the bottom temperature was found to be 32°·5 F. The existence of this exceptionally cold bottom water was discovered on the outward voyage in soundings near the Brazilian coast, so that the expedition was prepared to take up the study of it on the way home. This was done very thoroughly on a line from the mouth of the river Plate along the parallel of 35° to the meridian of Ascension. The depth of the water varied from 1900 to 2900 fathoms, and the distribution of temperature in the water was, roughly, a warm surface layer of perhaps 100 to 200 fathoms, then a thick layer of water of temperature about 36° F. down to 1600 fathoms near the coast, and to 2200 fathoms or thereabouts at sea. Here was a steep temperature-gradient falling away rapidly from 35° to 33° F., and more slowly to 32°·5 F. The occurrence of the steep gradient shows a renewal of the water, and therefore a current. The observations of the "Valdivia" show a similar distribution in lat. 60° to 63° S., with this difference—that the surface layer is colder than the intermediate one, which is itself colder than the former intermediate, being about 34° F. The bottom layer has as low a temperature as 31°·5 F. Unfortunately, there are not enough determinations of the temperature of the deeper

[1] 'On the Physical and Chemical Work of an Antarctic Expedition,' by J. Y. Buchanan, F.R.S. *Geographical Journal* for November 1899.

layers to indicate the gradient which separates the cold bottom water from the comparatively warm intermediate water. The recommendation, therefore, which the writer would make is, that in these regions temperature observations in the deeper layers should not be spared, and where there is water of exceptional coldness at the bottom, the position and steepness of the gradient which separates it from the overlying water should be accurately determined. Further, as the whole range of temperature to be dealt with in Antarctic water is at the most from 28° to 35° or 36° F., and therefore small differences of temperature are relatively of great importance, it is well to have the thermometers constructed specially for this work, the scale containing few degrees, but these wide apart. In the survey of the Gulf of Guinea in the "Buccaneer," the writer had such thermometers, and he regularly sounded with one thermometer at the end of the wire, and another usually 250 fathoms above it.

It is also of great importance to ascertain the density of this exceptionally cold bottom water. Near the coast of South America it was found in the "Challenger" to be very high, and this was confirmed by an observation of the "Gazelle" in the same locality. It is this density at constant temperature which decides whether a water can carry its surface temperature down to great depths, or whether it shall remain at the surface, and it is the annual range of temperature of such water which gives it its penetrating power. This was clearly set forth in a paper sent home during the first year of the voyage of the "Challenger," and published in the *Proceedings of the Royal Society*[1]. The highest surface densities are found in the North Atlantic, in the Trade-wind regions, and there the surface water has a higher density than any layer below it; consequently, when it is cooled in winter to the same temperature as the water immediately below it, it sinks through it, and in this way a high temperature is disseminated through the whole thickness of the water of the North Atlantic. In the eastern part of this ocean, the density and temperature

---

[1] 'Note on the Vertical Distribution of Temperature in the Ocean,' *Proc. R. S.* (1875), vol. clvii. p. 123.

of the bottom waters are sensibly increased by the "brining down" of the Mediterranean.

The common feature of Antarctic water found by all expeditions is the thick warm layer lying between a cold layer at the surface and another cold layer at the bottom. It is very important to trace these two cold layers Southwards, until they join and the warm intermediate layer has disappeared. Every particular connected with this will be of interest.

Freezing Temperature of average Sea-Water.—Sir James Ross, in his description of his voyage, frequently refers to 39° F. as the temperature of maximum density of all waters, and draws curious conclusions. Now it was well known before the date of his voyage that average sea-water continues to contract to a much lower temperature than 39° F. Indeed, its temperature of maximum density is below that of its freezing-point, which may be put at 29° F. A similar mistake is often made at the present day by geographical writers. Although everybody knows and recognises that sea-water freezes at a temperature below that of fresh water, and that this temperature is the lower the greater the quantity of salt contained in the sea-water, it is to some extent not known and to a great extent not recognised that pure ice, which when left to itself melts at 32° F., begins to melt in salt water at exactly the same temperature as that at which the same water begins to freeze. A piece of pure lake-ice immersed in average sea-water reduces its own temperature and that of the sea-water in its immediate neighbourhood to a temperature of roughly 29° F., which varies with the concentration of the resultant brine formed by the mixture of the sea-water with the pure water formed by the melting of the ice. An iceberg consists of pure land-ice, and if it is of sufficient thickness to reach the layer of warm intermediate water, its lower surface must be always melting at a temperature of about 29° F., and this temperature must in time be communicated to the body of the ice, if it did not have it before. But it must necessarily be at about this temperature, because it separates from the parent land ice after it has been pushed into the sea. If it had a temperature below 29° F., it would freeze the sea-water round it until it had got rid of its excessive cold; and if it had a temperature above 29° F., the sea-water round it would melt its ice until it had got rid of its excessive heat. The representative freezing temperature of sea-water is taken as 29° F., but it varies with the salinity. In this respect sea-water was found to agree closely with a solution of chloride of sodium containing the same percentage of chlorine. The subject was carefully investigated during the winter 1886–87, and the results were communicated to the Royal Society of Edinburgh, in a paper which was read on March 21, 1887[1]. The rule regulating the appearance or disappearance of ice in a solution of chloride of sodium is very simple. The number expressing the percentage by weight of chlorine in the solution expresses on Celsius' scale the depression of the freezing-point of the solution below that of distilled water, and by consequence the temperature at which pure ice begins to melt in the same solution. Thus the freezing-point of a solution of chloride of sodium, containing 1 per cent. of chlorine, is $-1°·0$ C.; if it contains 1·75 per cent. chlorine, its freezing-point is $-1°·75$ C.; and so on for concentrations not exceeding that of the saltest ocean water. Sea-water, the solid contents of which consist chiefly of chlorides, follows this rule approximately, but not exactly. The following table, from p. 133 of the memoir, is derived from twenty-five determinations made with the greatest care in sea-waters of different degrees of concentration and freezing at temperatures between $-0°·5$ C. and $-2°·22$ C.

Freezing temperature	$-2°·0$ C.	$-1°·5$ C.	$-1°·0$ C.	$-0°·5$ C.
Per cent. by weight of chlorine	1·940	1·445	0·963	0·475
Difference	0·060	0·055	0·037	0·025

From this we have the following approximate rule: *The number expressing on Celsius' scale the depression of the freezing-point of a sea-water below that of distilled water is found by adding 0·04 to the number expressing the percentage*

[1] 'On Ice and Brines,' by J. Y. Buchanan (1887). *Proc. R. S. E.*, vol. xiv. p. 129; also *Nature*, vol. xxxv. pp. 516, 608, and vol. xxxvi. p. 9. (Below, p. 130.)

by weight of chlorine in the same water. From these few
remarks, it will be seen that the mutual interaction between
ice, salt, and water must be taken into account in interpreting
the results of the sea-temperatures of the Antarctic.

Sea-Ice as met with in Polar Seas.—The above law refers
to ocean waters which contain not more than 4 per cent. of
dissolved matter. In the course of a Polar winter the sea
freezes to a thickness of 8 or 10 feet, and in proportion as the
ice gets thicker, the actions and reactions between ice and
brine and salt and water become more complex, and the law
of freezing is no longer so simple as that stated above.

Sea-ice, as it occurs in the Arctic ocean, has been
described in great detail by Weyprecht, in a work[1] which
should be included in the library of every Antarctic expedition. The "Tegetthoff," which was Weyprecht's ship, was
beset in the pack in lat. 76° 18′ N., long. 61° 17′ E., on
August 13, 1872. Twenty-one months later she was still
a prisoner in the pack, and had to be abandoned. During
all these months there was no lack of time or opportunity to
study sea-ice in all its forms and moods, and every line of
Weyprecht's book is of interest to the voyager in icy seas.
The matter is treated quite objectively. First the different
forms which the ice assumes and their origin are dealt with;
then ice-pressure, of which the "Tegetthoff" had sufficient
experience, and the nature of sea-ice in winter and in
summer, are described. In winter we have the formation
of the ice and its transformations under the combined
influence of cold and varying pressure; in summer, unfortunately for the "Tegetthoff," there was no opportunity
of studying the disappearance of the ice, but its transformations under the influence of melting and varying pressure are
described. After these detailed studies, we have a description
of the motions of the ice and of the water as observed in
North Polar regions, and speculations as to what may be
expected to take place in regions not then visited.

For the chemist and the physicist the following extracts,

[1] *Die Metamorphosen des Polareises*, von Karl Weyprecht. Wien: Moritz
Perles, 1879.

which describe the freezing of sea-water under severe cold, are of special interest:

Page 55.—The author is here describing the ice which surrounded the "Tegetthoff" during the winter. Referring to the openings which occur in it from time to time and without apparent reason, he says:

"If the cold is very intense, then such quantities of vapour rise from the water, so soon as it comes in contact with the air, that it looks as if a veil had been spread over the surface of the water. The masses of vapour which rise are so dense that it looks as if the opening of the ice were filled with hot water.

"This does not, however, last long. The evaporation of the water which furnishes the vapour removes heat from the water and assists the cold of the air in producing a covering of young ice. In a very short time the surface of the water begins to get thick, and threads like a spider's web run out from the edge of the old ice towards the middle. The covering which at first was thin and pasty acquires consistence and thickness; the production of vapour diminishes, and soon ceases.

"At this stage the salt-water ice is a pasty mass, which follows every surface movement of the water on which it floats. With increasing thickness this ice-mass acquires greater consistence and becomes tougher, but even with very intense cold it does not become sufficiently strong to bear the weight of a man with safety until after thirty to thirty-six hours. With a temperature of − 40° C. the new ice, even after twelve hours, is still so soft that, in spite of its thickness, a stick can be easily thrust through it. On December 13, 1872, the ice had in sixty hours attained a thickness of 20 centimetres, the temperature being − 35° C. But even with this thickness it is still in no way brittle, but is so pasty that it gives way under the weight of a man, without breaking. It gives the impression that one is walking on well-stretched leather, and it keeps this leathery character for a long time. Even after fourteen days, when its thickness is over half a metre, it does not break when exposed to moderate pressure, but crumples up with undulating folds. An expanse of young ice in this state looks as if the water, when in motion, had been surprised by the cold and every wave had suddenly been turned into ice.

"This persistent viscosity is caused by the large amount of salt which remains in the upper layers of the ice frozen by intense cold, and of the moisture which it attracts. In the formation of each ice-crystal, the salt is completely excluded. When the ice formation takes place very rapidly, under the influence of very low temperature, a large number of ice-crystals are formed in a very short time, and much of the salter brine remains entangled in them. Consequently the ice covering which first forms consists of loose ice-crystals, mixed with the brine from which they have been formed. As the cold continues, more ice is formed out of the brine, and the residual brine becomes more concentrated : but, however great the cold, the surface layers of sea-water ice never acquire the hardness or present the appearance of fresh-water ice. When the ice has acquired a certain thickness, the formation of ice at the lower surface takes place slowly, and the excluded brine disseminates itself at once in the water. The ice, so formed, is much freer from salt than that which was first formed on the water surface. The thicker the young ice becomes, the less influence does the comparatively warm sea-water have on the upper layers of the ice, and the lower does the temperature of the ice and the entangled brine fall.

" But as the brine is always in contact with ice, it is always at its freezing-point, which continually falls as the concentration of the brine increases. By this continual freezing process of brine, which is always getting more and more saturated, the liquid residue approaches more and more the point where it can sustain the greatest cold without freezing. On the surface there remains a very concentrated brine which keeps the ice moist day after day, and which gives it its pastiness. On walking over such a surface, so long as no fresh snow has fallen on it, one is astonished to find that every step one takes remains impressed on the white surface, and it is difficult, especially for a new-comer, to understand how what he takes for snow can be in a state of thaw at a temperature of $-40°$ C., and even lower. The moisture which collects in the foot-prints is, however, not water but a very concentrated saline solution, principally chloride of calcium, which in the course of time is absorbed into the ice.

" From the above description of the process of formation of sea-water ice, it is evident that when such ice is melted the saltness of the water produced will vary according as the ice has been taken from the surface layers or from lower down.

" The following determinations were made. The water formed by melting the white surface-ice which had taken thirty-six hours to form under a cold of $-33°·5$ C. had a specific gravity of $1·087$; and water from ice which had taken sixty hours to form under a cold of $-33°$ C. had a specific gravity of $1·076$, both measured at $+6°·2$ C. These measurements correspond to a salinity of $11·8$ and $10·0$ per cent. respectively. This ice was really the efflorescent surface skin.

" The specific gravity of the water produced by melting the uppermost 5 centimetres of the above ice along with the white surface skin was $1·017$ at $+19°·7$ C., that of the middle 9 centimetres was $1·009$ at $+11°·4$ C., and that of the lowest 5 centimetres was $1·008$ at $+16°·8$ C. These specific gravities correspond to salinities of $2·5$, $1·3$ and $1·2$ per cent. respectively. The average specific gravity of the sea-water is $1·025$."

Weyprecht assumed that the ice formed by freezing sea-water was pure ice, and that its saltness was derived from brine mechanically adhering to it, from which it was impossible to free it. There was no experimental evidence to prove this. Others thought that, as the ice formed by freezing salt water has a different melting-point from fresh-water ice, and it is impossible by melting it to produce water free from salt, the salt is an essential constituent of the ice. Experiments made by the writer[1] in the Antarctic area during the cruise of the "Challenger," supported the latter view, and he held it until subsequent experiments[2], which he made some twelve years later, enabled him to supply direct experimental evidence that the contrary is the case, that the ice-crystals are indeed pure ice, and that their saltness is due only to adhering brine from which it is impossible to free them.

Demonstration that the Ice produced by Freezing Sea-water and similar Solutions is pure Ice.—The principle guiding the research was, that if the ice, which forms when sea-water or a saline solution of similar concentration is partially frozen is pure ice, then pure ice of independent origin such as snow, must, when mixed with the sea-water or saline solution, melt at the same temperature as the ice which is formed by freezing the solution, when the concentration is the same. Sea-water and various saline solutions were experimented with on these lines, and the following was the scheme of experiment :—

The solution of a salt, for instance, chloride of sodium, was gradually cooled in a freezing mixture, and the temperature watched as more and more crystals separated out; at suitable intervals the temperature was accurately noted, and simultaneously a sample of the brine was taken; in this way a series of freezing temperatures t_1, t_2, t_3, etc., was obtained, and after analysis of the samples a corresponding series of salinities s_1, s_2, s_3, etc., was obtained. When the lowest temperature, say t_5, was obtained, the salinity was s_5; then the vessel containing the solution was removed from the freezing mixture, and it was exposed to the heat of the air of

---

[1] 'On Sea-water Ice.' *Proc. R. S.* (1874), vol. xxii. p. 431.

[2] 'On Ice and Brines,' by J. Y. Buchanan (1887). *Proc. R. S. E.*, vol. xiv. p. 129; also *Nature*, vol. xxxv. pp. 516, 608, and vol. xxxvi. p. 9. (Below, p. 130.)

the laboratory. The temperature of the mixture was observed to rise slowly, and when it arrived again at t_4, t_3, t_2 and t_1 respectively, samples of the brine were again taken and analysed. The resulting salinities s_4, s_3, s_2, s_1 were found to be identical with those observed at the same temperature during cooling.

Another solution of the same salt was now made, and of strength represented by salinity s_5. It was cooled down to close on t_5, and then snow was mixed with it. The vessel having been removed from the cooling bath was exposed to the heat of the laboratory, and the thermometer carefully observed as the temperature rose. When the temperature was exactly t_4, t_3, t_2, t_1, samples of brine were taken and analysed, and the resulting salinities s_4, s_3, s_2, s_1 were found to be identical with those observed in the two previous experiments.

It was thus shown that when a saline solution of concentration comparable with sea-water is gradually frozen, certain crystals which we call ice-crystals separate out, and during the process the temperature of the mixture gradually falls while its concentration increases. When the mixture of brine and ice-crystals is warmed the ice-crystals gradually melt, the temperature rises and the concentration diminishes; but when in the process of cooling and freezing the temperature has fallen to a certain point, say t, and the salinity is s, it was found that when the process was reversed and the same temperature t was reached during the process of warming and melting, the solution was found to have the salinity s. Therefore the substance, which forms the ice-crystals which separate out at a temperature t during cooling, melts again at the same temperature and concentration during warming. When the same solution, having the highest concentration which was used in the previous experiments, was cooled down and mixed with snow and then gradually warmed, it was found that the snow melted exactly as the ice-crystals formed in the solution itself had done, and at exactly the same temperature for the same salinity. But it is only a question of whether the salt is in the ice or in the brine. There is no salt in snow, and it behaves in a saline solution in exactly the same way as the crystals formed by freezing that solution; therefore the crystals formed by freezing the saline solution must be equally free from salt, and *it has thus been proved that the crystals formed in freezing saline solutions of moderate concentration are pure ice, and that the salt from which they cannot be freed does belong to the adhering brine.*

Analogy between Snow and Sea-Ice.—Snow is the result of the crystallisation of water-vapour dissolved in a gaseous mixture of nitrogen and oxygen. The freezing-point of this gaseous mixture is commonly called the "dew-point" of the air. The freezing of a saline solution is analogous. It is a homogeneous liquid, and the water is as much dissolved in the salt as the salt is in the water. When the solution is sufficiently cooled, ice separates in crystals from the liquid salt just as it did from the gaseous air, and *the freezing-point of a solution is in reality a dew-point.* The snow derived from these different sources has identical properties; and when real snow of atmospheric origin is mixed with a saline solution, it is as impossible to free it from salt and to get it to melt at 0° C. as it is to do so with the crystals formed by freezing a saline solution.

This matter has been dealt with at some length, not only because it is of importance for the Antarctic explorer, but because it is a matter of the highest importance in chemistry. The whole of that great branch of physical chemistry, which has the distinctive title of cryoscopy, depends on the fact that a saline solution, in freezing, yields pure ice; yet, in treatises on the subject, no adequate proof of this fundamental fact is offered.

Cryohydrates.—When a saline solution has been exposed to continued freezing, it finally acquires a concentration at which any further removal of water in the form of ice causes the precipitation of salt, because at the temperature attained the solution is saturated with the salt. If the cooling is continued the temperature remains constant, while ice and crystals of the salt separate out *pari passu*, and finally the whole solution may solidify to a porcelain-like mass, which has been called the cryohydrate of the particular salt. If warmed it will melt again at a constant temperature until all the salt has been dissolved, when the temperature will begin to rise. The case is quite analogous to the boiling mixtures[1] of constant temperature produced by blowing steam through salt, the temperature remains quite constant at that of the boiling saturated solution until nearly all the salt is dissolved, after which the temperature falls in proportion as the solution is diluted by the condensation of steam.

The temperature at which the cryohydrate forms has been called the cryohydric temperature. At this temperature crystals of ice and of the salt remain side by side without melting each other, and they behave with the same indifference to each other at lower temperatures. At temperatures above the cryohydric point they cannot be brought together without melting, when they produce the cryohydric temperature. The cryohydric temperatures of different salts differ. From the observations of Weyprecht, Nordenskjöld, and others, we learn that even at a temperature of $-40°$ C. there is liquid brine in the surface layers of the sea-ice; so that the cryohydric point of some of the salts in sea-water must be lower than this temperature. *The temperature of this mixture of ice and brine must be the freezing temperature of the brine and the melting temperature of ice in the brine.*

It has been said that above the cryohydric temperature crystals of the salt and of ice cannot be brought together without liquefaction. When air is saturated with moisture at temperatures below 0° C. the moisture is deposited as rime, which is ice. If the surface on which it is deposited is a soluble salt, and if the temperature is above the cryohydric temperature of the salt, liquefaction will take place at the point of deposit. Hence it follows that *every salt is deliquescent at temperatures between 0° C. and its cryohydric temperature.*

Numerical data regarding the freezing-points of saline

---

[1] 'On Steam and Brines,' by J. Y. Buchanan, F.R.S. *Trans. R. S. E.* (1899), vol. xxxix. p. 529. (Below, p. 151.)

solutions are to be found in collections of physical tables such as those published by the Smithsonian Institution.

The following table gives the freezing-points of saturated solutions of a number of salts, as recently determined with the greatest care by L. C. de Coppet[1], who was the first to prove and to clearly enunciate the relation which exists between the molecular weight of a salt and the freezing-point of its aqueous solution.

TABLE I. *De Coppet's Results.*

Salt dissolved	Formula	Amount of Salt dissolved in 100 grms. water	Freezing Temperature of saturated Solution
		grms.	°C.
Chloride of potassium	KCl	24·6	− 11·16
Chloride of sodium ...	NaCl	29·6	− 21·85
Chloride of ammonium	NH_4Cl	22·9	− 15·8
Nitrate of potassium ...	KNO_3	10·7	− 2·85
Nitrate of sodium ...	$NaNO_3$	58·5	− 18·5
Nitrate of ammonium	NH_4NO_3	70·0	− 17·35
Nitrate of barium ...	$Ba(NO_3)_2$	4·5	− 0·7
Nitrate of strontium ...	$Sr(NO_3)_2$	32·4	− 5·75
Nitrate of lead ...	$Pb(NO_3)_2$	35·2	− 2·7

The meaning of this table is that if, for example, we take 29·6 grms. of common salt (NaCl) and dissolve it in 100 grms. of water, which will take some little time, and we then expose it to an Antarctic temperature such as − 25° C. or − 30° C., the temperature of the solution will fall until it reaches − 21°·85 C., when it will remain stationary while crystals form in the solution. This crystallisation will proceed until the solution has become a white solid enamel-like mass. When it has completely solidified, its temperature will begin to fall again, and will continue to fall until it has reached the temperature of the air, − 25° C. or − 30° C., as the case may be. The white enamel-like mass is what was called by Guthrie the cryohydrate of chloride of sodium. Although

[1] *Zeits. f. Phys. Chem.*, vol. xxii. p. 239.

it is no longer believed to be a definite hydrate in the chemical sense of the term, still it is a mixture of NaCl and H_2O in a definite and constant proportion with a constant melting-point, and simulates a chemical compound so successfully that it is well entitled to retain its name of cryohydrate, all the more as it is convenient to have a special name to designate these mixtures. The cryohydrate of chloride of sodium is an intimate mixture of 29·6 grms. of chloride of sodium with 100 grms. of ice. It melts at the constant temperature − 21°·85 C., and when melted it will, if cooled, solidify again at the same temperature. Below this temperature chloride of sodium and ice are indifferent to each other; above this temperature they melt each other, and between this temperature and 0° C. chloride of sodium is a deliquescent salt.

Again, we see from de Coppet's table that the cryohydrate of chloride of ammonium (NH_4Cl) consists of 22·9 grms. of the salt and 100 grms. of ice, and this mixture can exist either in the liquid or the solid state at − 15°·8 C. In the case of chloride of sodium it was indicated that by making a solution saturated at ordinary temperatures (+ 15° C.) we obtained a solution of the concentration of the cryohydrate, which is a solution saturated at the cryohydric temperature (− 21°·85 C.). It is a remarkable property of common salt that its solubility in water is almost entirely unaffected by change of temperature. The law holds almost universally that salts are more soluble in warm than in cold water. This is the case with chloride of ammonium. At a barometric pressure of 742 mm., when steam is blown through the salt it forms a boiling mixture of salt and saturated solution having the constant temperature of 113°·8 C., the temperature of boiling distilled water under the same pressure being 99°·33 C. This boiling saturated solution contains 78·7 grms. of chloride of ammonium to 100 grms. of water, and on cooling to the ordinary temperature of 15° C. the amount of the salt remaining in solution is only 35·7 grms. to 100 grms. of water, the difference, 43 grms., has separated out as crystals. If the solution, saturated at + 15° C. and containing 35·7 grms. of NH_4Cl to 100 grms. of H_2O, be now exposed to a temperature of $-15°·8$ C., its temperature will fall gradually, until $-15°·8$ C. is reached. During the process of cooling a large amount of the chloride of ammonium crystallises out. It is well known—and the fact can be verified in a moment by making the experiment—that when crystals of chloride of ammonium are mixed with water of ordinary temperature, cold is produced. The cold experienced is the measure of the heat absorbed in the liquefaction of the salt by solution. When the process is reversed and the salt is deposited from the solution this heat is restored, and it is appropriated in the first instance by the solution which is cooling and crystallising, and *pro tanto* it diminishes the rate of cooling and of crystallising. At $-15°·8$ C. the saturated solution of chloride of ammonium contains 22·9 grms. of the salt to 100 grms. of water, so that 12·8 grms. of the salt have crystallised out. Had the solution been originally made to contain only 22·9 grms. of chloride of ammonium to 100 grms. of water at $+15°$ C., and it had then been exposed to the low temperature of $-15°·8$ C., the time in which the solution would have fallen to this temperature would have been much shorter, but the final result would have been the same—we should have an aqueous solution of chloride of ammonium of the cryohydric concentration and at the cryohydric temperature. If this solution were then exposed to a lower temperature, say $-20°$ C., it would lose heat, but its temperature would remain constant at $-15°·8$ C., and the equivalent of the heat removed would be apparent in the ice and salt which would separate out *pari passu* so long as heat was being removed and there remained anything liquid from which to remove it. So soon as the liquid has disappeared and the mass has become solid throughout, the further removal of heat is represented by a fall of temperature of the solid mass; the temperature will fall in time to $-20°$ C., and to any lower temperature to which the solid may be exposed. If it is then warmed by being placed, for instance, in a room having a temperature of $+15°$ C., its temperature will first rise to $-15°·8$ C., at which point it will remain stationary while the mass liquefies.

When the mass is all liquid the temperature of the liquid will rise rapidly until it finally assumes that of the room.

We see, then, that at temperatures below $-15°\cdot8$ C. chloride of ammonium and ice are indifferent to each other; above this temperature they melt each other, and at temperatures between $0°$ C. and $-15°\cdot8$ C. chloride of ammonium is a deliquescent salt.

It might be thought that, as the cryohydric temperature of chloride of sodium is $-21°\cdot85$ C. and that of chloride of ammonium is $-15°\cdot8$ C., the cryohydric temperature of a mixture of them in equal proportions would be somewhere near the middle, or about $-18°\cdot8$ C. But this is not so. The cryohydric temperature of a mixture of the two salts lies below that of either of the salts; and the same is the case even with the mixtures of the chlorides of sodium and of potassium, the cryohydric temperatures of which are so far apart.

The following observations (Table II) were made in

TABLE II. *Buchanan's Results.*

Snow Weight	Salt *a*		Salt *b*		Cryohydric Temperature
	Formula	Weight	Formula	Weight	
grms.		grms.		grms.	° C.
20	NaCl	32	—	—	− 20·7
32	KCl	38	—	—	− 10·7
34	NH₄Cl	22	—	—	−15·5
36	NaCl	11·7	KCl	14·9	− 22·7
42	NaCl	11·7	NH₄Cl	10·7	− 24·5
—	BaCl₂2H₂O	25	—	—	− 7·6
48	Ba(NO₃)₂	27	—	—	− 0·7
73	BaCl₂2H₂O	12·2	Ba(NO₃)₂	13·07	− 8·3

the Engadine on September 20, 1897, when a heavy snow-fall occurred. The temperatures were all observed with the same thermometer and no correction has been applied. The temperature marked $-0°\cdot3$ C. in melting snow, which makes the determined cryohydric temperature of nitrate of barium agree with that found by de Coppet, namely $-0°\cdot7$ C. The thermometer was one of ordinary German manufacture divided into whole degrees. The observations illustrate the fact that the cryohydric temperature of a mixture of two salts is lower than that of the salt which separately has the lower cryohydric temperature of the two. The mixtures experimented on were of equal molecular proportions, in fine powder, intimately mixed with *dry* powdery snow. Perhaps the most remarkable mixture in the table is that of the chloride and nitrate of barium : the cryohydric temperature of the mixture is nearly the sum of the separate cryohydric temperatures. In almost every case where accurate observations have been made, it has been found that the melting temperature of a mixture is lower than that of the most fusible of its components. The frequency of this observation has justified its enunciation as a law, and the recognition and elaboration of this law have done much for the advancement of physical chemistry. This is not the place to enlarge further on it in its general aspect, but in its particular application to the cryohydric and to the freezing temperatures of solutions of salts and of mixtures of salts, it is of great interest to the scientific members of an Arctic or Antarctic expedition. In recent years workers in physical chemistry have mainly limited themselves to the study of very dilute solutions, and the behaviour of concentrated and saturated solutions when exposed to great cold has received comparatively little attention. Consequently the chemist and physicist of the Expedition has a comparatively unoccupied field before him, and he is free to choose the simplest problems.

In our country the opportunities for the exact study of freezing mixtures are rare, because it seldom snows at all, and when it does, the snow is flaky and at or about its melting-point. Crystalline powdery snow, which has never experienced its melting temperature, is a rarity even on our mountains. This is the only form in which ice should be used for freezing or cryohydric mixtures, when these are to be studied exactly. When a freezing mixture is made with salt and ordinary moist snow, the salt forms little local freezing mixtures where it comes into actual contact with a snow-crystal, and the cold produced freezes the moisture adhering to contiguous snow-crystals, with the result that an altogether impracticable and heterogeneous mass of lumps of ice and masses of salt is formed. In the Antarctic regions cold powdery snow will be common enough, and exact experiments on the temperature and concentration of the freezing mixtures which it makes with different salts, and especially with definite mixtures of salts, will be easy, and cannot fail to be interesting.

The salts must be pure, dry, and in fine powder. When mixtures are used they should be in simple molecular proportions, as for instance, $NaCl + NH_4Cl$, $NaCl + 2NH_4Cl$, $3NaCl + 2NH_4Cl$, and so on. The weighed quantities of the two salts must be thoroughly mixed in a mortar before they are brought together with the snow. The experimenter must feel assured that there is no risk of his freezing mixture consisting of an indefinite association of local freezing mixtures, of, for instance, snow and $NaCl$ and of snow and NH_4Cl, but that it is certainly a homogeneous mixture, every element of which consists of three bodies.

The dry snow has necessarily a low temperature. The salt, or mixture of salts, when finely pulverised and intimately mixed, should be cooled in a stoppered bottle to a temperature below $0°$ C. When the cold dry salt is mixed with the cold dry snow, and the temperature, although low, is still above the cryohydric temperature of the mixture, and the temperature of the air when the mixture is made is also above the cryohydric temperature, the temperature of the mixture falls smartly to and stops abruptly at the cryohydric temperature. Even when exposed to the ordinary temperature of an inhabited room, such a mixture, if prepared with any care, will maintain a perfectly constant temperature for a length of time depending on the mass of the mixture made. The determination of the cryohydric temperature is thus quite simple, and its exactness depends on that of the thermometer, and on the certainty with which it may be

affirmed that the temperature of the thermometer is identical
with that of the mixture.

The determination of the concentration of the cryohydric
brine produced presents no difficulty, or only such as a
chemist knows how to deal with, and as it is an operation
which cannot profitably be attempted by any except a trained
chemist with considerable laboratory experience, it need not be
further described. The chemist may, however, be reminded
that he has the choice of two ways of approaching the subject
—the analytical and the synthetical. Further, he may either
take his freezing mixture and allow it to melt, or he may
make his presumed cryohydric solution, and allow it to freeze.

Cryohydrates of Salts forming Isomorphous Mixtures.—In
the study of the cryohydric constants of mixtures, a peculiar
interest attaches to mixtures of isomorphous salts which form
mixed crystals, such as the nitrates of barium, strontium and
lead, the nitrates of potassium and sodium.

We have seen that in the case of salts which, although
they crystallise in the same system, do not form mixed
crystals, as for instance NaCl and KCl or NH_4Cl, a mixture
of any two is more soluble than either separately, and consequently the cryohydric temperature is lowered. It is easy
to imagine why this should be. Looking to the analogy
between salts in solution and gases, a certain mass of water
when saturated with, say NaCl, is still virgin with regard to
KCl which dissolves in it with ease. But the introduction of
KCl interferes with the free meeting of the particles of NaCl,
which on the slightest lowering of temperature are prepared
to unite and fall out as crystals. Therefore the effect of
introducing this indifferent body KCl is to make the solution
of NaCl, which was saturated, no longer saturated : in other
words it depresses its temperature of saturation. But the
cryohydric temperature of a saline solution is the freezing
temperature of the saturated solution. This has been lowered :
therefore we see that the cryohydric temperature of a mixture
of NaCl and KCl must be lower than that of NaCl alone; and
it is quite independent of whether the cryohydric temperature
of KCl is high or low.

Imagine now that KCl and NaCl not only crystallise in the same system, but are isomorphous in the restricted sense, which is characterised by the formation of "mixed crystals." It is then obvious that in the saturated solution, the particle of NaCl when prevented from uniting with another particle of NaCl to form a crystal, will simply unite with the particle of KCl which stands in the way, and will crystallise. Consequently, if the solution of NaCl were originally not quite saturated, the introduction of the KCl, by increasing the amount of crystallisable material, would make it saturated. Hence, if KCl and NaCl were isomorphous and formed mixed crystals, the cryohydric temperature of a mixture of the two salts would be higher than that of the NaCl.

The nitrates of barium, strontium and lead are isomorphous salts which form mixed crystals. In boiling mixtures, it has been shown[1] that nitrate of strontium raises the condensing temperature of steam by $6°\cdot53$ C., and nitrate of lead raises it by $3°\cdot29$ C., while the mixture $\dfrac{Sr_2 + Pb}{3} (NO_3)_2$ raises it by $5°\cdot98$ C.; and the quantity of condensed steam required to produce the boiling saturated solution of the mixture is exactly the sum of the amounts required for the ingredients separately. In the case of the nitrates of strontium and barium, the elevation of the condensing point is not as great as with $Sr(NO_3)_2$ alone: the maximum temperature does not remain constant for a minute, and the condensed steam required to dissolve the mixture is about 25 per cent. more than is required to dissolve the salts separately.

Similarly in the case of freezing mixtures, the following cryohydric temperatures were observed: nitrate of barium $-0°\cdot7$ C., nitrate of strontium $-5°\cdot75$ C., and nitrate of lead $-2°\cdot7$ C.; and the cryohydric temperatures of pairs of these salts in equal molecular proportions: nitrates of strontium and barium $-5°\cdot73$ C., nitrates of strontium and lead $-5°\cdot23$ C., and nitrates of lead and barium $-2°\cdot53$ C.

The case of the isomorphism of salts which form mixed

[1] 'On Steam and Brines,' by J. Y. Buchanan, F.R.S. *Trans. R. S. E.* (1899), vol. xxxix. p. 547. (Below, p. 151.)

crystals is a feature of salts and solutions which has no analogy in the physics of gases.

The following table gives the temperature at which ice melts in solutions of various chlorides. The concentration of the brines is indicated by the percentage of chlorine by weight in the solution. It is taken from the writer's paper 'On Ice and Brines.'

TABLE III.

Temperature at which Ice melts in Brine	Name of Salt dissolved and Percentage of Chlorine in the Solution			
	HCl	KCl	NaCl	CaCl$_2$
°C.				
− 35	15·26	—	—	—
− 30	13·98	—	—	15·97
− 25	12·60	—	—	14·47
− 20	11·00	—	—	12·65
− 15	9·17	—	11·10	11·29
− 10	7·02	—	8·40	8·93
− 5	4·15	—	4·72	5·65
− 4	3·41	—	3·87	4·67
− 3	2·68	3·00	3·02	3·70
− 2	1·85	2·00	2·02	2·70
− 1	—	1·02	1·02	1·50

In dealing with sea-water ice, it is well for the chemist and physicist to confine his attention in winter-quarters to ice which he has seen freeze and with the whole of whose history he is himself acquainted. *Old* sea-ice is nothing but a curiosity. Every lump of it in a pack has a different composition, and when the composition of a hundred lumps is known, unless their history is also known, it does not really advance our knowledge.

Everything connected with the natural history of *young* ice—its birth, its growth, and its decay—is of interest, and its study, in the light of the foregoing remarks, will afford continual and interesting occupation.

Land-Ice and the Mechanics of Glaciers.—If ice is an important feature of the sea in Antarctic regions, it is a still more important feature of the land, and it should be the object of careful observation by the landing-party. The subject is a large one. The longer one studies ice the more one finds there is to learn about it, and the physicist or chemist who takes part in the Expedition should miss no opportunity of studying it in all directions. In order to do so with effect, he ought to have made preliminary studies of glaciers in Switzerland, where he finds every facility to his hand, and these studies should be made in winter as well as in summer.

Another educational preliminary for the members of a land expedition is to acquire as much skill as possible in ski-running and the method of travel adopted by Nansen in crossing Greenland. For this purpose a short visit to Norway in winter would be useful. The way into the Antarctic interior will almost certainly be over land-ice, and if it is the ice which is the parent of the great tabular bergs so well known from illustrations, it is probable that travelling over it will not be very difficult in so far as the nature of its surface is concerned. It has long been known that the glaciers of Greenland travel much more rapidly than those of Switzerland, but it is only since the publication of Drygalski's remarkable observations[1], carried on throughout the year on the glaciers of the west coast of Greenland, that we know that in some cases the motion of the ice reaches the astonishing rate of 18 metres in twenty-four hours, and that this rate of motion is very little affected by the change of season. According to Drygalski, what chiefly affects the motion of a glacier is its mass. Great as are the glaciers of Greenland, there can be little doubt that the parents of the Antarctic tabular icebergs are many times greater. If conditions such as these exist on the Antarctic land, it is little wonder that the supply of tabular icebergs is so abundant. Dr Arçtowski has described how, on the occasions when he landed on the rugged coasts visited by the " Belgica," and the weather was calm, the thunder of falling ice was continuous.

[1] *Grönland Expedition*, 1891–3, *unter Leitung von Erich von Drygalski.* Berlin, 1897.

On Heard Island, during the short visit which the writer was able to pay it from the "Challenger," the fall of ice from the western portion was also nearly continuous.

Since the days of Hugi and Agassiz, the intimate structure of glacier-ice has been the object of much study by Continental and chiefly Swiss naturalists. Englishmen, though they frequent glaciers as much as any other nation, have generally ignored it. Tyndall, to whom we owe so much of our knowledge about ice, recognised the existence of the *grain* of the glacier, as it is called, but made no use of it in his speculations with regard to the nature of the motion of glaciers. As his theories are independent of this fundamental feature of the constitution of glacier-ice, they must be *pro tanto* incomplete. The colour of the surface of a glacier, so dazzling in its whiteness that the inexperienced beholder is apt to suppose it covered with freshly fallen snow, is due to the disintegration of the compact blue glacier-ice into its constituent grains under the influence of the radiation of the sun. There is no more instructive or more impressive experiment than to expose a block of compact blue ice taken from the interior of the glacier to the direct rays of a powerful sun. Such a block is easily obtained by penetrating into the grotto, from which the glacier stream issues, to such a distance that direct sky-light is shut out. Any of the blocks found there will do, and it is to be brought out and exposed on a rock. In twenty minutes or half an hour the block falls down into a heap of irregularly shaped pieces of ice, each of which is a grain and a single crystalline individual. In higher latitudes or in dull weather, the power of the sun is not sufficiently strong to effect this complete and striking dissolution, but it loosens the block into its grains, which will rattle if the block be shaken.

The writer has analysed blocks of ice from many Swiss glaciers, and weighed the individual grains. They are of all weights up to a certain maximum, which varies with the glacier and the part of it furnishing the ice. The largest that he met with was from the Aletsch glacier, and it weighed 700 grams. It is the size or weight of the largest grains that it is important to determine. Small grains are abundant in every glacier, and are a necessity in order that the larger ones may pack close. The size of the largest grains is what is referred to when we read that the grain of this glacier is large or of that one small. The shape of the grain is irregular, and no two of them are alike, but they fit into each other like a puzzle. They resemble a collection of vertebræ more than anything else. Indeed, if the disarticulated vertebræ of an animal, especially one with a long tail, were carefully packed into a box of a suitable size, so as to occupy the least possible space, the boxful of vertebræ would resemble the block of ice which has been loosened by a moderate sun, and would rattle, when shaken, in much the same way. If gelatine were allowed to run into the box and set, we should have a model of the block of ice before exposure to the sun. If the box of vertebræ were exposed to the sun, the gelatine would be liquefied, and the mass would be loosened as in the case of the ice. What is it in the block of ice which corresponds to the gelatine in our illustration with the vertebræ? It is the slightly impure water which surrounds the grains and in which they *float* or try to float. Under the influence of cold, this impure water supplies pure ice to the grain with which it is in contact, while its freezing-point continually falls; finally its freezing-point and the temperature to which it is exposed reach a minimum, and the grain remains in contact, even in mid-winter, with a film of brine, which may be very minute. With rising temperature the grain begins to melt at the temperature at which it ceased to freeze, it dilutes the brine, and raises its own melting-point. We see, then, that the grain of the glacier may be surrounded in summer by a relatively considerable envelope of water of comparative purity and high freezing-point, while in winter it is surrounded by a mere film of brine of comparatively low freezing-point. The greater the amount of solid matter dissolved in the water, or the greater its salinity, the greater is the amount of liquid surrounding the ice-grain at a given temperature. The salinity of the water does not require to be very great for it to furnish, when frozen, an ice which is

not hard and brittle but soft and yielding, and this is seen
when the sea freezes in polar regions. *The difference between
land-ice and sea-ice is one of degree and not of kind. Sea-ice
is mixed with much brine and flows easily: land-ice contains
little brine and flows with difficulty.*

A ship floats in the smallest basin as perfectly as in the
largest ocean. We can imagine a dock being built round
a ship, and so exactly moulded to its shape, that between the
inner surface of the dock and the outer surface of the ship,
the clearance shall be so small that a pitcher of water poured
into it will float the ship. The floating of a grain in the
inside of a glacier is of this kind, and as it is enclosed on all
sides, it will press against the ice above it in preference to
that beneath it.

This feature of glacier-ice permits us to understand how
glaciers can move, and begin to move, even when their
temperature is very low. Before von Drygalski's work on
Greenland, we had no trustworthy information regarding the
temperature throughout the year of the inner mass of the ice
of any glacier. He carried out, at regular intervals of time,
a series of observations of the temperature at different depths
below the surface of one of the Greenland glaciers, and parallel
observations within the ice covering a neighbouring lake.
These observations showed that the temperature of the glacier
increases rapidly from the surface downward, and they render
it probable that the greater part of the thickness of a Greenland glacier is, even at the coldest time of the year, at or near
the ordinary temperature of melting ice[1]. The heat required
to support this temperature can only be supplied by the
friction of the grains of the ice, called into being by the
motion of the glacier. The *inland ice* which forms the great
reservoir for the supply of the glacier was found to have little
or no appreciable motion. Series of temperatures were not
taken in its thickness, but, in the absence of motion, we may
believe that the very low temperature at its surface penetrates
far into its interior, if not to the very bottom. If the motion

[1] Compare Hugi, *Ueber das Wesen der Gletscher und Winterreise in das
Eismeer*, p. 51.

B. 4

were dependent *only* on the lowering effect of pressure on the melting-point of pure ice, it would be difficult for such a mass of ice to *start* when it arrives at an outlet. The impurity of all natural water, and the effect which it has in lowering the melting-point of ice at ordinary pressures, remove this difficulty. However compact and solid the blue ice may look, there will always be *some* brine between its grains which will permit some yielding of its mass, which in its turn will produce a first generation of heat; this will produce a further yielding, and in due course a further generation of heat, and the effect of this initial agency, when combined with the powerful effect of fusion and regelation under conditions of very slight variations of pressure, is the extraordinary rate of motion observed in the glaciers of Greenland.

It has been pointed out that the whiteness of the surface of a glacier is due to what may correctly be termed sun burning or sun weathering. The icebergs which are met with at sea have an equally white surface; but where the interior is exposed, either in crevasses or in caves melted out by the waves, the deep blue colour of the fresh ice is visible. It is obvious that the whole of the surface of the glacier which is immersed in water at greater depth than that to which the sun's rays can penetrate must have the same blue colour, and it is equally obvious that when an iceberg turns completely over, it must stand out as an intensely deep blue mountain of ice among the multitude of sunburnt white ones. On one of the fine days during the sojourn of the "Challenger" in Antarctic waters, a striking and magnificent example of this was observed, but the cause of the blueness of the strange berg was quite unsuspected. If ice were collected by bombardment or otherwise, from such a berg, the grain would be large and well developed, and the ice would be quite compact and free from vesicles. In the region visited by Dr Arçtowski, the glaciers and the icebergs were comparatively small and of an Arctic character. The distribution of snow, *névé*, and ice he describes as being similar to that in the Alps at a height of 3000 to 4000 metres. There appeared to be very little melting, yet the glaciers advanced steadily towards the sea.

In the "Challenger" the writer observed at least one large tabular berg, which was melting freely on the top, and streams were cascading down the sides. In Spitzbergen the glacier streams often take large proportions; it will be interesting to know if in equally high Southern latitudes there is similar melting under the influence of the long polar day.

The Determination of the Temperature of Saturated Steam, and the Production of Higher Fixed Temperatures by the Condensation of Steam on Salts and in Saline Solutions[1].

The method of determining the pressure of the atmosphere by the temperature of saturated steam has for some time ceased to be in common use. Yet it is one which is very simple in its practice and accurate in its results. In the course of an investigation into the boiling-points of aqueous solutions, the writer had frequent occasion to determine the boiling-point of pure water, or, more properly, the temperature of saturated steam at the barometric pressure of the moment; and he has arrived at a form of apparatus which gives this with great exactness and constancy, no matter how delicate the thermometer used may be.

The instrument generally used for fixing or verifying the boiling-point of a thermometer is that of Regnault, which is described and figured in nearly every treatise on physics, e.g. Balfour Stewart's *Lessons in Elementary Physics*, 1870, p. 151. For thermometers having their boiling-point not too near the bulb, this instrument gives fairly trustworthy results. There are, however, several disadvantages. First, the stem of the thermometer is not *wholly* immersed in the steam, and there is uncertainty about the temperature of the mercury in the portion of the stem inside the cork and projecting beyond it. It is disadvantageous and unnecessary to make the steam space of the apparatus of metal, and to dispose it in the form of an inner space and surrounding jacket. The want of

---

[1] Chiefly a reprint of an article with the above title in the *Scottish Meteorological Journal*.

transparency of the metal is an obvious disadvantage, and the jacketing is unnecessary, because the latent heat of steam is so great that, with the supply of it which the boiler of Regnault's apparatus can furnish with any efficient lamp, a steam vessel of single envelope, under ordinary conditions, cannot be cooled by even a fraction of a degree below the temperature of saturated steam corresponding to the existing barometric pressure. Any cooling of the outside surface of the envelope is stopped at once and perfectly by the film of water continually descending along its inner surface, while the inner surface of the film is freely exposed to an abundant supply of saturated steam.

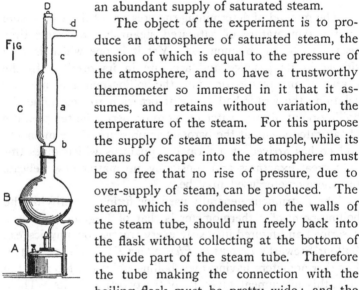

The object of the experiment is to produce an atmosphere of saturated steam, the tension of which is equal to the pressure of the atmosphere, and to have a trustworthy thermometer so immersed in it that it assumes, and retains without variation, the temperature of the steam. For this purpose the supply of steam must be ample, while its means of escape into the atmosphere must be so free that no rise of pressure, due to over-supply of steam, can be produced. The steam, which is condensed on the walls of the steam tube, should run freely back into the flask without collecting at the bottom of the wide part of the steam tube. Therefore the tube making the connection with the boiling flask must be pretty wide; and the exit tube from the steam vessel must be a trifle wider still. Of the steam which enters the tube, part is condensed on the walls and keeps them at constant temperature, and the remainder passes away in a stream of good volume through the exit tube. With very little attention to the construction and management of the apparatus, every risk of cooling from without or heating from within is completely avoided.

Description of the Apparatus.—A general view of the apparatus is seen in Fig. 1. It consists of four parts: the lamp *A*, the steam generator *B*, the steam vessel or distilling tube *C*, and the thermometer *D*. The lamp shown is one of a French pattern, sold with the *Réchaud à double flamme forcée* of smallest size. It holds about 250 c.c. of spirit, and gives a flame powerful enough to work a much larger flask. The steam generator *B* is a flask made of spun copper, and of 500 c.c. capacity. A suitable charge is 300 c.c. of water. With such a charge, and heated by the French lamp, the water boiled in six minutes at an expense of 12 grms. of spirit. While keeping steam at the rate suitable for the experiment the lamp consumed 21 grms. of spirit in fifteen minutes, and evaporated 92 grms. of water.

It is obvious that where gas is available a gas lamp may be used.

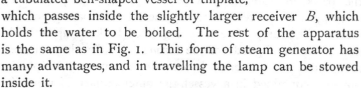

A very convenient lamp, especially for work out of doors, is the Swedish lamp for burning petroleum or paraffin oil under pressure, and with a blue flame. It is particularly useful when larger steam generators are used, such as the metal flasks made for the Napier's coffee machines, used in restaurants. One of the ordinary size holds 2 litres, and is a most useful article both in the camp and the laboratory.

Another and very convenient form of steam generator is shown in Fig. 3. *A* is a tubulated bell-shaped vessel of tinplate, which passes inside the slightly larger receiver *B*, which holds the water to be boiled. The rest of the apparatus is the same as in Fig. 1. This form of steam generator has many advantages, and in travelling the lamp can be stowed inside it.

The most important part of the apparatus is the steam tube or vessel *C*. The following are the dimensions of one which has been a good deal in use (Fig. 4).

The wide part (*a*) is 160 mm. long, and 41 mm. in diameter. The part (*b*) making connection with the steam generator is 60 mm. long, and 9·5 mm. wide. The upper portion (*c*) is

110 mm. long, and 15 mm. wide ; and the exit tube (*d*) is
50 mm. long, and 10 mm. wide. It is connected with the
flask by means of a well-fitting cork, which is
preferable to indiarubber as it does not adhere
to the hot metal. From the above measurements
it results that the internal volume of the whole
tube is 239 c.c., and its internal surface is
291·5 sq. cm. The weight of the tube is
109 grms. ; and if its specific heat be 0·2, the
weight of water thermally equivalent to it is
21·8 grms.

Fig 4

These dimensions are suitable for the thermometer which was used. The length of the tube
has to be suited to each thermometer. For
mountain work the thermometer should be
graduated from 85° C. to 101° C. and should
be divided into tenths of a degree, the space
between each division being about a millimetre,
so that the length of the scale would be 16 cm.,
and that of the whole thermometer about 20 cm.
It is convenient also to have a thermometer divided into
fiftieths of a degree. It is essential that the thermometer
be read entirely in the steam. If any part of the mercury
extends outside of the apparatus, it introduces an uncertainty
which stultifies the use of very delicate instruments. Whatever the size of the tube may be, it is essential to see that all
the parts are sufficiently wide and properly proportioned ;
there must be no resistance anywhere to the passage of the
steam. The lamp should be regulated to keep up a brisk
flow of steam of good volume through *d*. The thermometer
then finds itself immersed in saturated steam of atmospheric
pressure contained in a vessel, the outer surface of which is
glass, but the inner surface is a film of water continually
renewed by steam condensing on it. So long as the full
supply of steam is kept up, this continually renewed film of
water, in contact with saturated steam, is a perfect protection
against variation of temperature in the interior of the steam
tube. When the steam first passes through the tube there

is rapid condensation on the sides, and after that, on the thermometer, from which the water falls in a series of drops, which follow each other first slowly, then rapidly, then again more slowly, until, finally, a drop remains hanging from the lower extremity of the thermometer, and never falls. By that time the protecting film of water has established itself on the sides of the tube, and effectually guards the thermometer from external influence. If it were possible for the temperature of the thermometer to fall ever so little below that of the condensing temperature of the saturated steam in the tube, its temperature would be immediately restored by the condensation of some of the steam on the water which moistens its surface. If its temperature were to rise ever so little above the condensing temperature, it would be immediately brought back again by the evaporation of some of the moisture on it. Consequently, when the boiling is in full operation, the thermometer is exactly at the temperature when the smallest possible increase of heat will cause evaporation, and the smallest possible decrease of heat will cause condensation. *But the boiling-point of a substance is the temperature at which it, as a vapour, condenses on itself as a liquid, and as a liquid evaporates into itself as a vapour.* Therefore, the temperature of the thermometer is exactly the boiling temperature of the water. Further, the whole enclosure is guarded by a surface of water in contact with saturated steam, so that its walls are necessarily also at the boiling temperature of the liquid, and it is impossible that the thermometer can be at any other temperature than that of the boiling liquid and condensing vapour. It follows, therefore, that for this purpose we may graduate our thermometer into thousandths of a degree, if we choose, and it is quite certain that the temperature of the thermometer will not differ by that amount from that of the medium. During an operation the steam tube is the ideal enclosure at constant temperature. When the thermometer has taken the temperature of the steam, it remains perfectly steady so long as the barometric pressure remains the same.

As evidence of the efficiency of the steam tube, the rate of generation of steam was varied from the highest to the lowest which was possible with the lamp. The thermometer divided into fiftieths of a degree Centigrade was used, and the mercury stood at 100°·18 ; that is the top of the mercury was exactly in line with the centre of the division on the stem marking 100°·18 C. It occupied this position when steam was being generated at its highest rate of 8·4 grammes per minute, and never varied by a fraction of the width of a division-line until the rate of steam generation had been brought so low that it issued as an exhalation, and not as steam. The rate was then 2·5 grammes per minute, and the top of the mercury fell to the lower edge of the line marking 100°·18 C., which may be taken to represent a temperature of 100°·179 C.

In this experiment the rate of passage of steam was varied from 2·5 to 8·4 grammes per minute, or over three-fold, and it produced no effect on the thermometer ; therefore, the exit tube efficiently removed whatever amount of steam entered it, and offered no sensible resistance to it. No better evidence than this could be furnished of the perfect efficiency of the instrument for the purpose for which it was designed.

With regard to the relative advantages of the thermometer and the barometer in hypsometric work, this thermometer may be taken as an example. It is divided into fiftieths of a Celsius degree, the length of the degree being 35 mm. The tension of saturated vapour, of 100° C., is 760 mm., and of 99° C. it is 733·305 mm., giving a difference of 26·695 mm. pressure for a difference of 1° C. temperature. Thus, 26·7 mm. on the barometer are represented by 35 mm. on the thermo-meter. Converting the readings of the thermometer into the corresponding ones of the barometer, each division would correspond to about half a millimetre. At lower pressures the effect of change of pressure on the temperature of satura-tion is greater, and at higher pressures it is less. Thus, from Regnault's experiments, we have in Table IV the difference of pressure which causes a difference of 1° C. in the temperature of saturated steam at different parts of the scale.

This table shows how rapidly the tension of saturated steam rises, as compared with its temperature. It also shows

that the lower the tension of the steam, the more efficient
is the thermometer for indicating it. But, for hypsometric
purposes, it must be remembered that the higher we climb
on a mountain the greater is the height required to produce
a given fall of the barometer.

TABLE IV.—*Giving the Temperature (Celsius) of Saturated
Steam T, at the Pressure P, and the Difference of Pressure
D corresponding to 1° C. Difference of Temperature of
Saturation.*

T	P	D	T	P	D	T	P	D
°C.	mm.	mm.	°C.	mm.	mm.	°C.	mm.	mm.
0	4·600	0·335	50	91·982	4·483	140	2,717·63	76·19
10	9·165	0·591	60	148·791	6·776	160	4,651·62	117·26
20	17·391	1·045	80	354·643	14·155	180	7,546·39	171·87
30	31·548	1·766	100	760·000	26·695	200	11,688·96	241·50
40	54·906	2·867	120	1,491·280	46·730	220	17,390·36	327·07

That these two effects very nearly compensate each other
is shown in Tables V and VI[1], which give the temperature of
boiling water in Celsius degrees in Table V, and in Fahrenheit
degrees in Table VI, and the barometric pressure, in inches
and millimetres, which is equal to the tension of saturated
steam at the particular temperature, with the height above
the sea, in metres and feet, at which this barometric pressure
would be found in a still, dry atmosphere at the temperature
of melting ice. It will be seen that throughout a range of
4500 metres, or 15,000 feet, a depression of the boiling-point
by 1° C. corresponds very closely to an ascent of 300 metres,
or 1000 feet. If the column of differences of the barometer
be inspected, it will be seen that the diminution of pressure
for a given increase in altitude has fallen at 4500 metres to
nearly one-half of what it is at the sea-level. Hence the
efficiency of the thermometer as a hypsometric instrument,
compared with that of the barometer, increases steadily with

[1] These tables are compiled from Tables 20, 25, 33 and 34 of the Smithsonian
Meteorological Tables, 1893. No. 844 of the Smithsonian Miscellaneous Collections.

TABLE V.—*Giving the Tension of Saturated Steam in Milli-metres and Inches, and the Corresponding Heights above the Sea in Metres and Feet, for Temperatures in Celsius degrees.*

Temp. of Saturated Steam	Barometric Pressure				Height above Sea-Level			
	Millimetres		Inches		Metres		Feet	
°Celsius								
85	433·0	—	17·05	—	4495	—	14,750	—
86	450·3	17·3	17·73	0·68	4182	313	13,734	1016
87	468·2	17·9	18·43	0·70	3871	311	12,724	1010
88	486·6	18·4	19·16	0·73	3562	309	11,718	1006
89	505·7	19·1	19·91	0·75	3255	307	10,717	1001
90	525·4	19·7	20·69	0·78	2950	305	9,720	997
91	545·7	20·3	21·48	0·79	2647	303	8,728	992
92	566·7	21·0	22·31	0·83	2345	302	7,740	988
93	588·3	21·6	23·16	0·85	2045	300	6,757	983
94	610·7	22·4	24·04	0·88	1747	298	5,778	979
95	633·7	23·0	24·95	0·91	1451	296	4,804	974
96	657·4	23·7	25·88	0·93	1157	294	3,834	970
97	681·9	24·5	26·85	0·97	865	292	2,869	965
98	707·2	25·3	27·84	0·99	575	290	1,908	961
99	733·2	26·0	28·87	1·03	287	288	952	956
100	760·0	26·8	29·92	1·05	0	287	0	952

the altitude, so that for the greatest altitude the thermometer is the preferable instrument.

Fixed Temperatures produced by Steam in contact with Salt and Saline Solutions.—At the same atmospheric pressure, the tension of the vapour of water is reduced, not only by lowering its temperature, but also, while the temperature is kept constant, by dissolving any salt in it. The tension of the vapour of pure water at 100° C. is 760 mm. If a small quantity of common salt or chloride of sodium be dissolved in it, the tension of its vapour is no longer 760 mm. at 100° C.; it is necessary to raise it to a higher temperature in order that its vapour may attain this tension. In proportion as more salt is added to the water, the higher is it necessary to raise the temperature of the water, or rather the resulting saline solution, in order to attain a tension of 760 mm. But there is a limit to the amount of salt which water can dissolve when boiling under

TABLE VI.—*Giving the Tension of Saturated Steam in Milli-metres and Inches, and the Corresponding Heights above the Sea in Metres and Feet, for Temperatures in Fahrenheit degrees.*

Temp. of Saturated Steam	Barometric Pressure				Height above Sea-Level			
	Inches		Millimetres		Metres		Feet	
° Fahr.								
185	17·05	—	433·1	—	4653	—	15,267	—
186	17·42	0·37	442·5	9·4	4476	177	14,681	583
187	17·81	0·39	452·4	9·9	4292	134	14,082	602
188	18·20	0·39	462·3	9·9	4113	179	13,493	589
189	18·59	0·39	472·2	9·9	3937	176	12,917	576
190	19·00	0·41	482·6	10·4	3756	181	12,324	593
191	19·41	0·41	493·0	10·4	3580	176	11,744	580
192	19·82	0·41	503·4	10·4	3406	174	11,175	569
193	20·25	0·43	514·4	11·0	3228	178	10,592	583
194	20·68	0·43	525·3	10·9	3054	174	10,021	571
195	21·13	0·45	536·7	11·4	2876	178	9,436	585
196	21·58	0·45	548·1	11·4	2701	176	8,863	573
197	22·03	0·45	559·6	11·5	2530	171	8,302	561
198	22·50	0·47	571·5	11·9	2355	175	7,728	574
199	22·97	0·47	583·4	11·9	2184	171	7,166	562
200	23·45	0·48	595·6	12·2	2013	169	6,604	562
201	23·94	0·49	608·1	12·5	1842	171	6,042	562
202	24·44	0·50	620·8	12·7	1670	172	5,480	562
203	24·95	0·51	633·7	12·9	1499	171	4,919	561
204	25·46	0·51	646·7	13·0	1332	167	4,369	550
205	25·99	0·53	660·1	13·4	1161	171	3,809	560
206	26·52	0·53	673·6	13·5	994	167	3,260	549
207	27·07	0·55	687·3	13·7	824	170	2,703	557
208	27·62	0·55	701·5	14·2	657	167	2,156	547
209	28·18	0·56	715·8	14·3	491	166	1,610	546
210	28·75	0·57	730·3	14·5	325	166	1,066	544
211	29·33	0·58	745·0	14·7	160	165	523	432
212	29·92	0·59	760·0	15·0	0	160	0	523

a given pressure. When this amount has been added and dissolved, the solution is saturated and, so long as the atmospheric pressure remains the same, it is impossible to raise its boiling-point any higher. If heat is supplied, so as to keep the solution boiling, steam escapes from its surface, and crystals of salt separate out in the solution. The condition of the boiling solution is now precisely analogous with that of the freezing solution when, in the process of cooling, the cryohydric temperature and concentration have been reached. The temperature of the boiling solution remains constant, and steam and salt quit it *pari passu*. In the freezing solution the salt and ice are both solid, and remain associated in the cryohydrate. In the boiling solution the salt is solid and the steam is gaseous, and they part company of themselves.

The fact that steam produced by water boiling at 100° C., which under ordinary circumstances can produce by its condensation no higher temperature than that at which it was generated, should be able to raise the temperature of saline solutions many degrees above this temperature, appears at first sight paradoxical, and it was, in fact, largely disbelieved, in spite of the simplicity of the experiment to demonstrate it.

In a paper already referred to, the use of ice melting in saline solutions of definite nature and strength was strongly recommended as affording an absolute thermometric scale for such temperatures. Extending these researches to steam and brines, the writer found that the condensation of steam in saline solutions could be used for fixing, or verifying, temperatures above the boiling-point of water. Also steam, water and salt can be used to form *boiling mixtures*, exactly as ice, water, and salt are used to form freezing mixtures; and an independent and absolute thermometric scale is produced for high temperatures, just as with ice we have one for low temperatures.

These melting and condensing temperatures are intended to give fixed points of reference on the thermometer below the ordinary freezing- and above the ordinary boiling-point, and they can be chosen to suit the work in hand. Thus, if work is being done where temperatures about 108° or 109° C. are to be measured with the accuracy which is demanded of a thermometer on which the Centigrade degree occupies a length of one or perhaps two centimetres, it is not sufficient to verify the ordinary boiling-point, because the part of the scale of the thermometer used may be ten or fifteen centimetres distant from it, and to verify the scale by careful calibration entails great labour. It is, however, very simple to expose the thermometer in a vessel of the form about to be described to the action of steam condensing in a mixture of common salt and brine, which gives a perfectly fixed temperature at a given barometric pressure. And these conditions can be reproduced at any future time, and the readings of the thermometer in the immediate neighbourhood of this fixed temperature can thus be verified as accurately as if they had been as near to the ordinary boiling-point of distilled water.

For temperatures above the boiling-point of water it is convenient to use " boiling mixtures," where the dry salt is put in a suitable vessel holding a thermometer, and the steam is blown through it until the bulb of the thermometer is immersed in a boiling mixture of brine and solid salt, and the stem is immersed in its steam.

One of the most convenient salts for this purpose is chloride of sodium, on account of its almost uniform solubility at different temperatures. At ordinary atmospheric pressure it raises the condensing point of steam by 8°·4 C., or 15°·1 F. If a weighed quantity of salt has been used and the apparatus has been also weighed, then by continuing to blow in steam after all the salt has been dissolved, and weighing the apparatus when the temperature has fallen to certain definite degrees, a series is obtained of the temperatures at which steam condenses in solutions containing definite amounts of salt. This has been done for a number of salts. It is a much more accurate way of determining the boiling-point of a solution than by boiling it over a lamp flame. This holds generally. *To determine the boiling-point of a liquid it should be boiled by its own steam.*

The apparatus used is shown with the boiling flask and lamp in Fig. 2. The steam tube is U-shaped ; the one leg has a large body, *CB*, 15 cm. long and 4 cm. wide. This is continued upwards in the tube *AC*, which is 15 cm. long and 12 mm. wide. The exit tube *D* has an internal diameter of 7 mm., and the entry tube *E* has also 7 mm. diameter, or slightly less. The dry salt is introduced into *CB*; the thermometer passes through a cork at *A*, and the bulb is covered

by the salt at the bottom of *CB*. The tube is then weighed.
It is then connected with the steam generator and steam

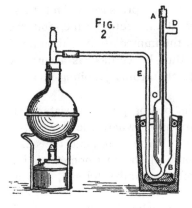

FIG. 2.

blown through, which in a
few minutes produces a magma of boiling brine mixed
with salt, while the thermometer takes the maximum
temperature, and retains it
until the solid salt approaches
complete solution.

The graduation of the
thermometer thus receives a
useful check at this part of
the scale. If it is wished to
verify the thermometer at
any temperature intermediate
between 100° C. and 108°·4 C., steam can be passed through
until the temperature of the boiling solution has fallen to the
degree wished, and by weighing the apparatus, or determining
the chlorine, the amount of steam condensed is found; and,
as that of the salt is known, the strength of the solution,
which boils at this particular temperature, under the observed
barometric pressure, is obtained, and that temperature can
always be again recovered independently of the thermometer
with which the original observation was made.

The elevation of the boiling-point of a saturated solution
of a salt above that of pure water is affected by the barometric
pressure. For chloride of sodium the rise may amount to as
much as 8°·45 C. at the sea-level; and in the course of
experiments made in Switzerland it has been found to fall
to 8°·0 C. at a height of 2770 metres above the sea. If,
however, we represent by t the boiling temperature of the
saturated solution, and T the boiling-point of pure water at
P, the barometric pressure at the moment: then, if p be the
barometric pressure under which pure water would boil at
temperature t, the ratio $p : P$ appears to be constant, and
equal to 1·345.

In Table VII we find the barometric pressure (P), given

TABLE VII.—*Giving the Boiling Temperature of Pure Water, and the Temperature of a Boiling Mixture of Steam and Chloride of Sodium, for Barometric Pressures between 550 and 770 Millimetres.*

Barometric Pressure	Temperature of Saturated Steam	Tension of Saturated Steam of Temperature t	Temperature of Boiling Mixture	Rise of Boiling-Point of Saturated NaCl Brine
P	T	$p = 1\cdot345 P$	t	$t - T$
Millimetres	° C.	Millimetres	° C.	° C.
770	100·37	1035·6	108·89	8·51
760	100·00	1022·15	108·50	8·50
750	99·63	1008·7	108·11	8·48
740	99·26	995·25	107·72	8·46
730	98·88	981·8	107·32	8·44
720	98·49	968·35	106·92	8·43
710	98·11	954·9	106·51	8·40
700	97·71	941·45	106·10	8·39
690	97·32	928·0	105·68	8·36
680	96·92	914·55	105·26	8·34
670	96·51	901·1	104·83	8·32
660	96·10	887·65	104·40	8·30
650	95·68	874·2	103·96	8·28
640	95·26	860·75	103·51	8·25
630	94·83	847·3	103·07	8·24
620	94·40	833·85	102·61	8·21
610	93·96	820·4	102·15	8·19
600	93·51	806·95	101·68	8 17
590	93·06	793·5	101·21	8·15
580	92·60	780·05	100·73	8·13
570	92·13	766·4	100·24	8·11
560	91·66	753·15	99·75	8·09
550	91·18	739·7	99·24	8·06

for every tenth millimetre from 770 to 550 mm., and the corresponding temperature of saturated steam (T); under (t), the temperature of a boiling mixture of steam and NaCl under the barometric pressure (P); under $p (= 1\cdot345 P)$, the calculated tension of saturated steam of temperature t; and under $t - T$, the excess of boiling-point of saturated NaCl solution above that of pure water under the given pressure. This table gives figures which agree closely with the results of the writer's observations in Switzerland at nine different altitudes, varying from 400 to 2773 metres. It appears that the alteration in the solubility of NaCl between the temperatures 91° C. and 100° C. is so slight that the relative depression of vapour tension is not sensibly affected.

The practical value of these results is that, if a thermometer be exposed to the action of pure saturated steam, and to a boiling mixture of steam and NaCl, and the temperature noted in each case, the difference of these temperatures, compared with the tabular value of $t - T$ for the existing barometric pressure, furnishes a direct and easily applied control of the graduation of the thermometer over the observed range.

TABLE VIII.—*Giving the Barometric Pressure at which a Boiling Mixture of Steam and Chloride of Sodium has a certain Temperature; the Limits of Temperature being* 109° *and* 99° C.

Temp. of Boiling Mixture	Barometric Pressure	Pressure corresponding to a difference of 0°·1 C.	Temp. of Boiling Mixture	Barometric Pressure	Pressure corresponding to a difference of 0°·1 C.
t	p	dp	t	p	dp
° C.			° C.		
109	773·4	—	104	650·9	—
108·5	760·2	2·60	103·5	639·8	2·22
108	747·4	2·40	103	628·7	2·22
107·5	734·9	2·50	102·5	617·8	2·18
107	722·4	2·50	102	607·0	2·16
106·5	710·2	2·44	101·5	596·3	2·14
106	698·0	2·44	101	585·7	2·12
105·5	686·1	2·38	100·5	575·3	2·08
105	674·2	2·38	100	565·1	2·04
104·5	662·5	2·34	99·5	555·1	2·00
104	650·9	2·32	99	545·1	2·00

For information about other salts in this respect, the reader is referred to the writer's paper on 'Steam and Brines.'

Before passing from this subject attention may be directed

to the following interesting connection between the freezing and the boiling of saline solutions.

The temperature of a saline solution at its freezing-point is *raised* by melting ice in it, and the rate at which it raises the temperature of the solution is the less the greater the amount of ice melted. The more nearly the temperature of the solution approaches to 0° C., the greater is the amount of ice which must be melted in order to produce a certain rise of temperature. When the temperature of the solution is ever so little below 0° C., the amount of ice which has to be melted in order to produce the smallest rise of temperature is ever so great. But the temperature at which ice melts in a solution is the temperature at which the solution freezes. Therefore, for solutions containing the same amount of a salt, the *depression of the freezing-point* is the less the greater the amount of water present. In some cases the depression of the freezing-point is exactly proportional to the reciprocal of the amount of water present; in other cases it deviates slightly from exact proportionality in one sense or in the other.

The temperature of a solution at its boiling-point is *lowered* by condensing steam in it; and the rate at which it lowers the temperature of the solution is the less the greater the amount of steam condensed. The more nearly the temperature of the solution approaches the temperature at which saturated steam condenses on pure water, the greater is the amount of steam which must be condensed in order to produce a certain fall of temperature. When the temperature of the solution is ever so little above the boiling-point of pure water, the amount of steam which has to be condensed in order to produce the smallest fall of temperature is ever so great. But the temperature at which steam condenses in a solution is the temperature at which the solution boils. Therefore, for solutions containing the same amount of a salt, the *elevation of the boiling-point* is the less the greater the amount of water present. In some cases the elevation of the boiling-point is exactly proportional to the reciprocal of the quantity of water present; in other cases it deviates slightly from exact proportionality in one sense or in the other.

If T_f be the freezing-point of water, and t_f be the freezing-point of a solution which for a certain quantity of dissolved salt contains a quantity W_f of water, then the value of the product $W_f(T_f - t_f)$ is constant in the case of some salts; in the case of others it increases as W_f, or the amount of water, increases; in the case of others again it decreases as the dilution increases. But the deviations from constancy are never great even in nearly saturated solutions.

If T_b be the boiling-point of water, and t_b be the boiling-point of a solution which, for a certain quantity of dissolved salt, contains a quantity W_b of water, then the value of the product $W_b(t_b - T_b)$ is constant for some salts; in the case of others it increases as W_b, or the amount of water, increases; in the case of others again it decreases as dilution increases.

The product of a mass into a temperature is a quantity of heat. Therefore, the expressions $W_f(T_f - t_f)$ and $W_b(t_b - T_b)$ represent quantities of heat[1]. The physical meaning or interpretation of the statement that either of these products is constant, when the proportion between the factors in it varies, is that *water can contain more heat without boiling, and less heat without freezing, in proportion to the amount of salt dissolved in it.*

This is a general law of which Blagden's law of freezing is a particular case.

The phenomena accompanying the freezing and the boiling of saline solutions have always been perplexing. Salt-water ice melts at a lower temperature than fresh-water ice, and it is impossible to prepare pure water from it; therefore it was held by some that the salt must belong to the ice. But by skilfully managing the melting, the temperature may approach very near to 0° C., and the water prepared from it may contain very little salt, and may be drinkable; therefore it was held by others that the salt belonged to the brine. The available experimental data did not conclusively prove either that the salt belongs exclusively to the ice or to the brine.

---

[1] 'Steam and Brines,' below, p. 173.

The proof that the salt does not belong to the ice was furnished by showing that snow or other pure ice, of independent origin, which contains no salt, behaves in a saline solution in exactly the same way as the alleged ice formed by freezing the solution.

Again, regarding the boiling of saline solutions, it was observed that the boiling temperature of a saline solution is higher than that of pure water, that the steam produced by it is pure steam, and that a thermometer in the steam above the boiling solution shows the same temperature as it would if immersed in steam above pure boiling water. The question which vexed many minds, was: When the steam is just quitting the solution, has it the temperature indicated by the thermometer immersed in the steam alone, or has it that of the thermometer immersed in the boiling solution? It was held by some that the temperature of the thermometer immersed in steam must be held to prove that the steam leaves the solution at the temperature of pure boiling water. Others contended that, although we have no means of knowing at what temperature steam would condense on *perfectly dry* glass, the moment there is the slightest moisture on it the steam is condensing on water, and we know exactly what that temperature is: it is and it must be the temperature of pure boiling water under the existing conditions, and it proves nothing as to the temperature at which the steam actually quitted the solution.

The proof that the steam must quit the solution at the temperature of the boiling solution is furnished by the following considerations. We observe that when we blow pure steam into a saline solution, it raises its temperature above that of pure boiling water, and until a certain maximum temperature is reached, which depends on the relation between the salt and the water present. As steam continues to pass and to condense in part, owing to the lower temperature of the air outside, the temperature of the boiling solution falls gradually as the amount of water present increases by the condensation of steam. We see then that pure steam of independent origin is condensed by a saline solution having a temperature higher than that of boiling water at the moment, and lower than that of its own boiling-point; but it is admitted that the boiling saline solution produces pure steam which must be in contact with the solution which produces it. If the steam so produced is at the temperature at which it would be produced by pure water boiling under the same conditions, then it must *by contact* cool some of the solution below its boiling-point. But it has been shown that pure steam in contact with a saline solution, the temperature of which is ever so little below that of its boiling-point, is condensed by it, and has its temperature raised by such condensation. But steam of a lower temperature than that of the boiling solution cannot be both condensed and generated under the same conditions. We see that it is condensed; therefore it cannot be liberated unless its temperature is at least as high as that of the boiling solution. There is no reason to suppose that it can be any higher; therefore, *the temperature of the steam leaving a boiling saline solution is the same as that of the boiling solution itself.*

Meteorological Observations and Instruments.

Full instructions for installation of instruments, and for methods of observation and recording, will be supplied with the instruments. It is unnecessary to repeat them here. But there are one or two points, seldom adequately insisted on in meteorological instructions, which may here be recalled.

Thermometers.—With the thermometers of the Expedition there will be supplied information as regards the position of the point of melting ice and as regards the agreement of other parts of the scale with the standard; but it is safe to affirm that no information will be supplied with regard to the *thermal mass* of the thermometers, or the rapidity with which they respond to changes of the temperature of the medium in which they are placed. The Rate of Cooling, or its reciprocal, the *Term of Cooling*, of a thermometer is as important a constant as the position of the ice-point or the length of a degree. The following considerations will show that the want of knowledge of this constant and of its application may, and in point of fact does, frequently introduce error and confusion into meteorological results.

When two unequal masses of the same substance have the same temperature and hang in the same medium, the temperature of which is constant and lower than their own, their temperature will fall at unequal rates. If, being at the same temperature, they are hung in a medium of constant but higher temperature, their temperature will rise at unequal rates. These unequal masses may be thermometers with similar but unequal bulbs. Under the same conditions they will lose or gain heat at unequal rates. If, at any epoch, they and the air in which they hang are at the same temperature, and the temperature of the air is changed to another, which is kept constant, it is clear that at equal intervals of time from the initial epoch the two thermometers will have and will show different temperatures, although their scales may be without error. Further, in the ordinary course of diurnal variations of temperature these two thermometers will never again show the same temperature except for a moment of time when, in the exigencies of their different rates of cooling and heating, their thermal paths cross.

The importance of self-recording thermometers for supplying a continuous record of the temperature of the air and of its variations is well understood. The form of instrument most commonly used is, on account of its compactness and comparative cheapness, that known as Richard's recorder. It is usual to control its indications by reading a standard mercurial thermometer hung alongside of it, and it is a very common practice to make this reading at 9 a.m., when the temperature of the air is changing most rapidly. When the paper is removed, the temperatures taken from the curve traced will be corrected so as to bring them into harmony with the 9 a.m. readings of the mercurial thermometer. Except in the very unlikely case of the mercurial and the recording thermometers having the same rates or terms of cooling, error will be introduced and not eliminated by this proceeding. The temperature of the air is as good as constant for a considerable time at the hour of the diurnal maximum, which is usually about 2 p.m.   This is the hour to compare the recorder with the mercurial thermometer.

Again, if the wet and dry bulb thermometers are not exactly equal and similar—and they never can be—their indications are not exactly comparable in air of changing temperature or humidity.

These few remarks will show the practical importance of a knowledge of the rate or term of cooling of every thermometer used in a meteorological observatory, and of the application of it to the correction of observations.   They also show the supreme technical importance of making such thermometers according to a perfectly uniform pattern.

When the thermometer was a novelty and philosophers studied its resources and applications in every direction, the importance of this constant was fully recognised, and the application of it to observations was insisted on.

With the prevalent dilettante character of meteorology its existence was forgotten, and its application fell into desuetude.

The subject is dealt with in great detail and quite exhaustively in the works of Newton, Lambert, Leslie, and others.   For profound but, at the same time, simple treatment of this interesting subject, the writings of these great men cannot be surpassed.

The method of determining the rate or term of cooling of a thermometer is simplicity itself; indeed, it can be carried out even in a shop, so that we need never buy a thermometer in ignorance of what may be termed its *thermal nimbleness*.

The necessary observations are best made in a dwelling-room of fair size, in which the air remains for at least a considerable time at a constant temperature.   The thermometer, which is not attached to any backing, is whirled in the air until it assumes a constant temperature.   This is noted as the temperature of the air at the time.   The thermometer is then hung up in the middle of the room, and a reading telescope should be set up at a little distance from it, so that its scale can be read without approaching too near.   Except in particular cases, as for persons who are short-sighted, the telescope is not absolutely indispensable. The bulb of the thermometer is then warmed in any convenient way, as by the heat of the hand through a fine cloth, to prevent soiling the glass. The temperature of the thermometer should be raised from fifteen to twenty degrees above that of the air. It is then allowed to cool while hanging quite motionless in the air. When its temperature has fallen a few degrees, time is taken to the nearest second when the mercury passes a given division. It is then observed as it falls, at regular equal intervals of time. The length of these intervals of time is regulated by the rapidity with which the thermometer cools, and may conveniently be 5, 10, 20, 30, or even 60 seconds. The observations should be continued until the temperature of the thermometer has fallen to about two degrees above that of the air. This should not have changed during the operation. If there has been any sensible variation the observations should be rejected, and the operation should be repeated when there is no variation.

The principle which should find interpretation in the observations is that *in equal intervals of time the bulb of the thermometer loses equal fractions of the heat which it possessed at the beginning of the interval.* Here heat is understood to mean excess of heat, or the heat corresponding to temperatures above that of the medium in which the thermometer is cooling.

If time has been taken when the temperature of the thermometer is exactly $10°$ above that of the air, and it is found that in the 1st second it falls to $9°·9$, then at the end of the 1st second the heat remaining is only $\frac{99}{100}$ of the initial amount which we may represent by unity, and in the interval it has lost $\frac{1}{100}$ of the amount that it had at the beginning of the interval. If the above principle holds good, it must lose in the 2nd second $\frac{1}{100}$ of the amount which it has at the beginning of that second, and at the end of the 2nd second the heat remaining will be $\frac{99}{100}$ of the amount which was present at the beginning of that second. But the amount present at the beginning of the 2nd second was $\frac{99}{100}$ of the

amount present at the beginning of the 1st second which is
represented by unity. Therefore, referred to the amount
present at the zero of reckoning as unity, the amount remaining at the end of the 2nd second should be $(\frac{99}{100})^2$.
Similarly, during the 3rd second the thermometer loses
$\frac{1}{100}$ of the heat which it had at the beginning of that second,
and the amount remaining at the end of the 3rd second will
obviously be $\frac{99}{100}$ of $(\frac{99}{100})^2$, or $(\frac{99}{100})^3$. Similarly, at the end of
the 4th, 5th, 6th, or nth second, the heat remaining will be
$(\frac{99}{100})^4$, $(\frac{99}{100})^5$, $(\frac{99}{100})^6$, ..., $(\frac{99}{100})^n$ of the original amount present at
the zero of reckoning represented as unity. It is obvious that
if the law holds good, by giving the suitable value to the
index n we can at once calculate the proportion of heat which
will remain after any number of seconds of cooling; and it is
also obvious that so long as the temperature of the medium
remains constant, the thermometer can never exactly reach
that temperature, although the difference of the temperatures
may be made as small as we like by making the duration of
cooling sufficiently long.

It has been said that by giving n the suitable value, we
can at once find the heat remaining after the lapse of any
time. But the computation of high powers of numbers by
ordinary arithmetic is very laborious. If, instead of simple
arithmetic, we use logarithmic arithmetic, the computation of
a high power is as easy and expeditious as the computation of
a low one. In the case which we have imagined the constant
fraction is $\frac{99}{100}$. Its logarithm is log 99 − log 100, that is

$$1\cdot9956352 - 2 = 0\cdot9956352 - 1,$$

which is usually written $\bar{1}\cdot9956352$.

This is the logarithm of the first power of $\frac{99}{100}$, which is
the heat remaining at the end of the 1st second of cooling.
If we multiply $\bar{1}\cdot9956352$ by 2 we have the logarithm of the
square of $\frac{99}{100}$, and if we multiply it by 3, 4, ..., n we have the
logarithms of the 3rd, 4th, ..., nth power of $\frac{99}{100}$, that is, of the
heat remaining at the end of the 3rd, 4th, ..., nth second.
These logarithms differ from each other by the same amount.
Therefore we have the following rule: If the initial excess of

the temperature of the thermometer above the temperature
of the air is t, and in any interval of time $d\theta$ it falls to
$(t - dt)$, then after any number n of such intervals $(d\theta)$ the
logarithm of the fraction of excess temperature remaining is
$n \log \left(\dfrac{t - dt}{t} \right)$; and by multiplying the number corresponding
to this logarithm by t, the excess of temperature of the
thermometer after n intervals of time, each equal to $d\theta$, is
given.

The law which has just been explained is Newton's law
of cooling, often called the logarithmic law. It is worthy of
remark that Newton looked upon this law as axiomatic and
self-evident the moment it is stated, and he did not think that
it required experimental demonstration. It did not escape
question, notably by Amontons; and Lambert, in vindication
of Newton, although he had held that the law was self-evident,
carried out a beautiful series of experiments, which are
detailed at length in the third part of his classical work on
Pyrometry[1] and the measurement of heat. They completely
bore out Newton's law.

The method of making the experiment has just been
described, and as it is important that every one should be
familiar with the practice of it, we give an actual example.
(See Table IX.)

The experiment here recorded was made under very
favourable circumstances in the month of September in
a large room the temperature of which was sensibly the
same as that of the air outside, namely 20°·2 C., and this
remained quite constant for a much longer time than was
required for the experiment; indeed, it hardly varied at all
during the day. The results are instructive, because they
give a good idea of the kind of agreement between observation and theory which we have a right to expect. The
temperature was observed at every ten seconds. The initial
excess of the temperature of the thermometer over that of the
air is 8°·0, and $\log 8·0 = 0·9031$. After the first interval of

[1] *Pyrometrie oder vom Maasse des Feuers und der Wärme*, von Johann
Heinrich Lambert. 4to. Berlin, 1779.

cooling the excess is $6°\cdot8$, and $\log 6\cdot8$ is $0\cdot8325$. Taking the initial excess as unity, the fractional excess, or the heat remaining after the first interval of ten seconds, is $\dfrac{6\cdot8}{8\cdot0}$, and its logarithm is $0\cdot8325 - 0\cdot9031 = \overline{1}\cdot9294 = \log y_1$. After the second interval of ten seconds the excess is $5°\cdot8$, and $\log 5\cdot8 = 0\cdot7634$. The fractional excess after the second interval is $\dfrac{5\cdot8}{8\cdot0}$, and its logarithm is $0\cdot7634 - 0\cdot9031 = \overline{1}\cdot8603$.

TABLE IX.

Number of Experiment	Epoch	Temperature of Thermometer	Difference between Temp. of Thermometer and that of Air	Logarithm of this difference	Logarithm of quotient of difference by initial difference	Difference of these Logarithms	Logarithm of quotients calculated for Mean value, $dl=0\cdot0669$	Logarithm of calculated difference of Temp.	Calculated Difference
n	secs.	t	$t-20\cdot2=s$	$\log s$	$\log s - 0\cdot9031 = \log y$	$\log y_n - \log y_{n+1} = dl$	$\log 1 - n(0\cdot0669) = \log y_c$	$\log y_c + \log 8 = \log s_t$	s_t
		°	°						°
0	0	28·2	8·0	0·9031	0	0·0706	0·0000	0·9031	8·00
1	10	27·0	6·8	0·8325	$\overline{1}$·9294	·0691	$\overline{1}$·9331	0·8362	6·86
2	20	26·0	5·8	0·7634	$\overline{1}$·8603	·0644	$\overline{1}$·8662	0·7693	5·88
3	30	25·2	5·0	0·6990	$\overline{1}$·7959	·0706	$\overline{1}$·7993	0·7024	5·04
4	40	24·45	4·25	0·6284	$\overline{1}$·7253	·0661	$\overline{1}$·7324	0·6355	4·32
5	50	23·85	3·65	0·5623	$\overline{1}$·6592	·0709	$\overline{1}$·6655	0·5686	3·70
6	60	23·3	3·1	0·4914	$\overline{1}$·5883	·0582	$\overline{1}$·5986	0·5017	3·17
7	70	22·85	2·65	0·4232	$\overline{1}$·5301	·0715	$\overline{1}$·5317	0·4348	2·72
8	80	22·5	2·3	0·3617	$\overline{1}$·4586	·0607	$\overline{1}$·4648	0·3679	2·33
9	90	22·2	2·0	0·3010	$\overline{1}$·3979	—	$\overline{1}$·3979	0·3010	2·00

According to the theory which has been explained above, if the thermometer in cooling loses in each equal interval of time exactly the same fraction of the excess heat which it held at the beginning of the interval, and if the observations are without error, then the logarithms of the fractional excess after the second interval ought to be $2 \times \overline{1}\cdot9294 = \overline{1}\cdot8588$ in place of $\overline{1}\cdot8603$, as above. The difference is obviously not great. In order to know what it amounts to in the observation of the temperature we have $\overline{1}\cdot8588 + 0\cdot9031 = 0\cdot7619 = 5°\cdot78$, which ought to be the excess of the temperature of the thermometer over that of the air, if the thermometer follows the law and if the first two observations are exact. The agreement with the observed difference, $5°\cdot8$, is quite satisfactory. But we know that no observations are free from error, which must affect the first observations as well as the others. In the table we have the observations made at the end of each of nine consecutive intervals of ten seconds. In the seventh column of the table we have the differences of the consecutive logarithms of the fractional excesses remaining. Theoretically these differences ought to be identical. They are not; and their variations are irregular. We may therefore take the mean difference, which is $0\cdot0669$, and with it calculate what ought to be the excess remaining after each interval of ten seconds. The initial fractional excess, y_0, is 1, and its logarithm is 0. Subtracting $0\cdot0669$ we get $\bar{1}\cdot9331 = \log y_1$; and again subtracting $0\cdot0669$ we get $\bar{1}\cdot8662 = \log y_2$; and so on. The logarithms obtained are found in the eighth column of the table. In the ninth column we have the sum of log $8\cdot0$, or $0\cdot9031$, and the respective numbers in the eighth column. They are the logarithms of the calculated thermometric excesses. These are given in the tenth column.

The first and the last entries in this column necessarily agree with the observed values in the fourth column. The greatest difference is $0°\cdot08$, so that the actual rate of cooling may be held to agree fairly well with the rate which, according to theory, we ought to observe if the bulb of the thermometer were a perfectly homogeneous body of infinite thermal conductivity and of symmetrical shape, cooling in a vacuum enclosed by walls having a definite and constant temperature. We know that this description fits neither the thermometer nor the room in which it was cooling. The shape of the bulb, whether it be cylindrical or spherical, is not symmetrical in the above sense, because, for purposes of observation, the thermometer must always have a stem, and the part of the bulb where it is united to the stem is exposed to different conditions, as regards cooling, from the other parts of it. Although the thermal conductivity of the bulb of a mercurial thermometer is not perfect, its degree of imperfection is not such as to introduce much error into observations of this kind. The temperature of the room, and no doubt that of its walls, was very constant, but of course there was no vacuum.

The instrumental deformity introduced by the necessity of a stem for the thermometer must always introduce some deviation from the normal rate of cooling, but it is, as thermometers are made, not practically of much importance. The disturbing element which takes precedence of all others is the air.

The conditions in which the experiments quoted in the table were made were as favourable as they could be, and it would not be possible to get air more motionless than it was. But however motionless the mass of the air may be, a thermometer, or any other object, suspended in it, and having a higher temperature, must produce convection currents in its immediate neighbourhood, which will be the more energetic the greater the difference of temperature. Hence the conditions under which a thermometer cools in air are complex. In the first place it cools by radiation to its surroundings, and, setting aside instrumental imperfections, this takes place independently, as it would in a vacuum, according to the logarithmic law, losing equal fractions of heat in equal times. In the second place it loses heat by contact with the air, and the rate at which this loss takes place depends on the rate of renewal of successive envelopes of fresh air, and this diminishes as the temperature of the thermometer approaches to that of the air in which it is cooling. This explains why the term of cooling when the thermometer is only one or two degrees warmer than the air is greater than when it is five to fifteen degrees warmer. If differences of temperature amounting to $10°$ or $15°$ C. are used, the terms of cooling found are very concordant.

In the case detailed in Table IX, p. 74, the observations were made at equal intervals of ten seconds, and the mean logarithmic difference (dl) was found to be 0·0669, and the logarithm of the fraction remaining after the lapse of the first interval ($\log y_1$) was $\bar{1}$·9331, whence $y_1 = 0·8572$. Now the fraction $\frac{6}{7}$ is expressed by the circulating decimal 0·$\dot{8}$5714$\dot{2}$, therefore $y_1 = \frac{6}{7}$, and in each interval of ten seconds the loss of heat is $\frac{1}{7}$ of the amount which was present at the beginning of it. Therefore, if in each succeeding interval of ten seconds the same *amount* of heat were lost, the whole of the excess of heat would disappear in seven such intervals, or in seventy seconds. Therefore, the arithmetical result which we arrive at from observations made at intervals of ten seconds is that the *term of cooling* of the thermometer is seventy seconds.

But if 0·0669 is the logarithmic difference for ten seconds, then 0·00669 is the logarithmic difference for one second, 0·000669 for one-tenth of a second, 0·0000669 for one-hundredth of a second, and so on. The resulting terms of cooling derived from these different intervals and logarithmic differences, and the method of arriving at them, will be apparent from the following table:

TABLE X.

	10	1	0·1	0·01
Length of interval (secs.) . $d\theta$	10	1	0·1	0·01
Logarithmic difference . dl	0·0669	0·00669	0·000669	0·0000669
Log. first fractional excess . . . $\log y_1$	$\bar{1}$·9331	$\bar{1}$·99331	$\bar{1}$·999331	$\bar{1}$·9999331
Fraction remaining at end of first interval . . y_1	0·8572	0·984714	0·99846	0·999846
Fraction lost in first interval . . . $1 - y_1$	0·1428	0·015286	0·00154	0·000154
Reciprocal of fraction lost . . . $\dfrac{1}{1-y_1}$	7	65·4193	649·35	6493·5
Term of cooling, in secs. . . $d\theta \dfrac{1}{1-y} = R$	70	65·4	64·9	64·9

The rule[1] for finding the term of cooling referred to the shortest possible intervals of time, and the smallest logarithmic differences, from observed values of $d\theta$ and dl is: *Divide the modulus of the system of logarithms,* 0·434295, *by the logarithmic difference dl, and multiply the quotient by the interval of time $d\theta$. The product is the term of cooling expressed in the same*

[1] *Experimental Inquiry into the Nature of Heat,* by John Leslie. Edinburgh, 1804. Page 265.

units of time that have been used in expressing the time interval $d\theta$.

For the above case we have for the true term of cooling :

$$R = \frac{0.434295}{0.0669} \times 10 = 64.917 \text{ seconds.}$$

In the application of this rule we may make the interval anything we please. We may, therefore, choose it so that the loss of heat during it is expressed by a simple fraction, such as one-half. The convenience of this method was first pointed out by Leslie, and it affords by far the best practical method. It is sufficient, for instance, to heat the thermometer to 10° or 11° above the temperature of the atmosphere, take time when it is exactly 8° warmer than the air, and again when its temperature has fallen to 4° above that of the air. The excess heat present at the beginning of the interval is 1, and that at the end of the interval is $\frac{1}{2}$. Then we have :

$$dl = \log 1 - \log \tfrac{1}{2} = 0.30103 \, ;$$

and the term of cooling is

$$R = \frac{0.434295}{0.30103} \, d\theta \, ;$$

or, very approximately,

$$R = \frac{101}{70} \, d\theta.$$

Hence the rule to find the term of cooling when the time in which one-half of the heat excess is lost is: *Multiply that interval $d\theta$ by* 101 *and divide it by* 70. *The quotient is the term of cooling.*

It is evident that there is no difficulty in making this observation in a shop, and the time in which the instrument loses half its heat is sufficient without further computation to give a good idea of what has been called its *thermal nimbleness.* The want of a term commonly used to express this important property shows how much the property itself has been neglected.

Table XI gives an example of the use of the method of

the "half-fall" in the case of a thermometer, first with its bulb plain, and secondly with its bulb silvered.

It will be seen that silvering the bulb has in this case increased the term of cooling in the proportion of about 3 to 4, and the rate of cooling is diminished in the inverse proportion. Also for initial excesses of temperature between 16° and 6° the terms of cooling are very concordant. In both cases they increase when the temperature excess falls to 4°. It is to be observed that not only is the effect of convection less powerful at low temperatures, but any slight change in the temperature of the air makes itself more felt when the difference between it and that of the thermometer is small, than when it is great.

TABLE XI.

Bulb Plain			Excess of Temp.	Bulb Silvered		
Term of Cooling	Time in which half the excess is lost	Epoch		Epoch	Time in which half the excess is lost	Term of Cooling
secs.	secs.	secs.	° C.	secs.	secs.	secs.
157	109	0	16	0	137	198
156	108	20	14	23·5	143·5	207
157	109	45	12	54	144	208
157	109	75	10	92	144	208
141	98	109	8	137	146	211
—	—	128	7	167	—	—
162	112	154	6	198	145	209
—	—	184	5	236	—	—
180	125	207	4	283	150	216
—	—	266	3	343	—	—
—	—	332	2	433	—	—

The Use of the Thermometer for Measuring the Velocity of Weak Currents of Air.—The difference between motionless air in a room and calm air outside is best shown and is accurately measured by the difference between the terms of cooling of the same thermometer as determined in the one medium and then in the other. This difference is due to the fact that calm air outside is not motionless, while in a room of constant temperature it is practically so. It is evident that if the difference is caused by the motion of the air, then that difference must also be a measure of the motion. As has been pointed out above, this was perceived by Leslie, and he gives formulæ[1] for calculating the velocity of the wind from the reduction of the term of cooling of a tin vessel holding about half a litre of water.

If R be the term of cooling in still air, or the *fundamental* term of cooling, and r be the *occasional* term of cooling when the air is moving with any velocity v, then Leslie gives the following expressions for this velocity :

$$v = \frac{20}{3} \cdot \frac{R-r}{r} \text{ in feet per second ;}$$

or,

$$v = 4\tfrac{1}{2} \frac{R-r}{r} \text{ in miles per hour.}$$

Converting into metrical units, we have

$$v = 2\text{·}032 \frac{R-r}{r} \text{ in metres per second.}$$

It is right to observe that R in Leslie's equations is the term of cooling of his flask of water when suspended "*out of doors, on a calm evening.*"

Difference between a Calm Indoors and a Calm Out-of-Doors.—The preceding equations give the diminution of the term of cooling produced by sensible wind as compared with a calm, both being out-of-doors. It does not appear that Leslie distinguished between a calm indoors and a calm out-of-doors.

Returning to Table IX at page 74, we find the term of cooling of the thermometer to be in round numbers 65 seconds in the still air of a room. A number of observations were made with the same instrument in the open air in very calm fine weather. The method of the "half-fall" as exemplified in Table XI was used, the excesses of temperature used being 12°, 10°, 8° and 6° C. The experiments were made in

[1] *Loc. cit.*, p. 283.

Edinburgh on September 16, 1894, during very fine anti-cyclonic weather. All the afternoon the air was perfectly calm. The smoke from chimneys went straight up and indicated no horizontal component of motion.

In the following table the "half-falls" for different initial excesses at different times during the afternoon are given, which show the extent of their agreement. The mean "half-fall" is then converted into the term of cooling by multiplying by $\frac{101}{70}$.

TABLE XII.

Excess	Time P.M.				
	12.45	1.10	1.40	2.10	6.30
	Duration of Half-fall in Seconds				
12° C.	26	—	33	22	36
10° C.	25	—	36	22	36
8° C.	27	31	36	25	35
6° C.	27	31	30	31	33
Mean	26·25	31	33·75	25	35
Term	37·9	44·7	48·7	36·0	50·5

This table shows well what great differences may exist in the calmness of calm air. Nobody doubts that such motions do occur; otherwise it would be impossible for the permanent difference of temperature which is found generally to exist at different elevations in the atmosphere to be maintained; and it is interesting to have a means of gauging them.

In order to obtain a standard of measurement, a number of observations were made by moving the thermometer at different velocities in the air of the room, and observing the rates of cooling. Dealing only with low velocities it was found that while in perfectly still air the term was 65 seconds; when

B.

6

moving through the air at the rate of 2 m. per second the term
was 17·2 seconds; and when the velocity was 1 m. per second
it was 24·6 seconds; while at 0·5 m. per second the term
was 30 seconds. Allowing that, whether moved or not, the
cooling of the thermometer goes on independently at the
still-air rate of $\frac{1}{65}$ per second, and subtracting this from the
reciprocals of the above numbers, we have the rates of cooling
due to motion of the air;—at the rate of 2 m. $\frac{1}{24}$, of 1 m. $\frac{1}{40}$,
and of $\frac{1}{2}$ m. per second $\frac{1}{56}$, the rate of cooling in calm air
being $\frac{1}{65}$. The reciprocals of these fractions, or the terms, are
24, 40, 56 and 65, which are in the proportion 3 : 5 : 7 : 8.

The importance of these figures is that they show that
thermometers in the open air, even when the air appears to
be calm, are in reality well ventilated. In a room, the
temperature of the dry bulb, and still more that of the wet
bulb, are very imperfectly given by a stationary thermometer;
both thermometers must be whirled in order to get anything
like exact observations. In the thermometer screens of a
meteorological station the instruments are certainly well
ventilated when there is a wind; and we see that, even in
a calm, the ventilation may be sufficient.

The Thermometer as a Calorimeter.—The term of cooling
of a thermometer and the method of its determination have
been dwelt on at considerable length, because the information
regarding it to be found in manuals is usually defective. The
same remark applies to the *thermal mass* of the bulb of the
thermometer and the method of determining it. *When we
have these two constants, namely, the term of cooling and the
thermal mass, the thermometer becomes a calorimeter, and its
radius of research is much increased.*

The thermal mass of the bulb of a thermometer is completely specified if we know (*a*) the weight of mercury which
it contains; (*b*) the weight of the glass which forms the
envelope; (*c*) the specific heat of mercury; and (*d*) the
specific heat of the particular glass used. As the mercury
used for thermometers must be perfectly pure, (*c*) is known.
Many different kinds of glass are used, and the information
regarding their specific heat is very defective. This constant

should therefore be determined for the particular sample of glass used in the construction of the thermometers. By working in association with the thermometer-maker there is no difficulty in ascertaining the exact weight of mercury in the thermometer, or that of the glass which goes to the bulb. Then if we multiply the weights of these substances used by their respective specific heats, the sum of the products is the thermal mass of the bulb expressed as the weight, in grammes, of water, which is thermally equivalent to it. This constant is usually, and conveniently, called the *water value*. Calorimetry is an important department of physics and physical chemistry, and the methods of determining water values are given in all treatises on the subject. Where the size of the thermometer is considerable the specific heat of its bulb can be determined directly by the old "method of mixtures," the thermometer itself being one party to the mixture. As the term of cooling of a thermometer increases in proportion to the size of the bulb, it is clear that thermometers intended for meteorological use should have as small bulbs as possible, and the method of mixtures is not applicable for the determination of their thermal masses. Again, it is out of the question to expect these thermometers to be constructed so that the respective weights of mercury and of glass in their bulbs shall be accurately known. The following method of determining this constant is very convenient. It depends on mensuration, and was published by the writer in 1894[1].

Estimation of the Thermal Mass of the Bulb of a Thermometer by Mensuration.—The method is shortly stated in the following rule : *Determine the external volume or displacement of the bulb in cubic centimetres; multiply it by* 0·475, *and the product is the water value of the bulb in grammes.* The factor 0·475 which occurs in this rule is arrived at as follows : The density of mercury is 13·596, and its specific heat is 0·033, therefore the capacity for heat of 1 c.c. is the same as that of

---

[1] 'On Rapid Variations of Atmospheric Temperature, especially during *Föhn*, and the methods of observing them,' by J. Y. Buchanan, F.R.S. *Proc. R. S.* (1894), vol. lvi. p. 126.

0·4486 grm. of water. The density of ordinary glass may be taken at 2·6 and its specific heat at 0·198, whence the capacity for heat of 1 c.c. of glass is the same as that of 0·5148 grm. of water. Therefore when referred to unit volume the specific heats of these two bodies are very nearly identical, and if we have the total volume and apply the mean of the above values, the result will be a very close approximation to the water value of the bulb, quite independently of the exact proportion in which the mercury and the glass enter into its construction. Suppose that the displacement of the bulb were 1 c.c. and that the glass were infinitely thin, so that the bulb were all mercury, then its water value would be 0·4486 grm. If, on the other hand, the internal volume of the bulb were infinitely small, so that it consisted entirely of glass, its water value would be 0·5148 grm. But the bulbs of thermometers of most ordinary patterns are very much alike in construction. It is not usual in the construction of even the best instruments to take into account the amount of mercury or of glass present in the bulb. Such data are available only in the case of thermometers especially constructed for use in calorimeters. As they are themselves parts of the calorimeter, their heat constants must be ascertained; and, as was pointed out above, this is most easily and most accurately done during their construction. Berthelot[1] gives the thermal constants of three thermometers which were determined during construction. These are embodied in columns 1, 2 and 3 of the following table. Column 4 refers to a thermometer belonging to the writer; it is by Chabaud, of Paris, and is constructed for use with Berthelot's calorimeter. The data were ascertained during construction, and are engraved upon the stem.

In the following table the numerical data in the first part are taken from Berthelot's *Mécanique chimique*, pp. 162, 167.

In this table the fundamental gravimetric data are—*b* the weight of mercury and *d* the weight of glass in the bulb—

[1] *Essai de Mécanique chimique, fondée sur la Thermochimie*, par M. Berthelot, Membre de l'Institut, Paris. Dunod, 1879.

both of which are furnished by the makers, whose names are given in line *a*. In columns 1, 2 and 3 the water values of the mercury and the glass, *c* and *e*, are given by Berthelot. In column 4 they were calculated, using 0·033 as the specific heat

TABLE XIII.—*Calorimetric Specification of Thermometers.*

Computation of the Water Value of the Bulb from Data supplied by the Maker					
Maker of thermometer	*a*	Fastré	Baudin	Tonnelot	Chabaud
Weight of mercury, grms. ...	*b*	18·003	30·20	3·781	25·09
Water value of mercury, grms.	*c*	0·60	1·01	0·126	0·828
Weight of glass, grms.	*d*	3·075	2·43	0·599	2·58
Water value of glass, grms. ...	*e*	0·61	0·49	0·120	0·511
Water value of bulb, grms. ... ... $c+e=$	*f*	1·21	1·50	0·246	1·339

Calculation of Water Value of Bulb and of its Specific Heat per Unit Volume					
Volume of mercury, c.c. $\dfrac{b}{13\cdot596}=$	*g*	1·37	2·221	0·278	1·845
Water value of mercury, grms. ... ... $0\cdot4486g=$	*h*	0·615	0·996	0·125	0·828
Volume of glass, c.c. ... $\dfrac{d}{2\cdot6}=$	*i*	1·18	0·935	0·230	0·990
Water value of glass, grms. ... ... $0\cdot5148i=$	*k*	0·607	0·481	0·118	0·510
Water value of bulb, grms. ... ... $h+k=$	*l*	1·222	1·477	0·243	1·338
Total volume of bulb, c.c. $g+i=$	*m*	2·55	3·156	0·508	2·835
Specific heat of bulb per unit vol. ... ... $\dfrac{l}{m}=$	*n*	0·474	0·475	0·484	0·472

Centesimal Composition of the Bulb by Volume					
Mercury volumes per cent. ...	—	53·75	70·3	54·7	65·1
Glass volumes per cent. ...	—	46·25	29·7	45·3	34·9

of mercury and 0·198 as that of glass, and the constant sought and arrived at in this part of the table is *f*, the water value of the bulb as a whole. The second part of the table is based on the same gravimetric data, namely the weight of mercury and of glass (*b* and *d*). From these the volumes are calculated, taking 13·596 as the density of mercury and 2·6 as that of glass, and they are given in lines g and i respectively. When the specific heat of a body is spoken of without further qualification, the water value of 1 grm. of the substance is meant. When we are dealing with volumes, we require to know the specific heat per unit volume, and that is the water value of 1 c.c. of the substance. Now the density of a substance, when the metrical system is used, is the weight of 1 c.c., therefore the weight of 1 c.c. of mercury is 13·596 grms. and that of the same volume of glass is 2·6 grms. Multiplying 13·596 by 0·033, the specific heat of mercury, we obtain 0·4486, which is the water value of 1 c.c. mercury, or its specific heat per unit volume. Multiplying 2·6 by 0·198, we obtain 0·5148 as the specific heat per unit volume of glass. If now we multiply the volume of mercury, g, by 0·4486 and the volume of glass, i, by 0·5148, we obtain h and k, the water values of the volumes of mercury and of glass respectively. The sum of these, l, is the water value of the bulb as a whole. If the values of l be compared with those of f in the first part of the table, the agreement will be found to be very close. With regard to the mercury, there is not room for much discrepancy, because it is an elementary body, and it can be used in the construction of thermometers only when it is in a state of purity. It is otherwise with the glass. Neither Berthelot nor the makers give its composition, its density, or its specific heat. Although the composition of the glass is of great importance, a knowledge of it is not necessary for thermometric or calorimetric purposes; on the other hand, a knowledge of both the density and the specific heat of the glass is essential. Yet it is very rarely furnished. The values used in the table are commonly occurring ones; and the agreement in the water values of the glass arrived at in the first and second parts of the table, lines e and k, shows that they apply to the glass used. Having obtained the total volume of the bulb, m, and its water value, l, we obtain at once $\dfrac{l}{m} = n$, the specific heat per unit volume of the bulb considered as a whole, or in other words it is the water value of 1 c.c. of bulb. The values of *n* agree very closely, the extremes being 0·484 and 0·472, and the mean 0·476. The factor used in the rule, page 83, is 0·475, which is a more convenient number than 0·476, and it has now been shown how it is arrived at.

The third part of the table gives the centesimal composition by volume of the bulbs of the four thermometers. The thermometers are by different makers, and the quantity of mercury in each shows how different they must be in pattern; yet there is great resemblance in their composition by volume. The mean composition is 61 volumes of mercury to 39 volumes of glass, and *we shall never be far wrong if we take the volumes of the mercury and glass in a thermometer bulb to be in the proportion* 3 : 2.

Method of Determining the External Volume or Displacement of the Bulb.—We have shown how the factor 0·475 is arrived at; it now remains to show how the volume or displacement of the bulb is determined. It can be roughly ascertained by actual displacement of water in a graduated vessel, but this method is not sufficiently delicate. When the thermometer has a spherical bulb, its diameter must be measured with calipers. The result is not usually satisfactory, because the length to be measured is very short, and it is never certain that the bulb is truly spherical. Fortunately it is usual nowadays to make thermometers with cylindrical bulbs, and their volume is easily measured. Instead of using calipers a fine thread is wound ten, twenty, thirty or more times round the bulb, then unwound and its length measured; this divided by the number of turns gives the circumference of the cylinder. The length of the cylinder is then measured, and its ends are assumed to be hemispherical. With the circumference thus determined, the diameter and sectional area are calculated, or, more conveniently, taken from tables. The length of the cylinder multiplied by the sectional area gives the volume of the cylindrical portion. The two hemispherical ends make up a sphere of the same diameter as that of the cylinder, and its volume is likewise calculated, or taken from tables. It is added to that of the cylinder, and the sum is the volume or displacement of the bulb. It is convenient to record the measurements in terms of the centimetre, and the volume is then given in cubic centimetres. The external cooling surface of the bulb is found by adding the hemispherical surface of the exposed end to the cylindrical surface, which is the product of the length of the cylinder into its circumference. The area of the external surface is given in square centimetres (cm.²). It is important to observe that, when cooling, a thermometer loses heat through the whole area of its external surface; when receiving heat from a particular direction, as, for instance, from the sun, the receiving area is, for the cylindrical part of the bulb, its length multiplied by its diameter, and for the hemispherical ends the area of a great circle of the sphere.

All these measurements have been carefully made on the bulb of the thermometer by Chabaud, col. 4, Table XIII. For the determination of the circumference, a fine thread was wound forty times round the cylindrical part. When unwound it measured 111 cm., whence we have:

Circumference	2·775 cm.
Diameter	0·883 cm.
Circular area of cylinder	0·6124 cm.²
Length of cylinder	4·08 cm.
Volume of cylinder ... ... 4·08 × 0·6124 = 2·498 c.c.	
Volume of sphere 0·883 cm. in diameter ...	0·377 c.c.
Whence total volume of bulb	2·875 c.c.
Water value of bulb ... ... 0·475 × 2·875 = 1·366 grms.	

It will be seen that by mensuration we arrive at 2·875 c.c. as the volume of the bulb in place of 2·835 c.c. as derived from the weights supplied by the maker. The difference, 0·04 c.c., is under 1½ per cent. of the whole volume. Applying 0·475 as the specific heat per unit volume of the bulb, we obtain for the water value 1·366 in place of 1·338 grms., a difference of 0·028, which is just 2 per cent. The agreement is quite satisfactory, and we can therefore use this method with perfect confidence in determining the water value of the bulbs of all our thermometers.

The area of the external cylindrical surface is 4·08 × 2·775 = 11·322 cm.²
Surface of one hemisphere ... ... = 1·225 cm.²
Whence the total external surface ... ... = 12·547 cm.²

The area of this surface is the area of the *outlet* for heat, and, other things being equal, it determines the rate and term of cooling. We found the total volume of the bulb to be 2·875 c.c., and its water value 1·366 grms. It is clear that, if the area of the external surface remains constant and the volume or the water value is increased, other things remaining the same, the term of cooling will be increased and, of course, the rate diminished. It is quite analogous to the case of a water-cistern. If the area of outlet remains the same the time which it will take for the cistern to empty itself will depend on the quantity of water in it, while, if the size of the cistern and the quantity of water in it remain the same, the time which it will take to empty itself will depend on the *smallness* of the outlet. Other things being equal, the term of cooling of a thermometer will depend on the magnitude of the ratio, volume of bulb : area of external surface.

In this case the ratio is 12·547 : 2·875 = 4·364. This may be called the *virtual cooling radius*. In the case of a cooling sphere it is the radius of the sphere.

The physical and thermal data relating to particular thermometers have been dwelt upon at great length because they convey a much better idea of the nature and propagation of heat than more voluminous general statements. It may be well here to remind the reader that in the department of science which has been called *geo-physics*, when it is a question of specific heat it is nearly always specific heat per unit volume that is required, and the thermal study of thermometers is helpful from this point of view.

Calorimetric Constants of a Thermograph.—It was interesting to know what could be obtained with a recording thermometer of ordinary type, and in Table XIV the results of some observations made at Cambridge with a Richard's recorder are given.

The figures in this table are taken from the curves drawn by the instrument on a drum revolving once in forty-eight minutes. The instrument was allowed to take the temperature of the room, then exposed in the shade in the open air when a fresh breeze was blowing, and allowed to remain there until it had taken the temperature of the air. It was then transferred to the room, and allowed to rise until it attained its temperature.

In this way two sets of curves were obtained, consisting of three curves in still air and three in a fresh breeze. The results are not very concordant, for, although the scale of time is very open—one minute occupying 5 mm.—the temperature scale was very close, 1° occupying only 1 mm. The object, however, of the table is to show what can be expected from

TABLE XIV.—*Giving the Time in Seconds required by a Richard's Recording Thermometer to change its Temperature by 1° C. for a given Difference of Temperature between it and the Air.*

Difference of temperature between thermometer and air at beginning of exposure	12°	11°	10°	9°	8°	7°	6°	5°	4°	3°	2°
In the open air and fresh breezes	20″	20″	25″	25″	30″	30″	30″	65″	90″	90″	240″
	—	—	—	—	—	35	45	120	130	150	300
	—	—	—	—	—	20	35	40	45	80	240
Mean from curve	20	22	24	26	28	30	35	52	84	140	250
In still air in a room	—	—	—	—	—	—	60	70	110	130	210
	—	—	—	—	—	—	—	90	100	300	450
	—	—	—	—	—	—	—	—	120	160	300
Mean ...	—	—	—	—	—	—	60	80	110	180	320

(Left-side label: Time in seconds required by thermometer to fall or rise 1° C. for above differences)

an instrument of the kind in the measurement of changes of temperature. The results obtained in the open air would necessarily vary somewhat, because, although a fresh breeze was blowing all the time, a fresh breeze varies in velocity.

In order to obtain the best results from a thermometer, it should be exposed to uniform ventilation. This can only be effected by artificial means, and they necessarily tend to efface sharp variations of temperature.

Departing from the mercurial thermometer the writer has

found the simple air thermometer very good for indicating and measuring quick variations of temperature. It has the advantage of lightness and cheapness. The form which I use is a glass bulb, of about 3 cm. diameter on a straight stem of about 10 cm. length. This can be attached to a U-tube of greater or less diameter, according as the differences of temperature to be observed are great or small. The U-tube has some coloured water as indicator, and the indications of the instrument are compared with those of a thermometer. As the instrument is only put together when it is wanted, the variations of barometric pressure do not affect it. It has the great advantage that it can be connected with a *tambour*, and thus be made to record. The sensitiveness of the glass air thermometer is about the same as that of a very fine mercurial thermometer made for me by Messrs Hicks. The air thermometer, however, would be more sensitive if the ball were made of thin metal instead of glass.

Air thermometers of this simple kind described are very easily made so as to give calorimetrical results. It is only necessary to weigh and measure the piece of glass tube before and after blowing the bulb. The shortening of the straight part of the tube after blowing gives the length of it which has been expanded into a ball, and from the known length and weight of the original piece of tube the weight of the bulb is found. By carefully gauging the diameter of the ball its surface can be obtained, and from that the thickness of the glass. When the specific heat of the glass is known, the water value of the bulb is given ; if the air contained is taken into account, the value is increased by from 1 to 2 per cent. The surface of the ball divided by the water value gives an expression for the sensitiveness of the instrument.

In Table XV the particulars of several air thermometers are given. As they are made of lead glass, both the density and the capacity for heat per unit of weight are higher than in the case of ordinary German glass, but the specific heat per unit of volume is probably very little affected.

Calorimetric Constants of Deep-sea Thermometers.—It is evident that a knowledge of the calorimetric constants of the deep-sea thermometer is necessary, if we are to have the conviction that the temperature indicated by it is in truth the temperature of the water in which it was immersed. This is all the more necessary because, in order to guard the bulb of the thermometer against the squeezing effect of the pressure of the column of water to which it is exposed when in use, it is hermetically enclosed in an outside bulb, the space between them being partially filled with mercury. This extra bulb increases greatly the term of cooling of the thermometer. The conditions are quite analogous to those regulating

TABLE XV.—*Particulars of Calorimetric Air Thermometers made of Lead Glass.*

Number of Instrument ...	1	2	3	4	5
Original weight of tube (grm.)	7·724	18·508	18·4186	18·8136	18·6169
,, length of tube (mm.)	225·7	193·0	192·1	196·0	194·25
Ditto after blowing	197·0	144·0	137·0	126·4	104·0
Difference	28·7	49·0	55·1	70·4	90·25
Weight of 10 mm. tube (grm.)	0·7853	0·9590	0·9590	0·9580	0·9580
Weight of bulb (grm.) ...	2·2538	4·6991	5·2841	6·7443	8·6550
Diameter of bulb (mm.) ...	24	32	38	45	51
Volume of bnlb (c.c.) ...	7·238	17·157	28·731	47·713	69·456
Surface of bulb (sq. cm.) ...	18·095	32·170	45·364	63·617	81·713
Volume of glass at sp. gr. = 3·0	0·7513	1·5664	1·7614	2·2481	2·8850
Thickness of glass (mm.) ...	0·415	0·487	0·388	0·353	0·353
Water value of bulb, sp. heat = 0·57	0·4282	0·8928	1·0040	1·2814	1·6445
Surface ÷ water value ...	42·26	36·03	45·18	37·25	42·24

the cooling of thermometers in air. The term is comparatively long when the thermometer is immersed in still water and kept motionless in it. When the water has relative motion with regard to the thermometer the term is reduced in proportion to that motion.

For practical purposes we require to know how long we must leave the thermometer at the particular depth in order to be sure that it has taken the temperature of the water. The experiments required in order to furnish this knowledge are extremely simple. The principle is exactly the same as

that which governs the behaviour of thermometers in air.
The thermometer loses equal fractions of its excessive heat
in equal intervals of time. These intervals are very much
shorter when the instrument is immersed in water than when
it is in air. When the difference of temperature is at all
considerable the thermometer falls very rapidly at first, and
more slowly as it approaches to the temperature of the water.
The divisions of the scale of a thermometer ought to be
about one millimetre apart. If in reading the thermometer
we estimate tenths of this amount, then it is important to
know how long the thermometer takes to assume the temperature of the water within one-tenth of one of its own divisions.
If we estimate only to one-half of a division, then it is
sufficient to know how long the thermometer takes to arrive
at that of the water within one-half of one of its own divisions.
It does not matter what the thermometric value of each
division is. The following imaginary case will illustrate this.
It is assumed that when immersed in still water the thermometer loses half its excessive heat in twenty seconds. When
immersed in the water the excessive temperature of the
thermometer is represented by 4·8 of its own divisions.

Thus we should have :

Excess temperature, divisions	4·8	2·4	1·2	0·6	0·3	0·15	0·075
Time elapsed, seconds ...	0	20	40	60	80	100	120

In this imaginary case the temperature of the thermometer
will have attained that of the water in 120 seconds, or two
minutes.

When the water touching the thermometer is moving,
whether the water runs past the thermometer or the thermometer runs through the water, then the time which the
thermometer requires to lose its excessive heat is very much
shorter. In actual sounding practice the thermometer arrives
at the depth at which it is to register the temperature, having
already very nearly the temperature of the water at that
depth. It is, therefore, generally speaking, quite safe to
despatch the messenger, which is to overturn the thermometer,
so that it may arrive at the required depth not later than one

or two minutes after the thermometer. This refers only to *oceanic* work. The physicist must himself test the actual thermometers which he has on board by observing them in still water and when moved at certain velocities in water, and from his own observations he will form his own opinion of how long the particular instruments with which he is supplied have to be left down, and then he will use his knowledge to guide his practice.

Application of Calorimetry to Hydraulics.—An interesting example of the application of calorimetry to problems of physical geography is afforded by some hydraulic experiments made by the writer in the Engadine in the summer of 1894. About a mile below the end of the Morteratsch glacier, the muddy stream which proceeds from the glacier is joined by the perfectly clear waters of the stream which descends from the summit of the Bernina pass, the waters of which drive the machinery for the supply of electricity to Pontresina. In the afternoon the glacier stream is running strongly, and its temperature was found to be 1° C. The temperature of the water of the Bernina stream was 11° C. Two or three hundred yards below the confluence, where the waters had been well mixed, the temperature of the water was 2°·5 C. If we represent by M the flow of the Morteratsch and by B that of the Bernina stream, we have the relation $M : B :: 10 : 1\cdot5$, whence we have $B = 0\cdot15 M$, or the stream below contains 85 per cent. of Morteratsch water and 15 per cent. of Bernina water.

Temperature of Insolation.—In a meteorological outfit it is usual to include a thermometer with blackened bulb enclosed in an external glass tube. This is to be exposed to the direct rays of the sun until it rises to a temperature which remains stationary. This temperature is to be observed. When taken in connection with the temperature in the shade this is to furnish information regarding the calorific action of the sun's rays. It is in a position to furnish reliable information of this kind only if the thermal mass or water value of the bulb and its term of cooling are known. These are not commonly supplied. Even if they were supplied it is doubtful if the instrument would furnish any trustworthy information; the external glass envelope introduces so much disturbance.

When a thermometer, whether its bulb be blackened or not, is exposed to the direct rays of the sun, its temperature rises at first rapidly, then more and more slowly until finally it reaches a temperature at or about which it remains steady, provided that it is in a place where the air is still and its temperature constant. While its temperature remains steady the thermometer is continually receiving heat from the sun's rays and giving it off to its surroundings, and the meaning of the constancy of its temperature is that the receipt and expenditure of heat per unit of time are equal.

If we know the thermal mass of the bulb of the thermometer and its term of cooling and the difference between its temperature and that of the air, we can calculate the rate at which it is losing heat, and this, at the stationary temperature, is the rate at which it is receiving heat.

If the thermometer is exposed with its axis perpendicular to the direction of the sun's rays, then it receives the rays which fall on a surface equal to the axial sectional area of the bulb. Consequently, the heating power of the bundle of solar rays having this sectional area is equal to the heating power of the thermometer when the excess of its temperature above that of the medium has become constant. This is rigorously true of that portion of the pencil of rays which penetrates the bulb of the thermometer. If the rays have had to traverse the windows of a room, some of the rays are absorbed by the glass, and a very large portion is dissipated by reflection from it. Further, even a blackened bulb does not transmit and absorb all the heat rays that strike it.

If two thermometers of different pattern are exposed side by side to the direct rays of the sun, they usually assume very different stationary temperatures, even although their graduation may be perfectly exact. This is to be expected, because thermometers of different pattern are sure to differ considerably both in term of cooling and in thermal mass. The nearest approach to equality in these particulars is found amongst common thermometers of the same pattern, which are turned out in large quantities at a time.

Example of the Method of Determining the Calorific Power of the Sun's Rays, which strike the bulb of a Thermometer and are absorbed by it; and of the difference between different thermometers in this respect.—The following observations were made with two thermometers of very different pattern. They were hung side by side in a room exposed to the powerful winter sun at St Moritz, in the Engadine, between 11 and 11.30 a.m. on February 26, 1901. The thermometers are designated A and B. A is an ordinary German thermometer, divided into whole degrees on a milk-glass scale and its bulb was coated with silver. B was an English thermometer with solid stem, and divided into tenths of a Centigrade degree, each degree occupying a length of 1 cm. Its bulb was not coated in any way. The particulars relating to the bulbs were obtained by mensuration as above described. The mean altitude of the sun was 33° and cos 33° = 0·84; the axial sectional area of each thermometer has to be multiplied by this factor in order to obtain the effective area exposed, or rather to obtain the sectional area of the bundle of rays which strikes the bulb of each thermometer.

TABLE XVI.—*Specifications of the Thermometers.*

Designation of thermometer			A	B
Diameter of bulb ... ... cm.	a		0·637	0·471
Length of equivalent cylinder ... cm.	b		2·3	2·4
Volume of bulb ... ... c.c.	c		0·7334	0·4181
Water value of bulb ... ... grms.	d		0·3375	0·1923
Area of external surface ... cm.^{2}	e		4·92	3·727
Area of axial section ... ... cm.^{2}	f		1·465	1·130
Effective area of ditto ... ... cm.^{2}	g		1·230	0·950
Term of cooling ... ... seconds	h		250	120

In Table XVI we have the specifications of the bulbs of the thermometers. It will be observed that the "length of the equivalent cylinder" b is given simply, in place of the cylinder and two hemispheres, as in the case of thermometers with large bulbs. The volume of the bulb c is found by multiplying

Table XVII.—*Example of Calorimetric Use of Thermometers for Determining the Radiant Heat which they absorb.*

Designation of thermometer		A	B	A	B	A	B	A	B	A	B
Stationary temperature reached in sun's rays ... ... °C.	k	21°·25	18°·5	22°·15	19°·1	22°·4	19°·2	23°·1	19°·6	22°·4	19°·2
Difference between stationary temperature and that of the air	l	6°·25	3°·5	7°·15	4°·1	7°·4	4°·2	8°·1	4°·6	7°·4	4°·2
Corresponding excess of heat in each bulb ... gr. °C.	m	2·110	0·673	2·406	0·788	2·498	0·808	2·734	0·885	2·498	0·808
Loss of heat per second ...	n	0·00844	0·00561	0·00962	0·00657	0·00999	0·00673	0·0109	0·00738	0·00999	0·00673
Ditto per square centimetre of effective axial sections ...	p	0·00686	0·00590	0·00782	0·00691	0·00812	0·00706	0·00889	0·00775	0·00812	0·00706
Ratio $\rho_A : \rho_B =$	q	0·860		0·896		0·869		0·872		0·869	

the circular area corresponding to the diameter a by the length of the equivalent cylinder b. The water value of the bulb is $d = 0.457c$. When Table XVI was calculated, 0.457 was used as the specific heat per unit volume of the bulb, instead of 0.475, which was adopted later. It has not been thought necessary to recalculate the table. The area of external surface is the circular area of the diameter, added to the product of the length of the equivalent cylinder and its circumference. The area of axial section is the product of the diameter of the cylinder and its length, and the effective axial area (g) is the axial section multiplied by 0.84; g is proportional to the quantity of the sun's rays, which strike the bulb in unit of time. The term of cooling, h, is given in seconds, and its reciprocal $\dfrac{1}{h}$ is the fraction of the excess of heat lost per second during cooling.

In Table XVII five separate observations of the stationary temperature reached by the thermometers side by side are given. The temperature of the air was $15°·0$ C. The excess of heat in each bulb is $m = ld$, or the water value of the bulb multiplied by the excess of its stationary temperature above that of the air. The loss of heat per second $n = \dfrac{m}{h}$ as was above described; $p = \dfrac{n}{g}$ is the loss of heat per second referred to unit area of effective axial section. It has been pointed out that g, the effective axial area, is the sectional area of the bundle of solar rays which strikes the bulb, and it is therefore proportional to the supply of radiant heat by the sun to the thermometer. But when the temperature of the thermometer in the sun has become stationary it is dissipating the whole of the heat which it is receiving. Now n gives the rate at which it is cooling, in gramme-degrees[1] (gr.° C.) per second; therefore the bundle of sun's rays of sectional area g is supplying

---

[1] It is convenient to give compound names to compound units; they then explain themselves. One gramme-degree (1 gr.° C.) is the heat required to raise the temperature of one gramme of water by one Centigrade degree. Names such as calorie or therm are indefinite, and may be confusing.

heat at the rate of n gr.° C. per second, and $p = \dfrac{n}{g}$ is this rate for a bundle of rays of 1 square centimetre section. It will be seen that all the values of p given by thermometer B are lower than those given by A. In line q we have the ratio of $p_A : p_B$, and it will be seen that the values of q agree very well with each other. The mean value is 0·873, the extremes being 0·860 and 0·896; that is to say, thermometer B indicates 13 per cent. less heat than thermometer A for 1 square centimetre of sun's rays. Yet A has a silvered glass bulb and B has a plain uncoated glass bulb. It is possible that to this difference is due the difference in the results obtained. The silvered bulb dissipates by reflection some of the heat which strikes its metallic surface, but whatever is not reflected by this surface passes into the bulb of the thermometer. In the case of the uncoated glass bulb there are two surfaces of reflection, namely that separating glass from air and that separating glass from mercury. The heat which has passed through the outer glass surface has still to pass the inner surface, where some of it is rejected by reflection.

Taking the maximum value of p, namely 0·00889 gr.° C. of heat supplied per square centimetre per second, and multiplying it by 60, we have 0·533 gr.° C. per square centimetre per minute. This is the heat actually taken up by the silvered bulb of a thermometer from 1 square centimetre of the rays of the winter sun, which have passed obliquely through a glass window. The sky was quite clear and cloudless, but the sun's zenith distance was 57°. In order to allow for this, we use the formula given by Sir John Herschel[1], from which we obtain the value of the *solar constant*

$$A = \frac{0\cdot533}{(\frac{2}{3})^{1\cdot84}} = 1\cdot1114 \text{ gr.° C. per square centimetre per minute,}$$

where $\frac{2}{3}$ is the *transmission coefficient* of the air for heat and $1\cdot84 = \sec 57°$.

Two-thirds of this, or 0·7409, would then be the heating power of the rays of the vertical sun at the surface of the earth

[1] *Meteorology*, by Sir John Herschel, Bart. Edinburgh, 1861, p. 10.

at a height of 1850 metres above the sea. This calculation
has been carried out in order to make the example complete,
and not as an experimental determination of the calorific
power of the sun's rays.

At the same time it illustrates the principle of the *actino-
metric method* of determining the heat of the sun's rays, which
has been much used.

Apart from the annual cycle of variation of the sun's
distance from the earth, there is no reason to expect any
measurable variation in its heating power. Therefore the
principal object to be attained is to find the maximum
heating power of the sun under the most favourable circumstances. This is perhaps best given by the calorimeter
depending on the rate of generation of steam[1], which was
designed by the writer, and used by him in Egypt in May
1882. The values of the solar constant obtained with the
actinometer by observers of the highest standing differ
greatly, and some of them are certainly exaggerated. What
we want to know in physical geography is how much heat
is quite certainly received from the sun by a given area of
the earth's surface exposed perpendicularly to its rays in
a given time. With the writer's steam calorimeter the highest
rate actually realised was 16·6 grms. of water converted into
saturated steam of the same temperature per minute on an
area of 1 square metre when the sun's zenith distance was
20°. This is equivalent to 17·04 grms. steam generated by
a vertical sun on the same area. The latent heat of steam
at 100° C. is 535 gr.° C. per gramme, therefore the generation
of 17·04 grms. of steam of 100° C. out of the same weight of
water of the same temperature requires 9116 gr.° C. of heat;
and this is the amount of heat which can with certainty be
extracted, per minute, from a bundle of sun's rays of 1 square
metre sectional area at the sea-level when the sun is at the
zenith. The mechanical equivalent of heat is taken at 425
kilogramme-metres (kgm.) of work per kilogramme-degree

---

[1] 'On a Solar Calorimeter used in Egypt in 1882,' by J. Y. Buchanan.
Proceedings of the Cambridge Philosophical Society (1901), vol. xi. pp. 37–74;
and *Nature* (1901), vol. lxiii. p. 548. (See below, p. 337.)

(kg.° C.) of heat, or per 1000 gr.° C. Converting 9116 gr.° C. at this rate we obtain as its equivalent in work 3875 kilogramme-metres (kgm.). This amount of work is done in one minute. One horse-power, or the rate at which the standard horse can work, is taken at 4500 kgm. per minute; *therefore the working value of the sun's rays at the sea-level is at least 0·87 horse-power per square metre for a vertical sun.* The total area of the bundle of sun's rays which is at all times being intercepted by the earth is the area contained by its great circle, and this is taken at 130×10^{12} square metres. Therefore the working value of the sun to the earth is at least 113×10^{12} horse-power. This figure depends on the amount of steam actually generated in a particular instrument; but no instrument is perfect, therefore the above figure falls short of the truth. One horse-power per square metre has been taken as a probable work-value of the sun's vertical rays at the level of the sea. This is equivalent to 1·06 gr.° C. per square centimetre per minute. In accepting these values of the solar heat constant at the sea-level we are assured that we are not exaggerating.

It is impossible to determine, or to estimate exactly, how much of the sun's heat is absorbed in its passage through the atmosphere. We have seen that Herschel estimates the amount transmitted to be two-thirds of the amount which arrives at the earth's orbit, leaving one-third to be absorbed. The true amount absorbed is probably rather under than over this figure.

Taking 1 horse-power per square metre as the total work-value of the sun's rays, and remembering that the mean distance of the earth from the sun is 212 times the length of the sun's radius, we find that the rays emitted by 1 square metre of the sun's surface are spread over 212^2, or in round numbers, 45,000 square metres of the earth's surface. Therefore, *the probable work-value of 1 square metre of the sun's surface is at least 45,000 horse-power.*

It is useful to note that the sun's heating power at the distance of the planet Mercury is $6\frac{1}{2}$ times, and at that of Venus it is twice, its value at the earth's distance.

The Necessity of the Knowledge of Calorimetric Factors in Connection with the Use of the Barometer.—It is an interesting historical fact that Fahrenheit, to whom we owe the thermometer as a physical instrument of precision, got the idea of filling his thermometer with mercury by observing and being troubled by the irregularities of the barometer due to change of temperature. It is also probably not altogether accidental that the length of one degree of his thermometric scale corresponds to one ten-thousandth of the volume of the mass of mercury used in the thermometer[1]. If mercury expands by $\frac{1}{10000}$ for every Fahrenheit's degree, then an error of 1° F. in the estimation of the temperature of the barometer introduces an error in the barometric pressure of $\frac{1}{10000}$ of the whole height of the column of mercury. If that height is 30 inches the error is three-thousandths of an inch. If the height is 760 mm. the error is 0·076 mm. It is obvious, therefore, that unless we can be perfectly certain that the temperature of the barometer is within $\frac{1}{3}$° F. of what we take it to be, there is no sense in reading the barometer to units in the third decimal place of the inch, and still less so to units in the second decimal place of the millimetre. But, it will be said, we have the "attached thermometer," and it can easily be read to $\frac{1}{10}$° F. Completing the suggested syllogism, we have :—Now the temperature of the attached thermometer is always identical with that of the mercury of the barometer to which it is attached; therefore, we know the temperature of the mercury of our barometer to $\frac{1}{10}$° F.; and by consequence we are justified in reading our barometer to 3 in the fourth decimal place of the inch, and to 7·6 in the third decimal place of the millimetre. If the premises are granted, the conclusion is necessary. But the second premise is true only in certain conditions which do not usually obtain. If the temperature of the room in which the barometer is hung is invariable, the premises may be granted and the conclusion is valid. If the temperature is not invariable, the premises cannot be granted, and the conclusion is false. Sufficient attention is not paid to this source

[1] This fact gives Fahrenheit's thermometer a genuine title to the name *centigrade* which Celsius' scale lacks.

of uncertainty in barometric readings. After what has been said about the term of cooling of a thermometer, the reader will easily see where the fault lies and how to remedy it. In so far as the effect of temperature on the length of the mercurial column is concerned, the barometer is a thermometer with a bulb of the volume and thermal mass of the column of mercury. The term of cooling of this mass must be very much greater than that of the attached thermometer; so that if at any epoch the two chanced to be at exactly the same temperature, and the temperature afterwards went through the usual diurnal changes, they would never again have the same temperature, except for a moment of time when their thermal paths happen to cross. The only adequate "attached thermometer" is one which has a bulb which is a copy of the barometer, has the same thermal mass and the same term of cooling, and is exposed to exactly the same condition as the barometer. In important central stations the standard barometer should be accompanied by such a thermometer. In standardising stations nothing but absolute uniformity of temperature should be admitted. The need of correction is the confession of imperfection. In exact work no correction should be admitted which it is mechanically possible to exclude.

In Europe and North America, and generally in countries where for a part of the year dwellings have to be artificially heated, the condition of uniformity of temperature is not secured of itself, and it may be troublesome to provide it. The Paris observatory has its classical cellars of invariable temperature. What could be artificially provided centuries ago can be provided now. For a central national standard-. ising institution such a chamber is indispensable, if its certificates are to have the value which ought unquestionably to belong to them.

In tropical, and still more in equatorial regions, where the diurnal variation of temperature is small and its rate slow, uniformity of temperature, sufficient for a first-class meteorological station, is easily obtained in any well-constructed building, and consequently perfection is more easily attained there than in regions more remote from the equator.

But the ideal conditions of temperature for a first-class physical laboratory are to be found on board a large wooden ship at sea between the tropics. The temperature of the surface-water does not vary by 1° F. in twenty-four hours. The temperature of a shaded thermometer on deck may vary by two or three, or even more, degrees, according to the greater or less efficiency of the shade, but *the true temperature of the air varies as little as does that of the sea surface.*

During the three years that the "Challenger" sojourned in tropical seas the writer had daily occasion to notice this fact, and to some extent, though inadequately, to take advantage of it. In particular, a series of experiments was made on the compression of deep-sea thermometers, which required practically absolute uniformity of the temperature of the air. This was found on the main-deck, where the compression apparatus was installed. The main-deck of the ship was protected by the spar-deck, and by an awning above that. It was ventilated by twenty-eight open gun-ports. All passages were made under sail; therefore, when under way, one side of the ship was always definitely a weather side and the other definitely a lee side, and the gun-ports afforded unobstructed passage to whatever wind was blowing. The main-deck of the "Challenger" was, therefore, a perfect "thermometer screen," and tested in these perfect conditions the air preserved as uniform a temperature as the water. While the compression experiments were being made the variation of temperature during the working part of any one day was not greater than one-tenth of a Centigrade degree.

It would be impossible on shore to provide such conditions of work. Hence, so far as temperature conditions are concerned, and other things being equal, observations of the barometer on board ship in tropical regions are entitled to more weight than those made elsewhere.

Giving effect to these considerations, we have the following :

Directions for Hanging a Barometer on Board Ship.—The barometer is to be hung in the inhabited part of the ship, in a saloon or cabin, not on deck, but with a substantial deck above it, and out of the radius of either sidelights or skylights. In such conditions the temperature of the barometer and that of its attached thermometer may agree.

The Effect of Change of the Force of Gravity on the Pressure of the Atmosphere and on the Height of the Barometer.—If we consider a siphon-barometer, and take the plane of the surface of the mercury in the open limb as *datum level*, all that is above it in one limb exercises the same pressure as all that is above it in the other. But all that is above it in the outer or open limb is the column of the atmosphere, and all that is above it in the inner or closed limb is the column of mercury: therefore, the weight of the column of mercury is the same as the weight of the column of air. The weight of a body is the product of its mass into the force of gravity at the place. The force of gravity is the same[1] in both limbs of the barometer, therefore the mass of the column of air is the same as the mass of the column of mercury. If, other things remaining the same, we reduce the density of the earth to one-half, there will still be equilibrium between the columns of air and of mercury in the barometer. The height of the barometer will be unaltered, yet the pressure of each of these columns on its base will now be only one-half of what it was before, because the pressure of the column depends on its weight, and the weight is proportional to the force of gravity. In both cases we shall have correctly the *height of the barometer*, and it will have remained unaltered, but with equal heights of the barometer the pressure of the air will, in the latter case, be one-half of what it was in the former.

The force of gravity is not the same at all points of the earth's surface. Hence arises the necessity for the *gravitation correction*, in order, from the observed height of the barometer, to ascertain the true pressure of the atmosphere expressed in

---

[1] Such slight differences as are due to the greater distance of the centre of the earth from the centre of gravity of the atmospheric column than from that of the mercurial column are here neglected.

standard units, that is, in millimetres of mercury having a temperature of 0° C., and subject to the attractive force of gravity which it would experience at the sea-level in latitude 45°. This force generates in one second a velocity of 980.6 centimetres per second.

If we consider the layer of air of 1 mm. in thickness which is contiguous to the surface of the mercury in the outer limb of the barometer, and is pressed down upon it by the weight of the whole atmospheric column above it, we see that the elasticity of this thin layer of air exactly balances the pressure of the air above it and that of the mercury below it. The tension of this layer of air is equal to the atmospheric pressure. Imagine the density of the earth to be reduced by one-half; the height of the barometer is still the same, but the pressure on both sides of the thin layer of air is reduced to one-half; therefore it expands to double its volume, and the thickness of the layer becomes 2 mm., and by consequence its tension has been halved.

Imagine now a thin layer of water on the surface of the mercury in the outer limb of the barometer. It is pressed down by the weight of the atmospheric column, and it is pressed up by that of the mercurial column. The tension of the air in contact with the water surface is equal to the pressure of the atmospheric column. Let the height of the barometer be 735.5 mm., then taking the mean force of terrestrial gravity at the sea-level in lat. 45°, the atmospheric pressure is 1 kilogramme per square centimetre, and the tension of the air in contact with the water is also 1 kilogramme per square centimetre. Let the layer of water be heated. When it arrives at a temperature of 99°.1 C. the tension of its vapour is exactly 1 kilogramme per square centimetre, and it is therefore equal to the atmospheric pressure, and any further supply of heat will cause the water to boil at the temperature of 99°.1 C. Let the water be cooled down again; and let the density of the earth be reduced by one-half. Then the height of the barometer will remain unaltered at 735.5 mm., but the pressure of the air will be halved and will be only 0.5 kilogramme per square centimetre. Let the water now be warmed. When the temperature of the water arrives at 80°·9 C. the tension of its vapour is exactly 0·5 kilogramme per square centimetre, and any further supply of heat will cause the water to boil at this temperature.

It is evident then that if we know the relation between the tension of aqueous vapour in kilogrammes per square centimetre and the temperature, we have in the boiling-point of water a means of measuring the true pressure of the atmosphere in which it is boiling.

But we have seen that the pressure of the atmosphere is proportional to the quantity of the air that is in the column which rests on the surface of the boiling water, and to the force of gravity or the earth's attraction which pulls it towards its centre. Suppose the force of gravity constant, the boiling-point of water will be the lower the less air there is in the column above it, or the greater the height of the locality above the level of the sea. If we know the law regulating the relation between pressure of the atmosphere and elevation above sea-level, we have in the boiling-point of water a means of determining the elevation, because the tension of the vapour of the boiling liquid is equal to the tension of the air on its surface, and the tension of the air at the bottom of a column of it is equal to the pressure of the column, and the relation between air pressure and altitude is known.

If the quantity of air in the column resting on the boiling water remain constant, and the force of gravity, or the attractive force of the earth on a particle at the surface of the water be altered, then the pressure of the column of air on the surface of the water and the tension of the air at the bottom of the column will be altered correspondingly.

If the force of gravity has been increased, the pressure of the air and the tension of the air at the bottom of the column will be increased ; and if the temperature of the water remain the same the tension of its vapour will be less than the tension of the air in contact with it, and it will cease to boil. If further heat be supplied to it, its temperature will rise to that at which the vapour tension of the water is equal to that of the increased tension of the air at its surface, and the water will again boil.

If the force of gravity has been reduced, the pressure of the atmosphere and the tension of the air in contact with the boiling water will be reduced. The vapour tension of the water will then be higher than that of the air in contact with it, the water will be super-heated and will boil for a time without any further supply of heat, the *super-heat* in the water being sufficient. When this is exhausted, the temperature of the water will be reduced to that at which its vapour tension is the same as the reduced tension of the air in contact with it, and with a further supply of heat it will continue to boil at this temperature.

During these changes of the force of gravity and consequent changes of the boiling-point of water, the height of the barometer has remained without change, because the changes of force of gravity act equally on both limbs of the siphon.

In the same way a balance which is loaded with a litre of water on the one pan and a kilogramme of platinum on the other remains in equilibrium, whatever changes may be made in the force of gravity. If, however, the litre of water were weighed on a *spring*-balance, its weight would vary in the same proportion as the force of gravity. Its weight on a spring-balance is the measure of the power which it has to overcome a certain constant resistance, the resilience of the spring. The whole of this power is conferred on it by the force of gravity. By virtue of its mass alone it has no such power. Hence, in the case considered, when the density of the earth is supposed to be halved, the weight of the litre of water would be halved also, and it would be registered as 500 grammes on the scale of the spring-balance. The same would apply to the kilogramme of platinum. The spring-balance would give its weight as only 500 grammes, and it is evident that under the altered conditions of gravity they would still balance each other in the opposite pans of a pair of scales. Like the hypsometer, the aneroid barometer measures the pressure of the air, not the height of the barometer. It is a spring-balance, whilst the mercurial barometer is a pair of scales.

The gravitation correction for the barometer, as supplied in meteorological instructions, is based on the assumption of a homogeneous earth, and refers only to varying distance of the station from the centre of the earth. The force of gravity varies inversely with the square of this distance. It is obvious that there will be some difference between the force of gravity at the level of the sea and that at the top of a high mountain in the neighbourhood. We have taken the mean radius of the sphere to be 6371 kilometres. At a height of 1000 metres above the sea the force of gravity is less than it is at the level of the sea in the proportions $6372^2 : 6371^2 = 0.99968$. At a height of 5000 metres the relative reduction is 0.99843, and at 10,000 metres it is 0.99687.

Owing to the spheroidal shape of the earth, places at the sea-level are at different distances from the centre of the earth if they are situated in different latitudes.

The most recent values of the dimensions of the earth's spheroid are :

Semi-axis major	...	...	...	6378 kilometres
Semi-axis minor	...	...	...	6356 ,,
Difference ...	...	...	22	,,

The greater radius is the equatorial and the smaller the polar, therefore the force of gravity at sea-level is a maximum at the poles and a minimum at the equator, and the relation between them is 0.99311. The difference between the equatorial and the polar radii is about three times the height of the highest mountain on the earth's surface.

In order to illustrate these remarks Table XVIII has been constructed. It is supposed that the barometer has been observed at a temperature of $0°$ C., and at the sea-level in the latitudes given in the first column, and that it stands always at a height of 735.5 millimetres. As the force of gravity at the sea-level increases from the equator to the pole, this constant barometric height corresponds to a true atmospheric pressure, which increases as the pole is approached. This is given in the second column in grammes per square centimetre. In the third column these pressures are given in terms of the height of a column of mercury at 0° C. at the sea-level, *and in latitude* 45°. In the fourth column the temperature of saturated steam or the boiling-point of water under these true pressures is given. It will be seen that, other things being the same, *the boiling-point of water at the sea-level is* 0°·147 *C. higher at the pole than it is at the equator.* This figure is based on the assumption that the mass of the earth is homogeneous, or if heterogeneous, that it is symmetrically heterogeneous with reference to its centre. It is also assumed throughout the whole of this argument, that there are no fluctuations in the atmosphere itself, and that the quantity of air in a column having unit area of base, at the sea-level, is everywhere the same.

TABLE XVIII.—*Giving the True Atmospheric Pressure and the Boiling-Point of Water Corresponding to a Barometric Height of* 735·5 *Millimetres at* 0° *C. and at the Sea-Level in Different Latitudes.*

| Latitude | Atmospheric Pressure | | Corresponding Temperature of Saturated Steam, or the Boiling-Point of Water |
	Grammes per Square Centimetre	Millimetres Mercury at 0° C. and Gravity as at Latitude 45°	
	gr. per cm.²	mm.	° C.
0°	1002·65	733·55	99·012
10°	1002·50	733·66	99·017
20°	1002·04	734·00	99·029
30°	1001·33	734·52	99·048
40°	1000·46	735·16	99·072
45°	1000·00	735·50	99·087
50°	999·54	735·84	99·099
60°	998·67	736·48	99·123
70°	997·80	737·00	99·142
80°	997·50	737·34	99·155
90°	997·35	737·45	99·159

In the tabular example it is imagined that we carry a standard barometer from the equator to the pole, and that we read it at the sea-level in certain latitudes, the barometer being entirely and exactly at the temperature of melting ice;

and it is further imagined that the height of the barometer at each of these stations is found to be the same, namely, 735·5 millimetres.

If we assume that the force of gravity is the same at the sea-level in all latitudes, then the fact that the atmosphere everywhere exercises the same pressure as a certain standard length of mercury, is evidence that the mass or the quantity of air in a column of the atmosphere having the same base is, at the sea-level, everywhere the same. If the height of the standard barometer had not been the same at each of these stations, and at the same time it was known that the force of gravity did not vary, then the conclusion would be necessary that the quantity of air in the atmospheric column at the sea-level is greater in one latitude than in another.

If the height of the standard barometer is, as postulated in the table, the same at each of the stations, and if it is known that the force of gravity is not constant, then the height of the standard barometer of itself gives us no information regarding the pressure of the atmosphere. The constancy of the height of the standard barometer justifies no other conclusion than that the pressure of the atmosphere is different in different latitudes.

If the law of the variation of the force of gravity with change of latitude is known, then the constant height of the barometer allows us to arrive at the *relative* pressure of the atmosphere at different latitudes. If, in addition, the absolute value of the force of gravity at any one latitude is known, the constant height of the standard barometer enables us to conclude what is the absolute pressure of the atmosphere at the different latitudes.

If the distribution of matter in the earth is homogeneous, and if the force of gravity at the sea-level varies, then the conclusion is necessary that different stations at the sea-level are at different distances from the centre of the earth, or they are at different *altitudes*, referred to this point as fundamental datum.

If the law which connects the variation of the force of gravity with distance from the attracting centre is known, the distance of the sea-level in any locality from the centre of the earth is at once ascertained by determining the force of gravity there; hence the gravitational method of determining the figure of the earth. Here it must be remarked that a perfect survey of the *figure* of the earth may be given without affording any information about its *size*.

To return to the barometric data in the table, we see that while the height of our standard barometer was the same, 735·5 mm. at every station, the indications of the hypsometer or the boiling-point varied. It has been shown that the *height of the barometer* is not affected by change in the force of gravity, but that *the pressure of the atmosphere* is so affected; we might conclude that in the temperature of saturated steam, or the boiling-point of water, at the sea-level, we have a method of determining the force of gravity at any place at that level. But we must be on our guard. Our knowledge of the relation existing between the boiling-point of water and the pressure on its surface is derived from observations of the barometer. Therefore, from the boiling-point of water we cannot get *absolute* data with regard to the force of gravity. But it may be that it can afford information about the *variations* of the force of gravity.

For instance, we observe our standard barometer having the temperature of melting ice at the sea-level, in lat. 20°, 45° and 70°, and at each of these places its height is 735·5 mm.; we apply the usual correction to reduce this barometric height to the equivalent height when the mercury is exposed to the same gravitational force of attraction. By convention, the standard force of terrestrial attraction is taken to be that exerted at the sea-level in lat. 45°. The barometric heights thus adjusted are now 734, 735·5 and 737 mm., and these columns of mercury under the influence of standard gravity exercise pressures of 1002·04, 1000 and 997·8 grms. per square centimetre. It results from experiments, principally by Regnault, on the relation between the boiling-point of water and the height of the standard barometer, that at these true pressures water boils at 99°·029, 99°·087 and 99°·142 C. respectively.

Supposing that we take the hypsometer to these latitudes, and that we determine the boiling-point of water with every care at each latitude at the sea-level when the standard barometer at 0° C. stands at 735·5 mm., and we find that the boiling-point is the same, namely 99°·087 C. at all these latitudes, then we conclude that the true pressure of the atmosphere at each of the localities is 1000 grms. per square centimetre. But we know from the geodetic determination of the figure of the earth that the sea-level is nearer the centre in lat. 70° than it is in lat. 45°, and nearer in lat. 45° than in lat. 10°. If the earth is homogeneous the force of gravity *must* be greater at lat. 70° than at lat. 45°, and at lat. 45° than at lat. 10°. But from the identity of the height of the barometer at the three localities we know that the *mass* of the atmospheric column is the same at the three places, and from the identity of the boiling-point of water we know that the *weight* of the three columns or the atmospheric pressure is the same in the three places. As the force of gravity depends only on the mass of the attracting body and on the distance of its centre from the attracted body, and we find that when this distance is varied in a certain measure no alteration is produced in the effective force of gravity, the conclusion is necessary, that the effect of variation of distance has been exactly compensated by variation of the effective attracting mass. In the example taken we find the force of gravity in lat. 10° greater than it should be, and in lat. 70° less than it should be.

As the attraction of the whole earth on a particle at its surface is the sum of the attractions of all the particles that make up its mass; and as the attractive force of each particle of the earth's mass on the particle at its surface is inversely proportional to the square of its distance from that particle, it is evident that the particles in the vicinity of the attracted body will exercise, mass for mass, a much greater attraction on the body than the particles that are more remote, for instance, at the opposite end of the diameter of the sphere. Therefore, the occurrence of rocks of relatively high density near the surface in a locality may easily cause an exaggerated force of gravity in the locality, and the occurrence of rocks of relatively low density may cause an analogous deficiency of local gravitational attraction.

Looking to the great variety of mineral substances which we meet with in the surface crust of the earth, and to the differences of their densities, it is not surprising that recent observations with the pendulum show that local peculiarities of the force of gravity are the rule and not the exception. The differences in neighbouring localities are not usually great, but the pendulum is an instrument of almost infinite delicacy, so that with patience the gravitational map of a district can be correctly constructed, no matter how slight the differences may be.

It will be evident from the examination of Table XVIII, that by the combined use of the barometer and the hypsometer we have in our hands the means of determining local deviations of the force of gravity from its normal value.

Rule for detecting the deviation of the local force of gravity from the normal by simultaneous observations of the standard height of the barometer and of the boiling-point of water.— Determine the height of the barometer at 0° C. Apply the gravitation correction for distance of the locality from the centre of the earth, so as to reduce it to its equivalent height when the mercury is exposed to the attraction of the standard force of gravity, as at the sea-level in latitude 45°. Determine the boiling-point of water at the same time and place. Refer to the tables connecting this boiling-point of water with standard barometric pressure, and from them take out the boiling-point corresponding to the observed standard barometric pressure. If the tabular and the observed boiling-points are identical, then the force of gravity at the place is normal. If the tabular boiling-point is higher than the observed boiling-point, then the local force of gravity is less than the normal; if it is lower than the observed boiling-point, then the local force of gravity is greater than the normal. In arriving at these conclusions, we postulate the complete exactness of the tables giving the relation between barometric pressure and the temperature of saturated steam, and of those giving the normal gravitational correction for latitude and for height above the sea.

The use of the hypsometer along with the barometer may give valuable results at sea. It may give us a means of *divining* chánges of depths of the ocean, especially in tropical latitudes where, when the sea is calm, all the conditions for such experiments are most favourable. On land, the subject has been taken up with great zeal by Professor Mohn[1], the distinguished head of the Norwegian Meteorological Service. He has already published a preliminary report, and further reports from him will be expected with interest. In his hands this method will be thoroughly tested in all directions.

It is not suggested that such observations should be made by the Antarctic Expedition. The margin of $0°\cdot147$ C. between the normal boiling-point at the poles and at the equator shows that very fine thermometers would be required, and many other refinements are necessary. It is quite evident that the height of the barometer at $0°$ C. can be arrived at with the requisite certainty only if the barometer has been hanging for at least a day in a room of uniform temperature. Of course the whole of the working part of the thermometer must be immersed in the steam. Correction for an exposed portion of the mercurial thread of the thermometer is quite inadmissible. The combined use of the barometer and the thermometer has been discussed at this length because it affords an opportunity of arriving at precise ideas of the fundamental principles of both instruments, and of the uses to which each may be legitimately put.

A legitimate, and at the same time an eminently practical, use of the hypsometer is to *replace* the mercurial barometer in the determination of the atmospheric pressure, and in the comparison of barometers at distant stations. On all camping excursions it is necessary to have the means of boiling water, and except under stress of circumstances, it is boiled night and morning. Saturated steam is therefore produced in the ordinary day's work; it is only necessary to provide the means of observing its temperature accurately.

[1] *Das Hypsometer als Luftdruckmesser*, von H. Mohn. Christiania, 1899.

The aneroid barometer has been mentioned as having, in some respects, the same advantages as the hypsometer, and it might be thought that the aneroid can, on such excursions, replace both the mercurial barometer and the hypsometer. But this is not so. The aneroid is a spring-balance, and its spring soon gets tired, so that, for instance, on an excursion in mountainous countries, it might, on arriving in the evening at the camping place, show a certain pressure of the atmosphere, and the next morning it might show a different pressure, even although the real pressure had not varied in the interval. If the temperature of boiling water varies in the same place, it is proof that the pressure of the atmosphere has varied. But although the aneroid cannot replace the hypsometer, it may with advantage be used in connection with it for determining local variations of height or of atmospheric pressure during one day. The night and morning comparisons with the hypsometer give its *error and rate* for that day.

Examples of Rapid Variations of Atmospheric Temperature, especially during Föhn.

The following pages contain the record of *peripatetic* meteorological observations made at different times. They are reprinted from a paper[1] on this subject published in the *Proceedings of the Royal Society* in 1894.

"The variation of the temperature of the air in the course of a day is a matter of familiar observation. It depends in the first instance on the relative positions of the locality and the sun. The temperature is generally highest a short time after the sun has attained its greatest altitude above the horizon, and it is lowest some time after it has attained its greatest depression below the horizon. Observations made at regular intervals over the twenty-four hours show a more or less regular rise of temperature during the early part of the day and a similar fall of temperature during the latter part of the day and the evening. When the interval between the observations is diminished the regularity of the march of temperature is found to diminish also, but the great variability of the

---

[1] 'On Rapid Variations of Atmospheric Temperature, especially during *Föhn*, and the methods of observing them,' by J. Y. Buchanan, F.R.S. *Proc. R. S.* (1894), vol. lvi. p. 108.

temperature of the air is best shown by the curve drawn by a recording thermometer of sufficient sensibility combined with a clock movement of suitable velocity. Such an instrument draws a sinuous line which is generally smooth during the night and serrated during the day. The shape and the crowdedness of the teeth on the serrated daylight portion of the line have a close connection with, and are to a certain extent an indication of, the character of the existing weather. In general the indented character of the daylight curve is an indication of the disturbing influence of the sun on the equilibrium of the atmosphere, and this continues just as long as he is above the horizon; after sunset, the atmosphere quickly reverts to a state of greater stability."

It is only necessary to watch a thermometer during one or two minutes to be convinced of the great variability of the temperature of the air not only from one minute to another, but almost from second to second. This is most easily and most briefly shown by quoting the series of observations made at St Moritz, in the Engadine, in February 1894.

"In the winter of this year I revisited the Engadine, and stayed for a fortnight at St Moritz. As the room which I occupied faced due north, the window of it was convenient for making observations of the temperature of the air. From the 25th February to the 3rd March I made every morning a series of observations of the temperature of the air, beginning when there was just light enough to read the thermometer, and continuing till between 8 and 9 o'clock in the morning. At first I took the temperature every minute, but finding the oscillations of temperature very great, I reduced the intervals to twenty seconds, and sometimes to fifteen seconds. To print the observations *in extenso* would occupy too much space, but the striking features can be easily summarised. They are given in Table XIX. Excepting on the 26th February, when it was snowing all the morning, the observations embrace the interval of an hour or an hour and a half after sunrise. The time was devoted entirely to this object, and observations were made at as close dates as possible. Working alone, an interval of twenty seconds is quite convenient; shorter intervals cause hurry. The time immediately following sunrise is when one would expect the temperature of the air to rise continuously, if not regularly; but we see that so far from rising continuously and regularly the thermometer rises, falls, and remains stationary quite irregularly. On some days, as on the 28th February, these irregularities are comparatively few; on others, as on the 1st and 2nd of March, they are numerous. The largest rise or fall in twenty seconds is $0°\cdot5$ C. From experiments in calm air outside and in still air in a room, we find that for this thermometer to rise or fall $0°\cdot5$ C. in twenty seconds the temperature of the air around it must be from $2°\cdot25$ C. to $4°\cdot65$ C. hotter or colder

TABLE XIX.—*Giving Results of Observations of the Temperature of the Air and its Variations at St Moritz.*

Date, 1894	Time of observation	Limits of temp.	Interval between observations	Number of intervals in which the temperature was observed to			Total number of intervals	Maximum rise or fall in any one interval	
				Rise	Fall	Remain constant		Rise	Fall
25 Feb.	6.24 a.m. to 7.32 ,,	°C. −8·0 −5·0	″ 60	32	22	13	67	°C. 0·51	°C. 0·50
26 ,,	11.10 ,, to 1.19 p.m.	+8·25 +3·5	20	32	33	61	126	0·13	0·50
27 ,,	6.55 a.m. to 8.40 ,,	+1·75 +5·35	20	80	43	45	168	0·37	0·47
28 ,,	7.0 ,, to 8.2 ,,	−6·48 −1·0	20	103	37	45	185	0·25	0·20
1 March	6.30 ,, to 7.30 ,,	−4·25 −2·1	15	93	68	80	241	0·25	0·20
2 ,,	6.38 ,, to 8.6 ,,	−6·23 −2·03	13	158	118	131	407	0·23	0·18
3 ,,	6.30 ,, to 8.6 ,,	−6·55 −1·68	20	120	89	77	286	0·28	0·20

than the thermometer. Taking even the lowest of these values, we see how great the possible error is in measuring the actual temperature of the air at any moment with a thermometer, and the error is the greater the more sluggish the instrument is. In Table XX the detailed observations are given for a few minutes on the 26th of February, when the temperature was changing very rapidly. In the third and fourth columns the rise or fall of the observed temperature is given. In the fifth and sixth columns the corresponding differences between the temperature of the air and that of the thermometer which would cause the observed rate of change of temperature are given; with these and the observed temperatures we obtain the amended temperatures of the seventh column. Although it was snowing on the 26th, the air was perfectly still, and the rate of cooling corresponding to the 'term' eighty seconds has been applied. Had the rate of cooling of the thermometer in the still air of a room been taken, the difference between amended and observed temperatures would have been nearly twice as great.

"*Characteristics of Föhn Weather.*—These observations show how rapidly the temperature of the air may vary even in ordinary weather.

TABLE XX.—*Temperature of the air at St Moritz, observed at Intervals of twenty seconds.*

Date, 26 February, 1894	Observed temp.	Difference		Corresponding difference of temperature of air		Amended temp. of air	Differences of amended temperatures	
		Fall	Rise				Fall	Rise
	$T°$ C.	−	+	$-t$	$+t$	$T' = T+t$	Fall	Rise
A.M. h. m. s.	°							
11 18 45	+5·88	—	—	—	—	6·48		
19 5	6·00	—	0·12	—	0·60	6·60	—	0·12
25	6·12	—	0·12	—	0·60	6·72	—	0·12
45	6·25	—	0·13	—	0·60	6·25	0·47	
20 5	6·25	—	—	—	—	6·25		
25	6·25	—	—	—	—	5·25	1·00	
45	6·00	0·25	—	1·00	—	4·30	0·95	
21 5	5·62	0·38	—	1·70	—	3·37	0·93	
25	5·12	0·50	—	2·25	—	4·12	—	0·75
45	4·88	0·24	—	1·00	—	2·63	1·49	
22 5	4·38	0·50	—	2·25	—	2·13	0·50	
25	3·88	0·50	—	2·25	—	3·88	—	1·75
45	3·88	—	—	—	—	3·28	0·60	
23 5	3·75	0·13	—	0·60	—	3·75	—	0·47
25	3·75	—	—	—	—	4·37	—	0·62
45	3·88	—	0·13	—	0·60	3·88	0·49	
24 5	3·88	—	—	—	—	3·28	0·60	
25	3·75	0·13	—	0·60	—	3·15	0·13	
45	3·62	0·13	—	0·60	—	3·12	0·03	
25 5	3·50	0·12	—	0·50	—	3·00	0·12	
25	3·62	—	0·12	—	0·50	4·00	—	1·00
45	3·50	0·12	—	0·50	—	3·12	0·88	
26 5	3·50	—	—	—	—	3·50	—	0·38
25	3·50							

There is a class of weather which is generally known by its Alpine name *Föhn*, the distinguishing feature of which is the rapidity with which the temperature of the air changes from moment to moment, and the exceptionally high average temperature of the air.

"It has been most observed in the valleys stretching in a northerly direction from the main summit line of the chain of the Alps and takes the form of an abnormally warm wind blowing from the mountains towards the plain. It has largely occupied the attention of continental meteorologists, and more particularly it has been the subject of exhaustive investigations by Hann, who has shown by very strong evidence that its high temperature must be due to its compression in descending from a great altitude. In the descriptions of the *Föhn*, attention is almost exclusively directed to the high average temperature of the air, and no mention is made of its extraordinary variations, although every observer must have noticed them. They are so great as to be recognised at once by the sensations and at the same time so rapid as to elude almost every other method of estimation or measurement. It has also, I believe, not been before remarked that the true *Föhn* occurs in our own country and with its characteristics quite as well marked as in Switzerland. It is sometimes supposed that a great absolute height of mountain chain is required for its production; but this is not so. A relative height of 1000 to 1200 metres is quite sufficient for its production; and this is equally available on the west coast of Scotland and on the northern slopes of the Alps.

"Some observations were made in the summer of 1893, which was abnormally warm all over the north of Europe. In the beginning of July I observed the *Föhn* at Fort William, and in the latter part of August in the upper Engadine, and more particularly in the valley occupied by the Morteratsch glacier. Besides the observation of the varying temperature of the air itself, the investigation of the temperature gradient set up between the melting ice surface of the glacier and the hot winds blowing over it presented considerable interest. The curious fact was observed that while the hot wind was blowing over the glacier and melting the surface in abundance, the temperature of the air, as close to the ice as a thermometer could be applied without touching the ice, was never lower than $5°\cdot5$ C.

"In the beginning of July at Fort William the weather was very warm, and in the midst of very warm air still hotter blasts made themselves felt from time to time. The sensation was much the same as is produced when, on the deck of a steamer, the air passing the funnel strikes the face. These hot blasts lasted only for one or two seconds, and repeated themselves every minute or two. Their effect on a thermometer, freely exposed in the shade, was to keep the mercury in a constant state of motion, the temperature rising often more than $1°$ C. in a minute, and falling again as much. The thermometers in the screens were also a good deal affected, though not nearly to the same extent as the freely exposed ones. The recording instruments, the clock motion of which was not sufficiently quick to draw the record out into an indented line, showed a broad band which measured the amplitude of the excursions of the instrument, though by no means the amplitude of the oscillations of the temperature of the air. This phenomenon was particularly observed on the 8th July, 1893. It was very warm, as the following observations (Table XXI) of the thermometers in the large observatory screens will show.

"It was during the heat of the day, from 10 a.m. to 2 p.m., that the hot puffs made themselves most felt; but I found it impossible to measure their temperature, owing to the thermal inertia of the thermometers. The puffs lasted not longer than one or two seconds, and their temperature, to judge by the sensation, was rather higher than that of the body. The thermometers had only begun to rise when the heating ceased, and they fell back again. From the figures in Table XXI, it will be seen that

TABLE XXI.

Hour	9 a.m.	10 a.m.	Noon	2 p.m.	4 p.m.
Dry bulb (° C.)	20·1	22·4	24·9	23·8	18·9
Wet bulb (° C.)	17·7	17·3	18·2	17·7	16·6
Vapour tension (mm.) ...	13·5	11·5	11·5	11·3	12·6
Relative humidity... ...	77	58	49	52	77

the temperature of the air at noon reached 24°·9 C., a very high figure for a station in nearly 57° north latitude. Along with the great rise of temperature there is a fall of absolute as well as of relative humidity, indicating that the air has come from a greater altitude. Attempts to measure the actual temperatures of the hot puffs gave no satisfactory result."

These few lines will give an idea of the nature of the weather called *Föhn*. The temperature of the air is abnormally high, and it is very unequally distributed through the mass of the air. The atmosphere seems rather to be made up of sheets of air of very different temperature; it is also very dry.

In the Antarctic regions there are plenty of high mountains and sharp gradients, and it is certain that conditions described as characterising *Föhn* must occur in some localities under certain meteorological conditions. Drygalski has called attention to the importance of *Föhn* as a climatic factor in Greenland. Where the heat disengaged by the contraction of the descending air is sufficient to raise its temperature above that of melting ice, its effect is at once apparent, and the fact that the West Coast of Greenland is free from land-ice at and near the sea-level, is attributed mainly to the heat of the *Föhn*. It is evident that, in the middle of winter with a general temperature of − 30° C. to − 50° C., *Föhn* may prevail without its being able to raise the temperature of the air to that of melting ice. Indeed, it may even raise it above this temperature, but the air will not melt any ice in the open until it has raised its temperature from − 30° or − 50° up to 0° C., and this will take some time, which may be longer than that during which the *Föhn* prevails. In summer in Greenland, the general temperature of the air is above that of melting ice, and the effect of *Föhn* is observed in the *increased* melting of the ice. From the observations of Mr Bernacchi, there seems to be reason to suppose that the surface of the ice which forms the great barrier does not usually rise to its melting-point even in the warmest month, February. Apart from the indications of the thermometer, in Greenland the occurrence of *Föhn* was made most evident in spring and autumn by its producing melting of the ice, which without it did not take place.

"Later in the year, in the middle of August, I visited the upper Engadine, and stayed for some weeks at Pontresina. Here, as elsewhere, the weather was very warm, and I was much struck by observing the same blasts of hot air as I had experienced in Scotland. The general characteristics of the weather were the same, and the temperature of the air in the valley rose nearly as high as it had done at Fort William.

"On the 18th August I went for an excursion on the Morteratsch glacier with a guide. On my remarking the hot puffs of air, which were much more striking on the ice than on the land, he said it was the *Föhn*, of which he considered them a characteristic. The sun and the hot wind were causing an enormous amount of surface melting of the ice, and having a thermometer with me, I took the temperature of the air by whirling at a height of about 1 m. from the ice, and found it 12°·0 C. ; the wet bulb was 5°·0, so that the vapour tension was 2·3 mm., the relative humidity 22, and the dew-point − 8°·6 C. The great dryness of the air will be remarked. I then swung the thermometer in a conical path as close to the ice as possible, and the temperature of the air was 10°·0 C. Being astonished to find this high temperature close to the ice, I put the bulb of the thermometer into a crack in the ice, so as to be below the level of the surface of the ice, and its temperature only went down to 7°·5 C.

"All the temperatures were taken with a mercurial thermometer, which was whirled at the end of a string so that its velocity was about 6 m. per second. It was not protected in any way, so that the temperatures observed with it are not free from a certain error due to radiation and reflection, although it was always shaded from the direct sun. These errors are not usually great with a whirled instrument, and most of my observations have to do with differences of temperatures observed with the same instrument and under similar circumstances. On the glacier the thermometer, when whirled, was not apparently affected by radiation or reflection from the ice, and only very slightly by that from the sun. On land I remarked that the greatest disturbing effect is produced by sunlight reflected from grass. If the thermometer was whirled in the shade of a north wall with a grass field or hill-side close by, the thermometer would be immediately affected to the extent of one or two degrees, according as the sun shone on the grass or was obscured by a cloud. The effect was immediate the moment the sun came out; sunlight reflected from rocks and light-coloured surfaces did not produce the same effect.

"On the 19th August I returned to the glacier. At 11 a.m. in the valley below the glacier I found the temperature of the air 22° C., and the wet bulb 12°·5, whence the vapour tension is 5·0 mm., and the relative humidity 26. In determining the temperature of the air by whirling the thermometer I found variations of as much as 2°. The hot puffs of air made themselves felt most markedly, and showed that the real variations of the temperature of the air were much greater than the thermometer showed. At 1 p.m., on the hill-side, to the west of the tongue of the glacier, and at a height of about 2100 m. above the sea, four good observations of the temperature were made, giving 17°·5, 18°·0, 19°·5, and 19°·0; they are all equally trustworthy, and represent the average temperatures of the air during the minute, or minute and a half, that the thermometer was whirled. The mean of these values, 18°·5, is taken as the temperature of the air. For determining the temperature of the wet bulb, the bulb of the thermometer was wrapped round with one thickness of Swedish filtering paper thoroughly moistened, and the thermometer was whirled as before and until the temperature ceased to fall; it then stood at 9°·5. Still higher up the hill at an altitude of 2250 m., the temperature of the air at 2 p.m. was 18°·5 C. Having returned to the spot where the observations had been made at 1 p.m. the following air temperatures were observed:—between 2.40 and 2.46 p.m., 17°·5, 18°·0, 17°·5, 17°·0, 17°·3, 17°·1 ; mean, 17°·4; and between 2.50 and 2.54 p.m., 16°·5, 16°·5, 16°·7, and 16°·5 ; mean, 16°·55. The mean of the two sets is 17°·06. It must be repeated that each of these individual observations is a faithful indication of the average temperature of the air in which the thermometer was whirled, and in so far as its sensibility enabled it to assume the same temperature as the air. From this spot I descended to the glacier and went up it until I got to a position which, judging by the eye, was at the same height as the station just left on the mountain side, and about one kilometre distant from it in a straight line. The weather was rapidly getting colder, the sky being covered with the characteristic *Föhn* cloud. The wind was fresh down the glacier, which made the exposure of the thermometer easy and good. The hot *Föhn* puffs were also very striking. The thermometer was first swung exposed to sun and wind, showing temperatures varying from 10°·5 to 11°·2, the mean being 10°·8 C. Swung in my own shadow, but exposed to the wind, the temperature was 9°·8. The wet bulb was 4°·7, showing a relative humidity of 37. The thermometer was now exposed, both wet and dry, in a horizontal position with the bulb at a distance of about 2 cm. from the ice, on the top of one of the superficial ridges of the glacier, and fully exposed to the wind, though shaded from the sun. The observed temperatures were: dry, 6°·6 C. ; wet, 3°·7 ; relative humidity, 58·5. The exposure of the thermometer was as good as could be desired, and, with the fresh breeze blowing, it was thoroughly ventilated. I was again much struck with the highness of the temperature of the air almost in actual contact with the ice. The observations at 1 m. and 2 cm. from the ice were repeated, giving substantially the same results—at 1 m., dry bulb $10°·2$, wet, $5°·1$; at 2 cm., dry bulb $6°·8$, and wet $3°·2$. The hot *Föhn* puffs were more striking on the ice than on the land, owing to the greater difference between their temperature and that of the surrounding air. At 4 p.m. I left the ice and returned to the station of 1 o'clock on the hill-side, and took the temperature at 4.35 p.m.—dry bulb $16°·0$, wet $8°·0$, relative humidity $24·5$. At the station in the valley below the glacier the temperature was at 5.45 p.m., dry bulb $16°·4$, wet $11°·8$, and relative humidity 56. These observations, besides showing the remarkable conditions of the air over the glacier, indicate the fineness and warmth of the weather which prevailed.

"On the 21st August another series of observations was made at the stations on the land and on the ice. The breeze on the ice was not so steady or so strong as on the 19th, and about 5 o'clock in the afternoon there was a heavy squall of rain and thunder. The same hot *Föhn* puffs made themselves felt as before, without there being any means of measuring their temperature. Their duration at their maximum temperature was never more than a few seconds, during which but little effect was produced on the thermometer. It occurred to me that the only way of gaining a knowledge of the temperature of these puffs of air would be by comparing the rapidity with which the thermometer moved when exposed to a known difference of temperature, with that observed in the puffs. A number of observations was made with this view, by warming the thermometer and noting its rate of cooling in air of known temperature. The reverse procedure was also followed on the ice. The thermometer was cooled by being laid close to, but not touching, the ice, it was then quickly raised to a height of 1 metre, and its rate of change of temperature observed. In this way it was found that for an initial difference of $4°$ the thermometer required 10 seconds to rise $1°$; for a difference of $3°$, 12 seconds; and for a difference of $2°·5$, 16 seconds. These ratios were observed in the open air, and under the circumstances where the hot puffs are observed. Unfortunately, owing to an accident to the thermometer, very little use could be made of them. Where the rate of change of temperature of the thermometer is used to determine the temperature of the air, the movement of the air must be measured or estimated. The observations made on the 19th and 21st August are given in Table XXII.

"For comparison with the temperatures on the ice on the 19th, the mean of the observations at the land station at 2.45 and 4.35 p.m. is taken, and on the ice the mean of the observations at 3.20 and 3.55 p.m. The altitudes of the two stations were as nearly as possible identical, and they were not more than 1 kilometre distant from each other. Considering the temperatures at a height of 1 m., there is a difference of 6°·5 between the land and the ice. The difference of vapour tension, 0·2 mm., is insignificant, and shows that substantially the air is the same. The dew-

TABLE XXII.—*Temperature Observations at Equal Altitudes on the Morteratsch Glacier, and on the Mountain West of it.*

	Thermometer		Diff.	Vapour tension	Rel. hum.	Dew-point
	Dry	Wet				
	° C.	° C.	° C.	mm.	p.c.	° C.
19*th August*, 1893 Land station, 2.45 p.m.	17·1	8·6	8·5	3·2	22	− 5·0
,, ,, 4.35 ,,	16·0	8·1	7·9	3·2	24	− 4·7
Mean ...	16·55	8·35	8·2	3·2	23	− 4·85
Ice station, 3.20 p.m.	9·8	4·7	5·2	3·26	36	− 4·4
Height 1 metre, 3.55 ,,	10·2	5·1	5·1	3·55	39	− 3·5
Mean ...	10·0	4·9	5·1	3·40	37·5	− 3·95
Ice station, 3.20 p.m.	6·7	3·7	3·0	4·2	57	− 1·4
Height 0·02 m., 3.55 ,,	6·6	3·2	4·4	4·0	56	− 3·0
Mean ...	6·65	3·45	3·2	4·1	56·5	− 2·2
21*st August*, 1893 Land station, 1 p.m.	14·5	7·5	7·0	3·5	29	− 3·5
,, ,, 3.45 ,,	14·3	8·0	6·3	4·2	35·0	− 1·3
Mean ...	14·4	7·75	6·65	3·8	32	− 2·4
Ice station, 2.22 p.m.	9·85	5·6	4·25	4·2	47	− 1·3
Height 1 metre, 2.54 ,,	11·0	7·0	4·0	5·1	52	+ 1·5
Mean ...	10·43	6·3	4·13	4·6	50	+ 0·1
Ice station, 2.15 p.m.	7·3	4·0	3·3	4·1	54	− 1·5
Close to ice, 2.40 ,,	5·5	3·2	2·3	4·2	65	− 0·7
Mean ...	6·4	3·6	2·8	4·2	59	− 1·1

point in both cases is several degrees below 0°, so that, on the air coming in contact with the ice, there would be evaporation from the latter. The evaporating power of the air may be represented by the difference between the tension of saturation and the actual vapour tension. It is very great on land, being 10·75 mm. at 16°·53 C., and it would rapidly evaporate water having that temperature. On coming in contact, however, with ice, the air actually in contact, which alone comes under consideration, is first cooled to 0° C., which reduces its saturation tension to 4·6 mm., and the difference is only 1·4 mm. We see, however, that this has been sufficient to increase the absolute humidity of the air in close proximity to the ice. At 1 m. above the ice the air had an average temperature of 10° C.; at 2 cm. from the ice its temperature was as high as 6°·65 C., and the air in actual contact with the ice must have been at 0° C. Many observations have been made of the temperature of the air at different heights above glaciers, and, as might be expected, considerable differences have been observed; but I am not aware that any observations have been made on the air almost but not quite in contact with the ice, as are those which have been made at 2 cm. from the ice. The bulb was perfectly shaded from the sun but freely exposed to the wind, it was also fully exposed to any cold radiations from the ice. There is, therefore, no doubt that 6°·65 was the temperature of the air passing the bulb of the thermometer. The vertical distribution of temperature shown by these figures is remarkable. From a height of 1 m. to within 2 cm. of the ice there is a gradient of 3°·4 per metre; in the remaining 2 cm. there is a gradient at the rate of 33° per metre; and, from various observations and considerations, it is probable that the moderate gradient is continued to within a millimetre of the ice, when it becomes precipitous. It is to be noted that the absolute humidity, as shown by the vapour tension of the air, has increased from 3·4 mm., at 1 m., to 4·1 mm., at 2 cm.; showing that ice is being evaporated and transferred from the glacier to the atmosphere. The wind was blowing freshly down the glacier, and its velocity was measured by noting the time which pieces of paper allowed to drift took to reach the ice, and then pacing the distance. The mean velocity was found to be from 8 to 10 kiloms. per hour.

"The observations made on the 21st and on the 22nd confirmed those of the 19th. The same variability of the air temperature at the land stations was noticed. Between 12.55 and 1.6 p.m. the following temperatures were observed by whirling:—16°·2, 16°·2, 16°·0, 15°·5, 16°·0, 15° 5, 15°·0, 14°·2, 13°·8, 14°·0, 13°·5, 13°·5. These are all good observations, and represent real variations of the temperature, or rather they indicate real variations of greater amount. Taking the mean of the last five observations, we have the temperature of the air 14°·0. The wet bulb was found at 1.15 p.m. to be 7°·5, giving a difference of 6°·5. On the glacier the air felt closer than on the previous occasion. The temperature at 1 m. was 11°·5, and at 2 cm. from the ice 7°·3. The difference 4°·2 is less than on the previous occasion. The wind was much less strong, and yet the temperature close to the ice is higher. The wet bulb under the same circumstances showed 4°·0. Five minutes later the dry bulb was observed at 1 m., 10°·2 and 9°·4, mean 9°·85. Another observation of the dry bulb at 2 cm. from the ice gave 6°·6. The interval between the bulb and the ice was now reduced to the smallest possible distance, about

2 mm. The wind fell very light, and the thermometer remained at 8°·0, when the wind returned it fell to 5°·8. The axis of the thermometer bulb would be about 5 mm. from the ice, and still the air is nearly 6° warmer than the ice. Another observation in the same conditions gave 5°·5. The wet bulb was now exposed, but it had to be kept about 5 mm. off the ice; it showed 3°·2. At 2.43 p.m. a great volume of warm air came down, and the wet bulb ran up to 4°·5 in three or four seconds. With the return of the breeze the wet bulb went back to 3°·0. The *Föhn* puffs were now very troublesome. At 2.52 p.m. the wet bulb at 1 m. was 7°·0; the dry bulb showed—at 2.54 p.m., 11°·0; at 2.55 p.m., 13°·5; and at 2.57 p.m., 14°·5. In one puff the thermometer was observed to rise one degree in eight seconds, which would make the true temperature of the air at the moment about 6°·0 higher, or 19°·5.

"At 3.30 p.m. I returned to the land stations, and again found the same variable temperatures. Between 3.35 and 3.45 p.m. the temperature varied between 16°·0 and 13°·5. The following averages were taken:

3.45 p.m., dry, 14°·3; wet, 8°·0; relative humidity, 35.
4.0 ,, ,, 14°·0; ,, 8°·5; ,,   ,, 42·5.

"Taking the first of these and the observations at 1 o'clock, we have for the mean temperature of the air 14°·15, and the wet bulb 7°·75. On the ice we have:

At 1 metre, dry bulb, 9°·85; wet, 5°·6, and
At 2 centimetres, ,, 7°·3; ,, 4°·0.

"The difference in the temperature of the air at 1 m. is only 4°·3, and that between 1 centimetre and 2 metres above the ice is only 2°·55, while the air at 2 centimetres is 7°·3 warmer than the ice.

"On the 22nd August the observations on the ice were repeated, with very much the same results. The temperature of the air ranged from 9°·0 to 9°·5 at 1 metre, and was 5°·5 at 1 centimetre from the ice.

"The result of the few observations here quoted is to show that the air, which over land has a temperature of 15° to 20°, or higher, in passing over a glacier is cooled to a comparatively slight degree. Although the air appears to be thoroughly mixed by its own motion, very sharp gradients of temperature are produced and maintained. The great and abnormal temperature of the air of the valley is kept up by the heat liberated by the compression accompanying the descent of local streams or striæ of air from high levels. These keep up an extra supply of heat over and above what is supplied by the direct radiation of the sun. The result is that the melting of the glacier in *Föhn* weather greatly exceeds that of even the hottest day of ordinary weather.

"In order to convey a general idea of the climate in the neighbourhood during the period when my observations were made, I subjoin a table of the air temperatures observed at the Pfarrhaus in Pontresina three times daily, and obligingly supplied to me by Herrn Pfarrer Falliopi.

TABLE XXIII.—*Temperature of the Air at Pontresina.*

Date	Temperature of the air observed at					
	7 a.m.		1 p.m.		9 p.m.	
	Temp.	Diff. from mean	Temp.	Diff. from mean	Temp.	Diff. from mean
1893	° C.		° C.		° C.	
August 15	4·7	− 2·92	19·2	− 1·26	10·0	− 1·36
,, 16	5·9	− 1·72	20·0	− 0·46	10·8	− 0·56
,, 17	7·2	− 0·42	20·8	+ 0·34	11·8	+ 0·44
,, 18	8·2	+ 0·58	21·8	+ 1·34	12·8	+ 1·44
,, 19	8·6	+ 0·98	21·2	+ 0·74	12·8	+ 1·44
,, 20	10·0	+ 2·38	19·8	− 0·66	12·6	+ 1·24
,, 21	7·6	− 0·02	22·2	+ 1·74	10·2	− 1·16
,, 22	8·2	+ 0·58	20·2	− 0·26	10·2	− 1·16
,, 23	6·9	− 0·72	19·2	− 1·26	12·8	+ 1·44
,, 24	8·9	+ 1·28	20·2	− 0·26	9·6	− 1·76
Mean	7·62	—	20·46	—	11·36	—

"In this table, the high temperature on the 18th, 19th, 20th and 21st, is very apparent. The *Föhn* prevailed during all these days.

"On the 23rd August, which was a very warm day, I made a series of observations between Pontresina and the top of the Piz Languard, which is the highest peak on the ridge immediately behind Pontresina, and is very easily accessible. It had been raining heavily in the night, so that in the early morning the air was rather cool; but the following observations, made before starting up the mountain, will show how rapidly the temperature was beginning to rise:

$$8.0 \text{ a.m., dry bulb, } 10°\cdot4; \text{ wet, } 9°\cdot2.$$
$$9.10 \text{ ,, ,, } 14°\cdot8; \text{ ,, } 11°\cdot4.$$
$$10.0 \text{ ,, ,, } 17°\cdot0.$$

"At 10 a.m. I started up the mountain, following the excellent path which leads to the summit.

"In Table XXIV the temperatures observed at various stations are entered, along with corresponding ones observed in the porch of the Hotel Reseg at Pontresina.

"Excepting in the first interval the rate of fall of temperature between Pontresina and the station on the mountain is less than 1° per hundred metres. At the summit the mean temperature of the dry bulb was 10°·75,

T ABLE XXIV.

	Height above sea	Time	Temperature		Difference
			On mountain	At hotel	
	m.		° C.	° C.	
Pontresina ...	1800	10.0	17·0	—	—
	2100	10.50	16·5	19·5	3·0
	2250	11.5	16·5	20·0	3·5
	2370	11.35	16·5	20·5	4·0
	2670	12.0	14·5	20·75	6·25
	2790	12.30	13·3	21·0	7·7
	2970	1.6	14·0	21·5	7·5
	3180	1.30	13·1	22·0	8·9
Summit ...	3266	2.10	11·0	—	—
	—	2.40	10·5	22·0	11·25

and of the wet bulb 6°·45, whence we have the vapour tension 4·5 mm. and the relative humidity 47. The weather was of the same kind as in the valley, abnormally warm, and the air very dry."

NOTE. The Föhn dealt with in this paper is fine-weather Föhn, with cloudless sky and strong sun. The Föhn occurring with overcast sky and rain is not treated. It is this Föhn particularly that wastes glaciers, because it persists through the twenty-four hours, while the melting by the fine-weather Föhn is limited to the four or five hours before and after noon. I have only once witnessed the overcast Föhn and it is a most powerful melting agent, but I am not prepared to furnish a satisfactory mechanical explanation of it.

Conclusion.—It will be observed that in this paper no attempt is made to give instructions to the Chemist and Physicist of the Expedition. Here and there a lead in one direction or another is suggested. He must rely on his own knowledge and experimental ability. Questions or problems which excite his curiosity should be worked out in his own way. If, in this way, he answers or solves them to his own satisfaction, it is probable that he has made a genuine addition to knowledge. In all his work it will be useful for him to remember that, primarily, Physics and Chemistry are branches of Natural History.

No. 3. [*From the Proceedings of the Royal Society of Edinburgh*, 1887, *Vol. XIV, p.* 129.]

ON ICE AND BRINES

THE composition of the ice produced in saline solutions, and more particularly in sea-water, has frequently been the object of investigation and of dispute. It might be thought that to a question of whether ice so formed does or does not contain salt, experiment would at once give a decisive answer. Yet, relying on experiment alone, competent authorities have given contradictory answers. All agree that ice, whether formed artificially in the laboratory by freezing sea-water, or found in nature as one of the varieties of sea-water ice, retains, in one form or another, and with great tenacity, some of the salt existing in solution in the water. The question at issue is whether this salt is to be attributed to the solid matter of the ice or to the liquor mechanically adhering to it, from which it is impossible to free it. Most bodies, and especially those which take a crystalline form, are easily purified and freed from all suspected foreign matter, with a view to analysis, by the simple operation of washing and drying. It is impossible to wash the crystals, formed by freezing a saline solution, with distilled water, because they melt at a temperature below that at which distilled water freezes. The effect of the addition of a small quantity of distilled water to a quantity of saline ice is at first the anomalous one, that what was a wet sludge is transformed into a dry crystalline powder. It is, of course, impossible to dry the ice by heat, and to do so by more intense freezing would be begging the question. The experimental difficulties therefore account for some of the divergence of opinion on the subject. The mixed character of the substances examined has also much to do with it. As a rule, it may be said that those investigators who have confined their observations to the laboratory have concluded that the ice formed when saline solutions of moderate concentration, including sea-water, are frozen, is pure ice, and the salt from which it is impossible to free it entirely belongs to the mother-liquor, while those who have collected and examined sea-water ice in high latitudes have come to the opposite conclusion.

During the Antarctic cruise of the "Challenger" I made a number of observations on the sea-water ice found in those regions, and, relying principally on the fact that the melting temperature of the ice was markedly lower than that of fresh-water ice, and that it was impossible by any of the ordinary means familiar to chemists for freeing crystals from adhering mother-liquor to materially reduce its salinity, I came to the conclusion that the ice formed in freezing sea-water is not a mixture of pure ice and brine, but that it contains the salt found in it in the solid state either as a crystalline hydrate or as the anhydrous salt, but most probably as a hydrate. In dealing with the subject, Dr Otto Pettersson (*Water and Ice*, p. 302) quotes my observations, and also rejects the view that "sea-ice is in itself wholly destitute of salts, and only mechanically incloses a certain quantity of unfrozen and concentrated sea-water." He founds his belief on the fact that numerous analyses of specimens of sea-water ice have shown that the constitution of the saline contents of different specimens of ice differs for each specimen, and is always different from that of the saline contents of sea-water. Were the salinity due to inclosed unfrozen and concentrated sea-water, we "ought to find by chemical analysis exactly the same proportion between Cl, MgO, CaO, SO_3, etc., in the ice and in the brine as in the sea-water itself." He quotes numerous analyses of specimens of sea-water ice from the Baltic and from the Arctic Seas to show that this is not the case. Calling the percentage of chlorine in each case 100, he found in various sea-waters the percentage of SO_3 to vary from 11·49 to 11·89. In specimens of sea-water ice it varied from 12·8 to 76·6, and in brines separating from the ice and remaining liquid at −30° C. it varied from 1·14 to 1·16.

This argument appears conclusive. In order to explain all the phenomena observed in connection with sea-water ice he cites Guthrie's investigations, which went to show that, in freezing saline solutions, under a certain concentration, pure ice is formed at a temperature which falls from 0° C., when the amount of salt dissolved is infinitely small, to a certain definite temperature when the solution contains a certain definite percentage of salt. Further abstraction of heat then produces solidification of the solution as a whole, in the form of a crystalline hydrate, of constant freezing- and melting-point. To such hydrates, Guthrie gave the name of cryohydrates. Pettersson quotes the following as being particularly applicable to the case of sea-water:

The cryohydrate of		Contains per cent. of water			Solidifies at ° C.
NaCl	...	76·39	...	...	− 22
KCl	...	80·00	...	...	− 11·4
CaCl$_2$	...	72·00	...	...	− 37·0
MgSO$_4$	...	78·14	...	...	− 5·0
Na$_2$SO$_4$	...	95·45	...	...	− 0·7

And he refers more particularly to the cryohydrate of Na$_2$SO$_4$ forming and melting at − 0°·7.

Now the bearing of Guthrie's experiments is to show that, while at sufficiently low temperatures, and with suitable concentration, the water will solidify along with one or other of the salts in solution, until this low temperature and high concentration are attained, pure ice must be the result of freezing.

The abnormal phenomena attending the formation and the melting of ice in saline solutions and sea-water find a natural explanation in an observation which I have frequently quoted, and which Dr Pettersson mentions in a footnote at p. 318, namely, that "a thermometer immersed in a mixture of snow and sea-water which is constantly stirred indicates − 1°·8 C." If this is true, it is clear that my melting-point observations proved nothing. On repeating the experiment I found it confirmed, and took the opportunity this winter of investigating the matter more closely. The paper now communicated to the Royal Society of Edinburgh contains the first portion of the results. It deals with the subject under two heads, namely, (*a*) the temperature at which sea-water and some other saline solutions freeze, and the chemical constitution of the solid and the liquid into which they are split by freezing; and (*b*) the temperature at which pure ice melts in sea-water and in a number of saline solutions of different strengths.

(*a*) The freezing experiments were limited to sea-water and solutions of NaCl comparable with sea-water.

Chloride of Sodium. Four solutions were used, and they were intended to contain 3, 2·5, 2, and 1·5 per cent. NaCl respectively. Forty grammes of this solution, in a suitable beaker, were immersed in a freezing mixture of such composition as to give a temperature from 2° to 2°·5 C. below the freezing temperature expected. The temperature at which ice began to form (if necessary after adding a minute splinter of ice) was noted, and the freezing was allowed to continue with constant stirring till the temperature had fallen 0°·2 C. A specimen of the mother-liquor was removed, and the chlorine in it determined; the chlorine in the original solution had been determined before. The beaker was then removed from the freezing-bath and the ice in it was allowed to melt. The temperature in all cases rose during melting exactly as it had fallen during freezing. In the following table are given the means of the temperature at which ice began to form in the original solution, and that of the liquid when the sample of brine was taken, and the means of the chlorine found in the original solution and in the brine sample:

| Mean freezing temperature | −1°·875 C. | −1°·63 | −1°·30 | −0°·975 |
| Mean per cent. Cl ... | 1·87 | 1·60 | 1·30 | 0·98 |

It will be seen that, in the dilute solutions experimented with, the percentage of chlorine expresses, in terms of the Centigrade scale, the lowering of the freezing-point of the solution.

Sea-Water. Similar experiments were made with sea-water of different degrees of concentration. In sea-water from the Firth of Clyde containing 1·84 per cent. of chlorine, ice forms at − 1°·9 C. The following results are from means of close-agreeing results :

Freezing temperature	−2·0	−1·5	−1·0	−0·5
Per cent. chlorine ...	1·94	1·445	0·963	0·475
Difference	0·06	0·055	0·037	0·025

Sea-water resembles a chloride of sodium solution containing the same percentage of chlorine, and the resemblance is closer the greater the dilution. When the beaker was removed from the freezing-bath, the temperature rose during melting as it had fallen during freezing. In these experiments, which had for their object the determination of the temperature at which the crystals melted, as well as that at which they began to form in the water, it was impossible to remove a sample for analysis large enough to enable the sulphuric acid to be determined in it.

For this purpose a series of observations were made, using quantities of 300 grammes of sea-water. Freezing was continued usually until the temperature had fallen 0°·3 C. below that at which crystals began to form. The mother-liquor was then separated from the crystals by means of a large pipette with fine orifice, before removing the beaker from the freezing-bath. The magma of crystals was then brought rapidly on a filter and drained by means of the jet pump. The ice, thus drained, was then melted, and the three fractions were analysed. In the following table (I) the results of four experiments are given. In the one column (W) will be found the weight of the original water taken and of the fractions into which it was split on freezing; in the other (R) will be found the ratio of SO_3 to Cl found by analysis, the chlorine being set down as 100; thus, in (I) the percentage of chlorine found in the crystals, melting at the lowest temperature, was 1·497, and that of the SO_3, 0·174; the ratio (R) is therefore 11·62.

It will be seen that the ratios (R) found for mother-liquor, drainings, and ice agree with one another quite as closely as those found in samples of pure sea-water from different localities. It is to be remembered that in these experiments the water was frozen *gently*—that is, the rate of abstraction of heat was low, the temperature of the freezing-bath being regulated so as to be about 2° C. below the freezing temperature of the solution. Much of the error and uncertainty about the freezing of saline solutions arises from the violence of the methods employed. Judging then by the constancy of the relation of the percentage of Cl to SO_3, we see that in

TABLE I.—*Freezing Sea-Water—Analyses of Fractions.*

No. of Experiment	I		II		III		IV	
Nature of Water	Forth 100°/₀		Mother-liquor		Clyde 100°/₀		Clyde 50°/₀	
	W	*R*	*W*	*R*	*W*	*R*	*W*	*R*
Original water	300	11·83	90	11·67	300	11·58	300	11·21
Mother-liquor	170·6	11·67	...	11·83	102	11·57	78	11·67
Drainings ...			...	...	94	11·56	109	...
Crystals ...	106	11·62	23	11·22	97	11·67	106	11·4
,, ...	22·5	11·11	...	...	...	...	...	...

sea-water, frozen at moderate temperatures, the composition of the saline contents of the original water, the mother-liquor, and the ice is identical; and we are justified in concluding that it is probable that the saltness of the ice is due to unfrozen and concentrated sea-water adhering to it. Ice formed in even very weak saline solutions closely resembles snow (which is ice formed in air), and has the same remarkable power of retaining mechanically several times its weight of water or brine.

If we assume that the ice formed in freezing sea-water is pure ice, and that the saline ingredients are retained by the portion remaining liquid, we can calculate the amount of ice which has been formed if we know the salinity of the original water and that of the residual brine. In the case of sea-water the salinity varies directly with the percentage of chlorine. The weight of the brine remaining after any freezing operation is found by multiplying the weight of the original water used by the ratio of the chlorine percentage found in the original water to that found in the brine. The difference between the weight so found and that of the original water is the weight of the ice formed. In Table II the results of this calculation are given for Experiment III on pure sea-water from the Clyde, and for Experiment IV on the same water diluted with an equal weight of distilled water.

TABLE II.—*Calculation of Ice formed, on the basis of the Salinity of the Original Water and of the Residual Brine.*

No. of Experiment		III	IV
Weight of original water (grammes)	W	300	300
Per cent. Cl in ditto	c	1·836	0·923
Per cent. Cl in mother-liquor ...	K	2·212	1·153
Weight of mother-liquor ... $W\frac{c}{K}=$	L	249·0	293·3
Weight of ice ... ... $W-L=$	I	51·0	60·7

Sea-water, like other saline solutions, is easily cooled several degrees below its freezing-point before crystals begin to form. While cooling down to and below what was known to be its freezing-point, simultaneous observations of the temperature of the sea-water and the freezing-bath were made from half-minute to half-minute. From these observations, the rate of abstraction of heat for different differences of temperature of sea-water and bath was found. At a given moment a minute splinter of ice (weighing much less than a drop of water) was introduced. Crystals immediately began to form, and the temperature rose in from ten to fifteen seconds to the freezing-point. During the freezing the temperatures of bath and sea-water were observed at regular intervals. The heat removed is thus made up of that eliminated during the few seconds when freezing began and the temperature rose to the freezing-point, which is found by multiplying the rise of temperature by the weight of the water in the liquid, and that removed during the subsequent cooling, which is found from the duration of the operation and the rate of loss of heat, deduced from observations made during the

cooling. The specific heat of the solution is taken as that of
the water which it contains, namely 0·965 for III, and 0·9825
for IV. The mean freezing temperature was − 2°·05 C. for III
and − 1°·05 for IV. The latent heat of water freezing at 0° is
79·25. The specific heat of ice being 0·5, the latent heat of
water freezing at − 2°·05 is 78·22, and that of water freezing
at − 1°·05 is 78·73. In Experiment III the total heat extracted during the freezing was 4230 × 0·965 = 4082 heat
units (gramme-degrees), and dividing this by 78·22 we find
52·2 grammes as the weight of pure ice formed at − 2°·05 C.,
equivalent to this abstraction of heat. In Experiment IV
the heat abstracted was found to be 5193 × 0·9825 = 5102.
Dividing this by 78·73 we find 64·8 grammes as the equivalent
weight of ice formed.

We have calculated the weight of ice which would be
found, first on the basis of the salinity of the solution; and
second, on the basis of the observed thermal exchange,
assuming in both cases that, in the act of freezing, pure ice
is formed. Thus:

TABLE III.

No. of Experiment	III	IV
	grammes	grammes
Calculated from thermal exchange ...	52·2	64·8
„ „ salinity	51·0	60·7
Difference	1·2	4·1

The agreement between the two quantities of ice formed
as calculated by the different methods is as close as could be
expected, and renders probable the truth of the common
assumption that the solid body formed is pure ice.

It was accepted by Guthrie, Rüdorff, and others that, in
solutions of the salts occurring in sea-water, ice separates
out at first, and continues to separate out until the concentration has become many times greater than that of
sea-water. Assuming that in sea-water all the chlorine is
united to sodium, 85 per cent. of the water would have to be

removed as ice before a cryohydrate would form, and if it contained nothing but sulphate of soda in the proportion corresponding to the sulphuric acid found in it, over 90 per cent. of the water would have to go as ice, before the cryohydrate would be formed.

In my experiments about 15 per cent. of the weight of the water was frozen out as ice, causing a lowering of freezing-point by 0°·3 C. In nature it is probable that the ice forming at the actual freezing surface does so at an almost uniform temperature, the local concentration produced by the formation of a crystal of ice being immediately eliminated by the mass of water below. In the interstices of the crystals there will be retained a weight of slightly concentrated sea-water at least as great as that of the ice crystals. These retain the brine in a meshwork of cells, and, as the thickness of the ice covering increases, and the freezing surface becomes more remote, the ice and the brine become more and more exposed to the atmospheric rigours of the Arctic winter. The brine will continue to deposit ice until its concentration is such that, for example, the cryohydrate of NaCl is ready to separate out. It probably will separate out until it comes in conflict with, for instance, the chloride of calcium or the chloride of magnesium, which will retain some of the water, without solidifying, even at the lowest temperatures. At the winter-quarters of the "Vega" brine was observed oozing out of sea-water ice and liquid at a temperature of − 30° C. It was very rich in calcium and especially magnesium chlorides. In fact, *it is probably quite impossible by any cold occurring in nature to solidify sea-water.*

(*b*) *Melting of pure ice in sea-water and other saline solutions.*—A large number of experiments were made with solutions of concentration comparable with that of sea-water, and in one or two cases the experiments were extended to low temperatures and strong solutions. As a rule, from 50 to 100 grammes of solution, cooled to 0° C., were mixed with an equal weight of pounded ice, also at 0° C. The thermometer used for all these determinations was one of Geissler's normal ones, divided into tenths of a degree Centigrade; and its zero-point was verified almost daily. Along with the thermometer, a pipette of suitable capacity was immersed in the beaker, and used with the thermometer for keeping the mass well mixed. Its upper aperture was closed with a small cork, which was removed from time to time to permit of some of the brine being sucked up and allowed to run back again. The inside of the pipette was thus kept constantly moistened with the slowly altering solution in the beaker. The temperature was read after very thorough mixing, and the sample thereupon immediately removed and preserved for analysis.

TABLE IV.—*Giving the percentage of Chlorine in Solutions of various Chlorides in which Ice melts at given Temperatures.*

Temp. of melting Ice	Chloride in Solution						
	HCl	NaCl	KCl	(Sea-Water)	MgCl$_2$	CaCl$_2$	BaCl$_2$
	Per cent. Chlorine in Solution						
°C.							
−3·5	3·06	3·30	...	...	4·12	...	...
−3·0	2·68	3·02	3·00	...	3·62	3·70	...
−2·5	2·28	2·53	2·50	...	3·12	3·20	...
−2·0	1·85	2·02	2·00	...	2·62	2·70	2·72
−1·5	...	1·50	1·50	1·500	1·19	2·15	2·10
−1·0	...	1·02	1·02	1·034	1·51	1·50	1·47
−0·5	...	0·50	0·52	0·588	0·87	...	...

As a rule, samples were taken for analysis at intervals of 0°·4 C. The results for three classes of salt in dilute solutions are arranged in Tables IV, V, and VI.

On considering them, it was at once evident that the lowering of the melting-point of ice followed the concentration of the solution, but the law deviated in all cases from that of strict proportionality to the amount of salt dissolved, in some cases to a greater extent than in others. In comparing the effects of different salts in solution on the melting-point

On Ice and Brines

TABLE V.—*Giving percentage of Potassium in Solutions of various Potassium Salts in which Ice melts at given Temperatures.*

Temperature of melting Ice	Salt in Solution			
	KCl	KI	$K\frac{Cl+I}{2}$	KOH
	Per cent. K in Solution			
°C.				
− 3·0	3·29	3·02	3·15	...
− 2·5	2·79	2·59	2·68	2·60
− 2·0	2·28	2·13	2·17	2·08
− 1·5	1·74	1·63	1·66	1·57
− 1·0	1·18	1·13	1·12	...
− 0·5	0·59	0·60	0·57	...

TABLE VI.—*Giving percentage of Hydrogen in Solutions of various Hydrogen Salts in which Ice melts at given Temperatures.*

Temperature of melting Ice	Salt in Solution			
	H_2SO_4	HCl	HNO_3	HKO
	Per cent. H in Solution			
°C.				
− 3·0	0·144	0·076	0·077	...
− 2·5	0·119	0·065	0·065	0·066
− 2·0	0·097	0·052	0·052	0·053
− 1·5	0·073	...	0·042	0·041
− 1·0	0·048	...	0·032	...

of ice, no simple connection could be traced between their absolute weights and the effects produced; but on comparing chemically equivalent weights, a very close connection was discovered. This will be evident from the inspection of the tables. In each the first column contains the temperatures at which pure ice melts; and in the parallel columns the percentages of chlorine, potassium, or hydrogen in the solutions of the salts indicated at the head of each column, when ice melts in them at the temperature indicated. The figures thus give numbers proportional in each table to the chemically equivalent weights of the different salts. They show at first that, whereas the presence of equal absolute weights in solution produces very different effects, the presence of chemically equivalent weights produces very similar effects. On closer inspection, it is seen that the effects are almost identical where the elements to which the common constituent is united belong to the same group of the periodic series, and differ sharply where these elements belong to different groups. In the case of the chlorides of sodium and potassium the number expressing the percentage[1] of chlorine in the solution expresses equally the depression of the melting-point of ice in terms of the Centigrade scale. The same depression of melting temperature is produced by 10 per cent. less of chlorine united to hydrogen, and by 30 to 35 per cent. more of chlorine when united to magnesium, calcium, or barium.

The results obtained with sea-water are also given, for comparison. It will be seen that it behaves very approximately as a solution of chloride of sodium containing the same amount of chlorine.

It is perhaps not very astonishing that unit weight of potassium in saline solution should produce the same effect in lowering the melting-point of ice, whether it is united to Cl or I; but it shows clearly how independent this action is of the general character of the body in solution when we find the effect produced by unit weight of hydrogen identical, whether it is united to such opposite radicals as Cl or OK. Table VI shows further the effect of valence. While a given weight of hydrogen produces the same effect in solution, whether it be united to the very different but both univalent radicals Cl and OK, its effect is reduced by one-half when united to the bivalent SO_4. That valence is not the only

[1] All percentages are *by weight*.

factor is shown by comparing the effects of hydrogen and potassium when united to the common element, chlorine. Hydrochloric acid in solution produces a markedly more powerful lowering effect on the melting-point of ice than the equivalent amount of chloride of potassium. Of all the substances that I have experimented on, hydrochloric acid is the most energetic in reducing the melting-point of ice, and with ordinary strong acid and pounded ice there is no difficulty in producing temperatures as low as the freezing-point of mercury. In the case of hydrochloric acid, sulphuric acid, chloride of sodium, and chloride of calcium, I have carried my experiments to low temperatures and great concentration. But before passing to them it is well to consider the more dilute solutions with regard to their density.

That the mere density of the solution in which the ice is melting has no direct connection with the lowering of its melting-point is shown by the following table, in which the specific gravities (at $15°$ C.) are given of the solutions of different salts which gave the same depressions of melting-point :

Temperature of melting Ice	Specific Gravity of Solutions of				
	NaCl	KCl	$MgCl_2$	$CaCl_2$	$BaCl_2$
$°$ C. $-2·86$ $-1·8$	1·03370 1·02174	1·03850 1·02535	1·03893 1·02715	1·04756 1·03262	... 1·06633

There are many similarities in the effects produced by greatly increasing the pressure upon pure water and by dissolving salts in it. First, there is an absolute diminution in the volume of the solution as compared with the sum of the volumes of its components; second, in virtue of this compression by molecular forces it has become less compressible by mechanical means; third, the temperature of maximum density and the freezing temperature are lowered ; and fourth, the former of these two temperatures is lowered more rapidly than the latter. All these effects are produced *in kind* by increasing the pressure on pure water. Whether, or in how far, they agree in degree must be decided by future experiments.

Experiments with Concentrated Solutions.—Several series of experiments have been made with hydrochloric acid, chloride of sodium, and chloride of calcium, and also with sulphuric acid. Table VII gives the results, in the same form as preceding tables, for the chlorides.

TABLE VII.

Temperature of melting Ice	Salt dissolved		
	HCl	NaCl	CaCl$_2$
	Per cent. Cl in Solution		
° C.			
− 35	15·26	...	...
− 30	13·98	...	15·97
− 25	12·60	...	14·47
− 20	11·00	...	12·65
− 15	9·17	11·10	11·29
− 10	7·02	8·40	8·93
− 5	4·15	4·72	5·65

It will be seen that, in proportion as the solution becomes more concentrated, further additions of salt produce a greater effect in lowering the melting-point of ice, and at a temperature of − 15° C. equivalent weights of NaCl and CaCl$_2$ produce identical results. In Table VIII the results for hydrochloric acid and sulphuric acid are given in terms of the percentage of hydrogen in the solution.

The temperatures given in these tables are all in terms of the same thermometer, which has not been verified for this part of its scale by comparison with a standard or with the air thermometer.

We have thus seen that, owing to its peculiar physical properties, it is impossible to prepare the crystalline solid which separates from sea-water and analogous saline solutions in a condition to enable the question, whether the salt does or does not form part of the solid matter of the crystals, to be solved directly by chemical analysis.

TABLE VIII.

Temperature of melting Ice	Acid dissolved	
	HCl	H_2SO_4
	Per cent. H in Solution	
°C.		
−25	0·355	0·538
−20	0·310	0·487
−15	0·258	0·418
−10	0·198	0·332
− 5	0·117	0·205

So far as chemical analysis is applicable, it is in favour of the salt belonging exclusively to the adhering brine. When sea-water is carefully frozen artificially, the ratio between the chlorine and the sulphuric acid is the same for the solid contents of the original water, the crystals, and the mother-liquor. It is exceedingly unlikely, if part of the salt went into the cystals, leaving the remainder in the brine, that there would be no selective separation of its constituents.

It has been shown that snow or pure lake ice, which, when melting by itself or immersed in pure water at atmospheric pressure, melts at the constant temperature called 0° C. or 32° Fahr., changes its melting temperature when immersed in a saline solution. The altered melting temperature, however, is the same for solutions of the same composition (no doubt with some allowance for pressure) and different for solutions of different composition.

The temperature at which pure ice melts in a solution is identical with that at which ice separates from the same solution on being sufficiently cooled.

When sea-water is frozen to the extent of 15 per cent. of its mass, and the crystals so formed are allowed to melt in the liquid in which they have been produced, they melt exactly as they have been formed. If snow or pure ice be immersed in the brine formed by partially freezing sea-water, it melts at the same temperature as the ice which had been formed by freezing the sea-water, so long as the chemical composition remains the same in each case.

The heat removed in freezing sea-water to the extent of 15 per cent. of its mass accounted for the production of the same amount of ice as was given by calculation on the basis of the chlorine found in the mother-liquor.

When some saline solutions are cooled for a sufficient length of time at a sufficiently low temperature, there arrives a certain concentration at a certain temperature, when further removal of heat causes solidification of the brine as a whole (cryohydrate).

The concentration necessary for the solidification of even the cryohydrate of highest melting temperature is such that in the *primary* freezing of the water of the sea no such body can be formed. It would follow from this consideration alone that the first ice formed on the sea in Arctic regions consists of pure ice, and it is also certain that it would retain a large quantity of the residual sea-water in its interstices. During the winter this inclosed liquor would solidify in the interstices of the crystals to ice and cryohydrates, in so far as the temperature and the nature of the salts in solution would permit. From my experiments with chloride of calcium, and the existence of brines observed to remain liquid at − 30° C. at the winter-quarters of the " Vega," it is unlikely that sea-water, as a whole, can ever be completely solidified in nature. The presence of unfreezable or difficultly freezable brine in freshly-formed sea-water ice, explains its eminently plastic character even at very low temperatures.

The fact that cryohydrates of different salts solidify and melt at different temperatures, sufficiently explains the various composition of different specimens of *old* sea-ice.

The apparent expansion, near the melting-point, of ice

formed by freezing water which contains any salt at all is perfectly explained on the hypothesis that in the act of freezing the water rigidly excludes all saline matter from participation in its solidification.

The residual and unfreezable brine which remains in considerable quantity liquid when sea-water is frozen, must also remain in greater or less quantity when fresh water is frozen. All natural waters, including rain-water, contain some foreign and usually saline ingredients. If we take chloride of sodium as the type of such ingredients, and suppose a water to contain a quantity of this salt equivalent to one part by weight of chlorine in a million parts of water, then we should have a solution containing 0·0001 per cent. of chlorine, and it would begin to freeze and to deposit pure ice at a temperature of − 0°·0001 C.; and it would continue to do so until, say, 999,000 parts of water had been deposited as ice. There would then remain 1000 parts of residual water, which would retain the salt, and would contain, therefore, 0·1 per cent. of chlorine, and would not freeze until the temperature had fallen to − 0°·1 C. This water would then deposit ice at temperatures becoming progressively lower, until, when 900 more parts of ice had been deposited, we should have 100 parts residual water, or brine as it might now be called, containing 1 per cent. of chlorine, and remaining liquid at temperatures above − 1°·0 C. When 90 more parts of ice had been deposited, we should have 10 parts of concentrated brine containing 10 per cent. chlorine and remaining liquid as low as − 13° C. In the case imagined, we assume the saline contents to consist of NaCl only, and with further concentration the cryohydrate would no doubt separate out and the mass become really solid. On reversing the operations, that is, warming the ice just formed, we should, when the temperature has risen to about − 13° C., have 999,990 parts ice and 10 brine containing 10 per cent. chlorine. Now, owing to the remarkable fact that pure ice, in contact with a saline solution, melts at a temperature which depends on the nature and the amount of the salt in the solution, and is identical with the temperature at which ice separates from a solution of the same composition on cooling, the brine liquefies more and more ice at progressively rising temperatures, until, as before, when the temperature of the mass has risen to $-0°·1$ C., it consists of 999,000 parts of ice and 1000 parts of liquid water, containing 1 part of chlorine. The remainder of the ice will melt at a temperature gradually rising from $-0°·1$ to $0°$ C.

The consideration of this example furnishes an easy explanation of the anomalous behaviour of ice formed from anything but the very purest distilled water, in the neighbourhood of its melting-point. This subject has been studied with great care and thoroughness by Pettersson. The apparent expansion of all but the very purest ice, when cooled below $0°$ C., is ascribed by him in part to solid saline contents of the ice which exercise a disturbing and unexplained influence on its physical properties. Viewed in the light of the fact that the presence of even the smallest quantity of saline matter in solution prevents the formation of ice at $0°$ C., and promotes its liquefaction at temperatures below $0°$ C., we see that this apparent expansion of the ice on cooling is probably due to the fact that we are dealing, not with homogeneous solid ice, but with a mixture of ice and saline solution. As the temperature falls this solution deposits more and more ice, and its volume increases. But the increase of volume is due to the formation of ice out of water, and not to the expansion of a crystalline solid already formed.

TABLE IX.—*Water containing 7 parts Cl in* 1,000,000.

Temp. $°$ C.	Water frozen c.c.	Ice formed c.c.	Brine remaining c.c.	Ice and Brine c.c.	Pettersson III Vol. of Ice at $T°$ c.c.	Diff. c.c.
T	V_1	v_1	V_2	v_2	P	$P \cdot v_2$
$-0·07$	99000	107979	1000	108979	108980	1
$-0·10$	99300	108306	700	109006	109007	1
$-0·15$	99533	108561	467	109028	109038	10
$-0·20$	99650	108687	350	109037	109048	11
$-0·40$	99825	108879	175	109054	109057	3

In Table IX are given the volumes occupied by the ice
(with inclosed brine) formed by freezing 100,000 c.c. (at 0° C.)
of a water containing chloride of sodium equivalent to
7 grammes chlorine in 1,000,000 cubic centimetres (at 0° C.).

The volume (v_2) of the ice and brine formed on freezing
this water is compared with that (P) observed by Pettersson
in freezing a sample of the distilled water in ordinary use in
the laboratory.

It will be seen that the volumes observed by Pettersson
agree very closely with those calculated for a water containing
7 parts of chlorine in a million, on the assumption that
the saline matter is contained entirely in adhering liquid
brine.

The irregularities in the melting-points of bodies like
acetic acid, to which Pettersson refers, are without doubt due
to a perfectly similar cause.

Also the very low latent heat observed by Pettersson for
sea-water is to be explained by the fact that the salt retains
a considerable proportion of the water in the liquid state even
at temperatures many degrees below the freezing-point of
distilled water.

Thus, he made two determinations of the latent heat of
sea-water containing 1·927 per cent. Cl and 3·53 per cent. salt.
The freezing took place in the one case between the temperatures −9°·0 and −7°·47 C., and in the other between −8°·35
and −6°·94 C., and the results he found were 52·7 and 51·5.
The mean initial temperature in these two experiments is
−8°·7 C., and the mean final temperature −7°·2 C. At
−7°·2 C. ice would form on cooling, and would melt on
warming a solution of chloride of sodium containing 6·48 per
cent. Cl, which represents 11·87 per cent. of the sea salt. In
order to concentrate a brine containing 3·53 per cent. salt to
one containing 11·87 per cent., 70 per cent. of the water in it
must be removed. Hence in sea-water freezing at a final
temperature of −7°·2 C. there is formed 70 per cent. of ice,
and there remains liquid 30 per cent. of brine. Freezing
began at the mean temperature −8°·7 C., and the latent heat
of pure ice at this temperature is 75. Calculating the latent

heat of this mixture from the heat liberated in the calorimeter during freezing, and assuming that the whole mass had solidified, Pettersson's results give the mean latent heat of this sea-water as 52·1. Calculating the apparent latent heat on the assumption that 70 per cent. of the mass solidifies into pure ice and that 30 per cent. remains liquid, we get the number 51·5. On all grounds therefore we must conclude that pure ice is the primary product in freezing sea-water and saline solutions of moderate concentration.

The *plasticity* of ice and the motion of glaciers receive a simple and natural explanation when we see, as in Table IX, that, if the water from which this ice is produced contains no more than 7 parts of chlorine per million, it will, in the process of thawing, when the temperature has risen to $-0°·07$ C., consist to the extent of 1 per cent. of its mass of liquid brine or water. The water considered in Table IX is certainly not less free from foreign ingredients than rain or snow. It follows, therefore, that a glacier, in a climate where the temperature is for the greater part of the year above 0° C., must have a tendency to *flow*, owing to the power of saline solutions to deposit ice and to dissolve it at temperatures below 0° C.

The *verification of thermometers* by comparison with the air thermometer is always troublesome. It results from the above investigations that, if the temperature at which ice melts in solutions of a salt, such as chloride of calcium of different degrees of concentration, were once and for all carefully determined by means of a standard air thermometer, a thermometer could be indirectly but satisfactorily compared with the air thermometer at temperatures below 0° C. by immersing it in a mixture of ice and chloride of calcium solution, and taking a series of readings of the thermometer and samples of the brine simultaneously. By determining the chlorine in the samples the concentration of the brine is ascertained, and the comparison with the standard effected.

Freezing Mixtures.—The results obtained in examining the melting-point of ice in saline solutions afford data for mixing freezing-baths of any degree of cooling power. With chloride of sodium, for instance, a rough rule is to have such an amount of salt dissolved in the brine that the percentage of chlorine shall give the desired temperature in Centigrade degrees below the freezing-point. In my experiments in freezing sea-water in quantities of 300 grammes, I usually made up the bath of 500 grammes pounded ice, 400 grammes water, and 45 grammes common salt. When mixed, the liquid contained about 4 per cent. Cl, and gave a temperature a little below $-4°$ C. In the course of an hour the liquid would contain 3 per cent. to 3·25 per cent. Cl, and the temperature have risen to $-3°$ C. By using such baths freezing operations can always be kept completely in hand.

No. 4. [*From the Transactions of the Royal Society of Edinburgh,* 1899, *Vol. XXXIX, Part III, No.* 18.]

ON STEAM AND BRINES

THE immediate purpose of the present research was the investigation of the temperature at different pressures of *boiling mixtures* of steam and salts, analogous to the well-known *freezing* mixtures of ice and salt.

When steam is blown through common salt in coarse powder, it condenses to water, which dissolves some of the salt, and the resulting brine is kept boiling by the arrival of more steam. The temperature of this boiling mixture is quite constant so long as there is an abundant supply both of steam and of salt, and as the atmospheric pressure does not change, it is about 8°·5 C. above the temperature of boiling water when the barometric pressure is the normal of 760 mm. When the barometric pressure is 560 mm. this excess has fallen to 8°·0 C. Most other salts behave in a similar way.

The investigation was not confined to the determination of the temperature of the boiling mixture of steam, brine, and salt; the experiments were continued after all the salt was dissolved, and the passage of the steam was interrupted from time to time, and the weight of steam condensed and the temperature of the liquid observed, so as to give the concentration and the boiling temperature of solutions of the salt of different strengths. This part of the work, namely, passing steam through non-saturated solutions, was in continuation of work begun immediately after the publication of my paper, 'On Ice and Brines,' in the *Proceedings of the Royal Society of Edinburgh,* 1887 (see above, p. 130). It is well known to all those who have tried it that it is impossible to produce pure ice by freezing a saline solution, and the purpose of the research reported in the above paper was to ascertain whether the salt, from which the ice could not be entirely freed, was an essential part of the ice, or only an inseparable adherent to it. The latter was shown to be the case, and the ice, or crystalline body which freezes out of a saline solution, to be really ice and nothing else, by showing the identity of the temperature at which snow or pure pounded ice of independent source melts in a saline solution, and that at which ice forms on freezing a saline solution of the same nature and concentration.

It was natural to pass from the freezing to the boiling temperature of solutions, and to use the temperature at which steam condenses in the solutions as a measure of their boiling temperature in the same way as the melting temperature of ice had been used for indicating the freezing temperature. It had also the advantage that no preliminary investigation was necessary. There was no doubt that the steam produced by a boiling saline solution was pure steam, there was no doubt about the temperature of the boiling solution from which the steam proceeded, and there was equally little doubt that the steam so proceeding had the same temperature as the boiling solution, because when steam is blown into such a solution it is condensed until the solution has been heated to its boiling-point. If the solution can condense outside steam, it necessarily must be able to condense its own steam, for the two substances are identical.

These conditions are expressed in the following law:—*The temperature at which steam condenses depends on the nature of the medium in which it condenses; and the temperature at which it is generated depends on the medium in which it is generated; and the temperature at which steam condenses in a given medium is that at which it is generated in the same medium, the pressure in all cases being the same.* (1916.)

When the naked flame is used as source of heat it is very difficult to obtain the exact boiling temperature of a brine, or indeed of pure water.

On the other hand, it is very easy to get results of any desired degree of exactitude by using an abundant current of saturated steam to boil the water or brine. Overheating is impossible, and bumping equally so. A thermometer of any degree of delicacy may be used ; there is never any room for doubt that its temperature is, to the minutest fraction of a degree, the same as that of the boiling liquid, and the value of the observation thus depends only on the trustworthiness of the instrument. In the paper on 'Ice and Brines' above referred to, it was pointed out that the melting-point of ice in brines of determined nature and strength could be used as fixed points on the thermometric scale, in the same way as the melting temperature of ice in pure water is habitually used. Similarly the condensing temperature of steam in saline solutions forms a ready means of fixing exactly certain points above the ordinary boiling-point of water. Also the minimum temperature, or cryohydric point, of freezing mixtures is very useful, many of them being as well defined as the melting-point of ice in water.

Our boiling mixtures occupy a similar position with respect to the boiling-point of water that the freezing mixtures do to its freezing-point. In freezing mixtures the dry salt is mixed with pounded ice or snow ; if the mixture is properly made, the temperature falls at once to the true minimum, and remains quite steady for a great length of time. In boiling mixtures the dry salt is placed in a U-tube of special dimensions, to be described presently, and steam is passed into it by one leg, while the other leg carries the thermometer, and the surplus steam escapes through a side tube. The supply of steam must be abundant, while the exit tube for the steam must be sufficiently wide to make it impossible for the steam, after it has passed through the mixture of salt and brine, to have a pressure above that of the atmosphere into which it exhausts. From the time when enough steam has condensed to form a more or less liquid magma through which the steam bubbles, the mass is kept thoroughly well mixed, and the thermometer keeps the temperature with absolute steadiness until so much steam has condensed and so little solid salt remains that it cannot be assumed that every particle of steam condensed can immediately find a particle of salt to dissolve. Then the temperature begins to fall in the same way as that of a freezing mixture begins to rise. Operating in this way it is possible to obtain definite and perfectly constant temperatures with most of the ordinary soluble salts of the laboratory. They are, however, not all equally good, and there are some that are of no use for boiling mixtures.

The effect produced when steam meets a salt depends on the properties of the salt, especially on its solubility in boiling water and on the thermal effect of its solution. Generally speaking, the greater the solubility of the salt the greater is the elevation of the condensing point of steam which it produces. At ordinary temperatures salts commonly dissolve in water with absorption of heat, which tends to increase the condensation of steam. The absorption of heat may be so great that the brine produced by the salt dissolving in the condensed steam is unable to rise to its boiling-point until all the solid salt has disappeared.

Amongst anhydrous salts nitrate of ammonium is a good example of this. An experiment was made with 60 grammes of the salt. In $2\frac{1}{4}$ minutes after the steam reached the salt it was all dissolved, and the temperature barely rose to that of boiling water. Twelve grammes of steam had been condensed.

Of salts containing water of crystallisation, acetate of soda is a good example amongst those which dissolve with great absorption of heat. Sixty-two grammes of this salt treated in the same way as the nitrate of ammonium were completely dissolved in $2\frac{1}{2}$ minutes after the steam reached the salt, and the temperature had only reached 60° C. These and similar salts are not suitable for boiling mixtures.

The following are one or two examples of the solubilities observed at the boiling-point of the saturated solution for one-fifth gramme-molecule of each salt:

Salt used		NaCl	KCl	$BaCl_2$	$(NH_4)_2SO_4$
Temperature of Conden-{	water	100°·44 C.	100°·44	99°·95	99°·40
sation of Steam on {	salt	108°·98	108°·94	104°·46	107°·03
Weight of Steam Condensed					
for complete solution grms.		29·9	25·9	64·0	25

The least soluble of these salts is chloride of barium, requiring 64 grms. of condensed steam for solution, and the most soluble is sulphate of ammonium, which requires only 25 grms.

Apparatus and Method of Experimenting.—The apparatus, Fig. 1, consists of the lamp, *A*, the steam generator or boiler, *B*, the U-tube or receiver, *D*, and the connecting tube, *G*. In the laboratory a gas lamp was used, except when it was wished to check the results of the fuel consumed, when a spirit lamp was used, and it was weighed before and after the experiment. In the experiments at high levels, where gas was not available, spirit was used. The lamp employed was one of a French pattern, and forming a part of the *Réchaud à double flamme forcée* of smallest size, which has a large sale for domestic purposes. It is the most efficient pattern of spirit lamp with which I am acquainted, and as it is especially constructed for use in travelling, it was very suitable for my high-level work. It holds about 250 c.c. of spirit.

The steam generator or boiler, *B*, used at high levels, was a flask made of spun copper and of 500 c.c. capacity. A suitable charge is 300 c.c. of water. With such a charge, and heated by the French lamp, the water boiled in six minutes, with a consumption of 12 grammes of spirit. While keeping steam at the rate suitable for the experiment, the lamp consumed 21 grammes of spirit in fifteen minutes, and evaporated 92 grammes of water. For experiments in the laboratory I use a large copper flask of 2 litres capacity. This is a very convenient laboratory vessel. They are manufactured to replace the glass flasks of Napier's coffee machines for use in restaurants, and they can be had in larger sizes. A copper flask can always be obtained at short notice from a plumber, by getting him to fit a neck to one of the copper balls which are used as controlling floats for cistern taps. These are to be had of all sizes up to 10 inches diameter, and are very cheap.

The receiver, *D*, has the appearance and shape shown in the Fig. The actual dimensions are variable, according to the quantity of salt on which it is proposed to operate, and according to the length of the thermometer. The working part of the thermometer must be entirely within the receiver. The uncertainty caused by the exposure of any part of the stem occupied by mercury stultifies the use of very delicate

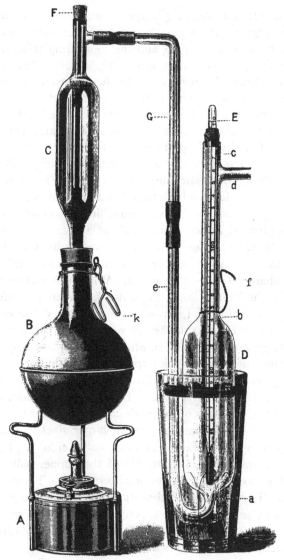

Fig. 1.

instruments. If the whole range of the thermometer is to be utilised, then the length of the receiver must be such as to take the whole thermometer. The two thermometers which I used in all my experiments differed slightly in length, and it was convenient to have a receiver made for each. Of course the longer thermometer could be used with the shorter receiver, so long as the temperature to be observed was not too high. One receiver which I have used much has the following dimensions :—Total length of the body of the tube, from the entrance, *a*, of the steam tube at the bottom to the top, *c*, where the thermometer, *E*, is retained by a perforated cork, is 30 cm.; the length, *ab*, of the body for the reception of the salts and brine is 16 cm., and its diameter 42 mm. The length of the neck, *bc*, is 14 cm. and diameter 13 mm. The steam exit tube, *d*, has a diameter of 7 mm., and the entry or connecting tube, *e*, has the same or a slightly less diameter. The entry tube is bent up parallel to the main body of the instrument, and is connected with the boiler by a tube, *G*, as shown in the Fig. The steam tube, *C*, on the boiler is important. The straight portion which enters the boiler should have a diameter of 8 to 9 mm. Its upper part, which is fitted with a cork, should have a diameter of not more than 7 or 8 mm. Steam is kept constantly in the boiler, and connection with the receiver is made or broken instantly by inserting or removing the cork. The most convenient support for the receiver, whether it be used in or out of the laboratory, is an ordinary tumbler or drinking glass. The tube rests on a piece of cork, grooved to take the bend of the tube, and it is steadied by a ring made of a piece of india-rubber tubing. This form of support has many advantages. In the first place, the apparatus has great stability; then the glass is transparent, and it is essential to be able to see the boiling mixture during the whole course of the experiment; also, while being transparent, the glass protects the receiver from excessive loss of heat.

In the experiments to be reported, the principal thing to be observed is the difference between the temperature of pure saturated steam and the temperature produced by the condensation of this steam in the mixture of salt and brine. It is, therefore, of equal importance to observe accurately the temperature of pure saturated steam as to observe that of the boiling mixture. The same apparatus suffices for both purposes. Before charging, and being clean and dry, the receiver is connected with the boiler, and steam blown through it. Some of it condenses and collects at the bottom of the receiver, forming a pool of distilled water boiled by steam, the steam produced by which is perfect for the purpose. As the division marking 100° C. is usually some way down the stem, the thermometer is preferably pulled up, so that the bulb is entirely in the steam. The steam so produced in an apparatus of the proportions described, must be truly saturated steam of the tension equal to the actual barometric pressure, and must, therefore, have exactly the temperature which corresponds to this pressure. This is the temperature of steam condensing on pure water.

The supply of steam is abundant, and the latent heat of steam is very great ; the thermometer, therefore, must take, in a very short time, exactly the temperature of the steam with which it is completely surrounded. The thermometer may be constructed so that a millimetre on the stem corresponds to a hundredth or a thousandth of a degree. Its indication will be perfectly steady, provided that the conditions remain unchanged. The slightest change in the barometric pressure makes itself at once apparent, and at all times, especially during unsettled weather, the temperature of saturated steam must be observed at frequent intervals.

It is convenient to have a separate apparatus for this purpose. If we imagine the receiver, *D*, with the entry tube at the bottom straight instead of bent, so that it can take the place of the T-tube in the boiler, we have a perfect apparatus for determining the temperature of saturated steam and consequently also for fixing the point corresponding to 100° C. on the scale of the thermometer. It is assumed that the steam tube is large enough to take the whole working part of the thermometer. It is unsuitable in itself or to the steam generator used if the steam makes its exit with an audible sound. Tenths of a Centigrade degree are then uncertain. Yet it is essential that there should be an abundant flow of steam through the tube ; therefore, the exit tube must be sufficiently wide to allow the steam to issue in a stream of good volume, and must not be so wide as to incur any risk of regurgitation of air. Again, the entry tube for the steam must not be too narrow. It is not essential, but it is very convenient, that it should be so wide that the steam condensed on the walls and flowing back into the boiler should continue to flow down the sides of the entry tube and not collect at the bottom of the wide part of the tube. When the steam proceeds along a short straight passage from the boiler to the steam tube, it throws about any water in a violent and inconvenient way. In a tube which I use for this purpose the entrance tube has a diameter of 9 mm. and the exit tube 10 mm. The entry tube is thus wide enough to permit the condensed steam to flow back along its sides ; at the same time, it is smaller than the exit tube, so that, apart from the continual condensation of a portion of the steam, there is no danger of the tube receiving more steam than it can freely get rid of.

That an apparatus such as that here described does, in fact, exclude the possibility of the steam supplied having a tension which differs at all from the pressure of the atmosphere with which it exhausts, will be evident from the following experiments. The small copper flask, the spirit lamp, and the steam tubes above described were used. The thermometer was divided into fiftieths of a degree Centigrade, the length of one degree being 35 mm. The atmospheric pressure happened to be pretty high, and when the thermometer had taken the temperature of the steam the top of the mercury was exactly even with the centre of the line on the scale marking 100°·18 C. It occupied this position when steam was being generated at its highest rate of 8·4 grammes per minute. The flame of the lamp was reduced by degrees until it reached its lowest point, when steam was being generated at the rate of 2·4 grammes per minute. It was then issuing continuously, that is, there was no regurgitation, but it partook more of the nature of an exhalation than of a stream; yet the mercury remained exactly on the centre of the line of 100°·18 C., and it was only when the lamp had been reduced to its very lowest that it could be said to have fallen to the lower edge of the line. The temperature then indicated by the thermometer was not lower than 180°·179 C. When the reading of such a thermometer remains unaltered while the supply of steam is varied in the proportion of nearly four to one, the efficiency of the steam tube may be said to be perfect.

A further condition affecting the usefulness of these tubes is that they and the thermometer shall be perfectly clean. The thermometer is the most liable to contamination, and I generally found it convenient to wash it with soap and water before every experiment. The steam which condenses on the thermometer and on the inside of the tube should do so in a film and not as a dew, and it does so if the surfaces are perfectly clean. The inside of the tube is more difficult to deal with than the outside of the thermometer. On the other hand, when once cleaned it remains clean much longer. Soap is used here also, and the best way of using it is to smear the inside of the upper part of the tube with soap, preferably soft soap. Steam is raised, and so soon as it reaches the soap it condenses and forms a uniform film of solution which drains down back into the boiler and by continuing to boil the steam condensing washes the inside walls quite clean from everything, and for a considerable time afterwards there is no trouble about the steam condensing in the tube as dew. If the boiler has been charged for this operation with distilled water, there is the disadvantage that it immediately primes, and the steam, instead of washing down the sides of the tube, continues to blow soap bubbles at its upper end, without washing the tube. I always use ordinary tap water, which supplies equally pure steam with distilled water, but it has this advantage, that when the soap solution drains back into it, being in comparatively small quantity, it is immediately precipitated by the earthy ingredients of the water, which continues to supply pure steam without priming.

Attention to these small matters is all-important, not only in order to secure accuracy but also comfort in experimenting.

Thermometers.—Two thermometers were used for these experiments, both made especially for me by Mr Hicks, of Hatton Garden. One of them, A, was intended for use at ordinary levels; the scale was in Fahrenheit's degrees, and ranged from 210° F. to 240° F. Each degree was divided into tenths, and had a length of 7·2 mm. The length of a Centigrade degree in this thermometer would thus have been 13 mm., and in order that the single divisions should not be too far apart it would have had to be divided into twentieths of a degree.

The other thermometer, B, was especially made with a view to experimenting on chloride of sodium at all available heights above the sea. It was graduated into Centigrade degrees and tenths, the length of one degree being 10 mm., and the range was from 85° C. to 110° C.

By the kindness of Dr Bilwiller, director of the Central Meteorological Office of Switzerland, I was able to compare the indications of this thermometer in saturated steam at different heights with the temperatures which it ought to have shown, on the basis of the atmospheric pressure as given by the barometers of the central bureau at different stations. For higher temperatures it was verified in terms of the standard barometer of the Scottish Meteorological Society in Edinburgh, which was obligingly put at my disposal by Dr Buchan. The following are the readings in the order of height:

Locality ...	Julier	Sils	Zürich	Edinburgh
Height above sea metres	2244			
Barometer, at 0° C. mm.	581·84	616·4	720·3	764·18
Observed temperatures	92°·86 C.	94°·38	98°·62	100°·23
Calculated „	92°·69	94°·24	98°·51	100°·16
Correction	−0·17	−0·14	−0·11	−0·07

Thermometer B was compared at Kew, giving corrections

amounting in the extreme to 0·2° F.   All the observed temperatures have been corrected accordingly.

The General Order of the Experiment.—The temperature of saturated steam was determined in the straight steam tube. The U-shaped receiver being clean and dry, was weighed. The portion of salt, usually one-fifth of a gramme-molecule, was weighed out carefully into the receiver, which was then again weighed for the purpose of afterwards arriving at the weight of the condensed steam.   The weight of the receiver, both empty and charged, includes that of the thermometer and its attachment.   When steam is issuing from the boiler at the top of the T-tube, the receiver is connected with it, as shown.   The top of the T-tube is now closed with the cork, and the passage of steam through the salt begins.   The time is noted when the steam reaches the salt, and this is the beginning of the experiment, and the time is logged as a part of every entry in the note-book.   It is particularly noted when the mass in the receiver forms a liquid magma through which the steam bubbles, when the thermometer attains its maximum, when it begins to be unsteady, and when steam is shut off.   Before reaching the maximum the temperature is noted every half-minute, afterwards every minute, and at the end every half or quarter-minute.   When it is judged by the fall of temperature and the quantity of salt undissolved that these two small quantities compensate each other, the cork is removed from the T-tube and the operation is interrupted.   The receiver is then quickly disconnected from the india-rubber tube, and suspended from the scale, which is within arm's reach from the working bench, and the weight ascertained to the nearest decigramme.   The receiver is then immediately reconnected with the boiler, the cork inserted, and boiling recommenced.   Steam is passed until the temperature has fallen to the first attainable whole number of degrees above the temperature of saturated steam, when it is interrupted, the weight observed and the receiver reconnected, to be again weighed when the next whole degree is reached, and so on until so much water has collected in the receiver that the steam can no longer be passed through it at a suitable rate without risk of throwing out some of its contents. The temperature of saturated steam is now again determined with the thermometer used in the experiments. This has been determined several times during the experiment in another apparatus and with another thermometer. This enables the effect of any change in the barometric pressure to be spread correctly over the time occupied by the experiment. This series of observations gives the concentration of solutions whose boiling-points are higher than that of pure water by certain definite amounts. The small uncertainty which attaches to the determination of the concentration of the boiling saturated solution does not affect that of the less concentrated solution. When the saturated solution has been weighed and reconnected with the steam generator, it often happens that, however expeditiously the operation may be performed, some of the salt has crystallised out, and this generally requires an extra amount of heat, or steam condensed, to redissolve it. Then the boiling temperature of the solution when saturated is lowered very much by a small dilution, an effect which diminishes rapidly with increasing dilution.

As result of the series we have the temperature of the saturated boiling solution and approximately its concentration, also the boiling temperature and exact concentration of a series of more dilute solutions. When the boiling tube has been emptied and washed, steam is blown through it until the whole tube is heated up to the temperature of the steam ; it is then quickly disconnected, the water ejected from it, and air blown through it from the lungs, which in a few seconds dries the inside of the receiver completely. This is the easiest way to dry the inside of all complicated glass apparatus. The glass of the apparatus is always sufficiently massive that when it has been heated to 100° C. it has more than sufficient immediately available heat to evaporate all the water that will adhere to its surface, and still not fall to such a temperature as to condense moisture from the air of the lungs.

The receiver is immediately ready for another experiment

As above described, each experiment must be expected to take from an hour and a half to two hours.

Steam Condensed in Heating the Apparatus.—When the thermometer is in its usual experimental position, that is, with its bulb in the middle of the salt and so low down that it will be immersed in the brine or water whenever enough steam has condensed to make this possible, and the whole of the working part of the stem is in the steam space, it arrives at the maximum temperature almost simultaneously with the first exit of steam from the apparatus ; but that part of the steam tube situated above the exit tube is not yet thoroughly warmed through, and a little time must be allowed during which the strength of the steam current increases until it becomes steady. In ordinary circumstances the apparatus cannot be held to be warmed through in less than 90 seconds, and, for purposes of heat calculation, we take the initial period of heating to be two minutes, during which it may be said always to be complete. Experiments made with boiling mixture tube, weighing with thermometer 239·65 grms., showed the following results :

Steam condensed	...	grms.	8·3	8·3	8·5	8·35
In time ...	... ...	seconds	85	90	100	90

From these we find the mean amount of steam condensed in the first 90 seconds 8·25 grms. When the steam was passed through for exactly two minutes before weighing, the following weights of steam were condensed :—9·1, 8·9, and 9·0 grms., or a mean of 9·0 grms.

For vessels of the same pattern and nearly the same size the quantity of steam required to heat them or to keep them hot depends simply on the amount of glass.

Thus, our apparatus weighs 239·65 grms., and may be considered to be all glass. If the specific heat of the glass be 0·2, then the amount of water thermally equivalent to it is 48 grms. In order to raise the temperature of 48 grms. water from 15° C. to 100° C., we require 4080 g.° C. (grammedegrees-Celsius), and this can be supplied by 7·61 grms. steam saturated at 100° C., and condensing at 100° C. But the mean temperature of the apparatus during warming may be taken to be 57°·5 C., and the 7·61 grms. water formed would, in cooling from 100° C. to 57°·5 C., give out 323·4 g.° C., a quantity which is furnished by the condensation of 0·603 grm. steam at 100° C. Deducting this from 7·61, we have 7·007, or 7 grammes as the least amount of steam required to raise the temperature of the apparatus instantaneously to 100° C. In practice, the operation takes a minute and a half, during which the apparatus is losing heat at an increasing rate. This is supplied by additional steam condensed. We have seen that the steam condensed in two minutes is 9·0 grms., and in 1½ minutes 8·25 grms., giving 0·75 grm. of steam condensed in half a minute, or 1·5 grms. in one minute after the whole apparatus has taken the temperature of 100° C. The mean temperature during heating has been taken as 57°·5 C., so we may take the rate of cooling at $\frac{6}{10}$ of the above rate, or equivalent to the condensation of 0·9 grm. steam per minute, or of 1·35 grms. for 1½ minutes. Adding this to 7 grms., the weight of steam required for instantaneous heating, we obtain 8·35 grms. as the theoretical weight of steam required to warm the apparatus under the above conditions. The mean observed amount is 8·25 grms., which may therefore be accepted with confidence. Also, we may confidently calculate the amount of steam required to warm another apparatus of the same type on the basis of its weight.

In order to determine more carefully the amount of steam required to keep the apparatus at a temperature of 100° C. during some time, two experiments were made, the apparatus being dry and cold to begin with. In the first, steam was passed through for 12 minutes, when 22·2 grms. were condensed; in the second, the steam was passed for 32 minutes, when 50·8 grms. were condensed. Allowing that in each case 9 grms. of steam were condensed in the first two minutes, we have 13·2 grms. condensed in 10 minutes, and 41·8 grms. in 30 minutes. The first is at the rate of 1·32 grms. per minute, and the second at the rate of 1·39 grms. per minute, or a mean of 1·35 grms. per minute.

Experiments of a similar kind were made with ⅕ NaCl, or

11·7 grms. of this salt in the tube to begin with. Two minutes were sufficient for heating up to 100° C. In two experiments the amounts of steam condensed were 9·8 and 9·6 respectively, giving a mean of 9·7 grms. In a similar experiment, where the passage of steam was not stopped until the salt was all dissolved, which took 16⅓ minutes, the steam condensed was 31·5 grms. Deducting 9·7 grms. we have 21·8 grms. condensed in 14⅓ minutes, or 1·5 grms. per minute. The rate of condensation is naturally higher, because salt is being dissolved. With the apparatus empty at the start, 9·0 grms. steam are condensed in the first two minutes; with a charge of 11·7 grms. chloride of sodium, 9·7 grms. of steam are required; the excess, or 0·7 grm., may be taken as the steam condensed by the NaCl in the two minutes, and as constant for the same amount of NaCl in other apparatus.

The boiling tube used in all the experiments up to 26th October 1897 weighed, with thermometer, 157·3 grms. This would require 5·5 grms. of steam in order to raise it to 100° C., and the heating would be complete in one minute instead of in one and a half minutes as with the apparatus weighing 239 grms. Allowing 1·0 grm. for the amount of steam condensed in the next minute, we should have, after two minutes, 6·5 grms. steam condensed. Where ⅕ NaCl was used, we should have to add 0·7, and the amount thus condensed at the end of the first two minutes would be 7·2 grms. The rate of condensation per minute, after the first two minutes, would be ⅔ 1·35 or 0·9 grm., and adding 0·15 for the chloride of sodium, we have 1·05 grms. per minute, taking 30·4 grms. as the amount of steam required to be condensed for ⅕ NaCl.

Localities where Experiments were made.—The lowest station, and the one representing the sea-level, was my laboratory in Edinburgh: its elevation is about 85 metres, or 279 feet, above the sea. Although a large number of experiments in this field had been made in the course of previous years, those utilised for this paper were all made after my return from Switzerland, with the same apparatus and the same thermometers that were used there, and with the experiments arranged on the same plan. Being the last series, it is also the most symmetrical.

The stations at higher levels are all in Switzerland, and they are as follows in order of elevation:

Place	Height above the Sea	
	Metres	Feet
Zürich	410	1345
Fiesch	1054	3458
Andermatt	1444	4738
Pontresina	1820	5970
St Moritz	1860	6102
Eggischorn	2193	7195
Julier Hospiz	2244	7362
Schafberg (Engadine) ...	2733	8966

Of these places Pontresina, with St Moritz, was the most important. The most complete series of observations on mixtures, as well as on single salts, was made there. A corresponding series of observations on single salts was made on the Schafberg, which rises immediately behind Pontresina, and about 900 metres above it. Shelter is obtained at the top in the chalet which does duty as a restaurant, and it is approached by a well-made path. Experiments on mixtures of salts could not be made at this station, because the weather became so persistently bad that the chalet was closed for the season early in September.

Of the other places on the list, St Moritz is taken as one with Pontresina, because the difference of level is less than that corresponding to ordinary fluctuations of the barometer.

Zürich and Julier Hospiz were visited for the purpose of verifying the thermometers by observing the temperatures of saturated steam, and the barometric pressure by standard barometers at the same time and place. One or two observations with salts were made at the same time, but they have only the value of isolated observations. The boiling mixture of chloride of sodium was observed at all the stations, and it was the only salt experimented with at Eggischorn, Fiesch, and Andermatt, as at that date I intended to confine my observations to chloride of sodium alone.

The salts used in this research are the chlorides of sodium, potassium, ammonium, barium ; the chlorate of potassium ; the nitrates of sodium, barium, strontium, and lead, and the sulphates of potassium and ammonium.   These salts were used singly, and also in mixtures of not more than two salts each.   The charge of the apparatus was usually one-fifth of a gramme-molecule, but with sparingly soluble salts, such as nitrate of barium or sulphate of potassium, one-tenth and sometimes one-twentieth of a molecule were used.   A watch, giving minutes and seconds accurately, was observed during all the experiments.   In the case of simple salts the time was noted when the steam reached the salt, when the salt formed a magma, with the steam condensed, when the maximum temperature was reached, when it began to fall, and when the passage of steam was stopped previous to making the first weighing.   These particular epochs were always noted, but, as a matter of fact, a complete time-log was kept of every experiment.   In experiments with mixtures the temperature was noted every minute, and often every half-minute, so long as salt remained undissolved.   A complete time record of this kind often furnishes valuable incidental information, and is often useful in detecting and rectifying errors of observation.

Chloride of Sodium.—The series with this salt is very complete, including sixteen independent experiments ; the atmospheric pressure varied from 550·4 to 772 mm., and the temperature of saturated steam from 91°·2 C. to 100°·44 C.   The corresponding temperatures of the boiling mixture ranged from 99°·3 C. to 108°·98 C., so that the elevation of boiling-point caused by saturation with NaCl ranges from 8°·1 C. to 8°·54 C., or nearly half a degree Centigrade of increase for a rise of boiling-point of the salt solution of 9°·68 C.   Roughly, it diminishes 0°·05 C. for every degree that the boiling temperature of the boiling solution falls.   The results of the observations in different localities are collected in Table I, page 186.

If we consider the relation between atmospheric pressure

and the vapour tension of water at the temperature of the boiling mixture, we see that it is practically constant. The mean of the sixteen values is 0·7435. The average deviation from the mean is 0·0004, and the maximum deviation 0·0009. The mean of the observations made at Edinburgh is 0·0001 below this. The five observations made at Pontresina and St Moritz give a mean of 0·0004 above, and the three observations on the Schafberg a mean of 0·0005 below 0·7435. Taking 0·7439 for the value at Pontresina, the value of $t - T$ would be 0°·015 C. less than when the general mean 0·7435 is used, and on the Schafberg, using the factor 0·7430, the value of $t - T$ comes out 0°·025 C. higher than with the general mean.

Table II (page 187) has been constructed on the basis that $\dfrac{P}{p} = 0\cdot7435$, and therefore $\dfrac{p}{P} = 1\cdot345$. The barometric pressure, P, is given for intervals of 10 mm. from 790 mm. to 550 mm. The temperature of saturated steam at pressure, P, is given under T. Under p the vapour tension of water at the temperature of a boiling mixture of steam and NaCl at barometric pressure, P, is given, where $p = 1\cdot345\,P$. The temperature of this boiling mixture is found from Regnault's tables connecting the temperature and pressure of saturated steam, and it is given under t. The difference $(t - T)$ gives the elevation of the boiling-point of saturated NaCl brine above that of pure water at barometric pressure P.

The figures in this table show that a boiling mixture of steam and NaCl at a known barometric pressure gives the means of obtaining an independent fixed point on a thermometer about 8° to 8°·5 C. above that furnished by the boiling-point of pure water at the same pressure. In most cases the normal pressure of 760 mm. would be used, but as the mean pressure in inhabited countries is less than 760 mm., it is convenient to be able to use directly the values observed at the existing pressure, and our table affords the means of doing so, assuming that the readings of our thermometers as corrected are exact.

In Table III are given the saturation values for the simple salts. They are arranged in order of the temperature of the boiling mixture (t), and this temperature indicates quite clearly the locality where the observation was made. The second column contains the values of $t - T$, from which, with the first column, the values of T are at once obtained. The third column contains the relative reduction of vapour tension $\left(\dfrac{p - P}{p}\right)$ produced by saturating the water with the salt at its boiling temperature.

The means of the observations at the same heights above the sea are inserted between lines.

The case of NaCl has been already discussed. The observations with KCl show a diminution of the value of $\dfrac{p - P}{p}$ with a fall of barometric pressure, and consequent fall of boiling temperature. This is observed in all the other salts experimented with, and depends chiefly, if not wholly, on the diminished solubility of the salt at the lower temperature. Although, in kind, the effect is the same in all the salts, it varies much in amount. It is most pronounced in the case of $KClO_3$, for which the mean values are :

Locality:—	...	Edinburgh	Pontresina	Schafberg
$t - T$...	...	$3°\cdot80$ C.	$3°\cdot30$ C.	$3°\cdot06$ C.
$\dfrac{p - P}{p}$	...	0·1250	0·1138	0·1078

It is also well marked in the case of NH_4Cl. This salt also crystallises with great promptitude so soon as the temperature falls at all. As is well known, salts differ much in this respect. A large number of cooling observations were made with NH_4Cl, and with some of the other salts, and *eutectic* points were observed, but they are not of sufficient importance for the present research to justify their being printed. The salt which appears from the value of $\dfrac{p - P}{p}$ to vary least in solubility at its boiling-point is $(NH_4)_2SO_4$, and in this respect it closely resembles NaCl. Nitrate of sodium was observed only at Pontresina and Schafberg. It could not be observed near sea-level with either of the thermometers used in the investigation. Nitrate of potassium was excluded altogether from boiling mixtures both because of the great elevation of boiling-point, and on account of the readiness with which it solidifies to a crystalline mass the moment the temperature begins to fall.

The chief part of the research is contained in Table IV. Three columns are devoted to each experiment ; namely, $(t - T)$ the elevation of the boiling temperature of the mixture or brine above that of pure water at the same time and place ; W the weight of steam condensed in the time from the beginning of the experiment until the value of $(t - T)$ has become as tabulated ; and $W(t - T)$ the product of the corresponding pairs of numbers. The experiments are numbered consecutively in the first headline (N), while under n each line in the table is numbered consecutively from o upwards. In this way any entry in the table can be referred to at once by its co-ordinates (N, n). The second headline gives the name and quantity of salt taken expressed in gramme-molecules. In the third headline will be found the temperature of saturated steam, or that of pure water boiling at the same time and place. The temperature of the boiling mixture or brine is obtained at once from the values of T and $(t - T)$.

The figure o is always used as a suffix when the boiling mixture of steam and salt or saturated brine is being dealt with. Thus t_0, p_0, W_0 always represent the temperature of the boiling saturated mixture, the steam tension of pure water of that temperature and the *dilution* of the mixture ; that is, the weight of water exactly saturated by the amount of salt at temperature t_0. The temperature of the mixture when the steam was stopped for the first time, and the first weight of condensed steam, W_1, ascertained, is always t_1. It has already been pointed out that it was the custom to stop the steam while there were still some particles of solid salt present, and while the temperature of the mixture showed that there was also already unsaturated water present. The idea was that the moment might be correctly judged when the amount of free salt present would be just enough to saturate the amount of free water if time were given. As a matter of fact, this was in most cases very nearly attained, as will be seen by comparing the observed values of W_1 with the computed values of W_2 in the cases where $t_0 - t_1$ is not more than $0°\cdot 1$ to $0°\cdot 3$ C. When this difference is larger, then the passage of steam has not been interrupted until all the salt has disappeared. This is the preferable practice. When any solid salt is present, and the temperature has fallen below the maximum, we know exactly the temperature of the boiling brine and the weight of water in it, but as there is an uncertain proportion of the salt originally taken which has not passed into solution, the concentration or dilution of the brine is uncertain. For this reason it has been impossible in the majority of cases to use t_1 and W_1 in the computation of W_0. The values of W_0 have been arrived at in the following way:—The difference $W_2(t - T)_2 - W_3(t - T)_3$ is found, also the differences $t_2 - t_3$ and $t_0 - t_2$, then we assume that

$$W_0(t - T)_0 = \frac{t_0 - t_2}{t_2 - t_3} \{ W_2(t - T)_2 - W_3(t - T)_3 \} + W_2(t - T)_2$$

and this value divided by $(t_0 - T)$ gives W_0. Thus in experiment No. 1 on $0\cdot 2$ KCl, when $T = 100°\cdot 44$ C., we have $W_2(t - T)_2 = 217\cdot 6$, and $W_3(t - T)_3 = 212\cdot 9$, their difference being $4\cdot 7$. Also $t_0 - t_2 = 0°\cdot 81$ C., and $t_2 - t_3 = 1°\cdot 05$ C., whence we have $\qquad W_0(t - T)_0 = \frac{81}{105} 4\cdot 7 + 217\cdot 6 = 220\cdot 6,$ whence $\qquad\qquad\qquad W_0 = 26\cdot 30.$

It will be seen that the value of W_1 is $25\cdot 91$, and $t_1 - T = 8°\cdot 28$, only $0°\cdot 11$ C. below $t_0 - T$. The passage of steam had therefore been stopped too soon; there was solid KCl present in greater quantity than could in any length of time be dissolved in the amount of water present. Again, if we look at experiment No. 4, we find the observed value of W_1, $27\cdot 25$, almost identical with the computed value of W_0, $27\cdot 21$. Here the difference, $t_0 - t_1$, is $0°\cdot 2$ C. So that, working in the way described, with chloride of potassium, when the temperature of condensation of the steam has fallen by $0°\cdot 2$ C., we should expect to have the amounts of free water and free salt present in compensating amounts. No. 6 represents a case in which the steam was passed until all the salt was dissolved, and until the temperature of condensation had fallen by a whole degree. Here we use $W_1(t - T)_1$ and $W_2(t - T)_2$ for finding $W_0(t - T)_0$

and W_0. If we consider experiments Nos. 1 to 9, it will be seen that the values of W_0 in experiments 1 to 3, which were made in Edinburgh, are lower than those found in Nos. 4 to 9, which were made at high levels, and therefore at lower temperatures, in Switzerland.

It must also be noted that the weight of the salt taken was less exactly ascertained in Switzerland than in my laboratory in Edinburgh. The Swiss weighings were made with a pair of hand scales, and were exact to the nearest 0·05 grm., that is to say, generally to ± 0·025 grm. In the Edinburgh experiments quoted in Table IV the weights are exact to the nearest 0·01 grm., or to ± 0·005 grm. The quantity of salt usually taken was one-fifth of a molecule in grammes. In the case of KCl, which has a medium molecular weight, this represents 14·92 grms., and we see that even the roughest of the weighings would be exact to within less than one-half per cent.

The earlier experiments in Switzerland were not always made with equivalent weights of the salts; all such cases have been recalculated for this table. While the usual quantity taken is one-fifth of a gramme-molecule, on some occasions two-fifths have been taken; and in the case of sparingly soluble salts as little as one-tenth or one-twentieth has been taken. The values of W are given for the quantity of salt quoted in the headline (M). The products have been all reduced to their value for one-fifth of a molecule of salt.

The physical meaning of the expression $W(t-T)$ is important. W is a weight of water expressed in grammes, and $(t-T)$ is the excess of the boiling temperature in degrees Celsius of that water, when it holds in solution a certain amount of a given salt, above its boiling temperature when in a state of purity; therefore $W(t-T)$ expresses, in gramme-degrees (g.° C.), the quantity of heat required to be in the water when it is boiling with salt dissolved in it above what is required when it is pure. If the values of $W(t-T)$ were constant for each salt at all dilutions, then the law connecting the dilution of a saline solution and the elevation of its boiling-point would be graphically expressed by a hyperbola, like the law connecting the volume and pressure of a gas at constant temperature. If we look over Table IV we see that for some salts, and mixtures of salts, the values of $W(t-T)$ are very nearly quite constant, while for the others, some deviate from constancy in the one sense, and some in the other. In the case of the chlorides of potassium and of sodium, the values of $W(t-T)$ diminish very considerably as the value of W increases. The case of ammonium chloride is peculiar, because at the sea-level $W(t-T)$ diminishes as W increases; at a height of 3000 metres it increases with W, and at a height of 2000 metres it is sensibly constant. In the case of barium chloride, $W(t-T)$ diminishes with dilution; the same is the case with strontium nitrate and ammonium sulphate. Potassium chlorate, barium nitrate, and lead nitrate show $W(t-T)$ increasing with W, while sodium nitrate and potassium sulphate show almost constant values of $W(t-T)$. Mixtures of salts follow the rule of their components. There are several examples in the table of pairs of salts which individually differ in the sense in which the values of $W(t-T)$ depart from constancy, and in mixture give constant values of $W(t-T)$. Examples are Nos. 63, 70, 71, 72, 73, 78, and 79.

At the beginning of an experiment, when the steam reaches the salt, it condenses very rapidly owing to abstraction of heat by the glass and by the salt, then it condenses at a very regular rate, the salt dissolving in proportion as steam is condensed. After a certain time the exact amount of steam has condensed which is necessary to form a boiling saturated solution of the salt taken; having observed (t_0-T) and W_0 we have the value of $W_0(t_0-T)$. If the elevation of the boiling point were proportional to the concentration, the factor $W(t-T)$ would remain constant while the solution was diluted by further condensation of steam. But if we deny thermal importance to the salt and consider only the water, then $W_0(t_0-T)$ is the heat in the boiling *saturated water*, counting from the temperature of pure boiling water. If we prevent it from losing heat, and provide for dilution by furnishing water of exactly the temperature of pure steam condensing on pure water at the time and place, the saturated water will mix with the free water, producing a resultant temperature depending on the relative quantities of the saturated and the free water. The case, then, of sodium nitrate, for instance, in which the value of $W(t - T)$ is nearly constant, could be represented by imagining the steam to condense at the temperature of the boiling saturated solution of the salt so long as solid salt is present, and the condensation temperature of the steam remains constantly at the maximum. When the solid salt has all disappeared, then the steam condenses at the temperature at which it condenses in pure water. The two portions of water, the saturated and the free, then mix, giving the resultant temperature, depending on the relative quantities and on the assumption that the heat of the saturated solution is that which the water present in it would have if it had the same temperature.

In the cases where $W(t - T)$ is constant, we have Blagden's law of the lowering of freezing-point applied to the raising of the boiling-point of saline solutions ; both vary directly with the concentration or inversely with the dilution. But in the case of a saline solution following Blagden's law, when ice is melted in the saturated solution already cooled to its freezing-point, the solution is diluted and its temperature rises. The rise of temperature is thus the same as would have been produced if the quantity of ice which has melted had been added as pure water of 0° C. to the saturated solution at the initial temperature of its freezing-point, and the two had been mixed.

The same is the case, *mutatis mutandis*, in the condensation of steam by a saline solution. If the boiling temperature of the stronger solution be t_1, and steam be passed through it until this temperature has fallen to t_2, the temperature of steam condensing on pure water being T, and if the quantity of water in the stronger solution be W_1, and in the weaker W_2, then we should have

$$W_1 t_1 + (W_2 - W_1) T = W_2 t_2,$$

whence
$$W_1 (t_1 - T) = W_2 (t_2 - T),$$

the values of $W(t - T)$ are constant, which is the characteristic of Blagden's law, whether applied to freezing or boiling. The deviations from Blagden's law in freezing and in boiling solutions may be attributable to some deviation from the law that the capacity for heat of a saline solution is the capacity for heat of the water which it contains. It is easy to obtain from the values of $W(t-T)$ in Table IV what must be the specific heat of the saturated water which would make $W(t-T)$ constant[1].

Table V gives for dilute solutions what Table IV gives for strong solutions. As the values of $(t-T)$ are not exactly the same for each salt, the relations of $(t-T)$ and $W(t-T)$ were expressed by curves and the values of $W(t-T)$ taken from them for the same values of $(t-T)$ for every salt. These values are given in Table VI. In all the experiments of Table V, exactly one-twentieth of a gramme-molecule of salts was taken, and the weight was exact to the nearest milligramme, and the experiments were made in every way alike. Table V contains no mixtures; on the other hand, some simple salts are included which are not found in Table IV. They are LiCl, RbCl, CsCl, KI, KBr, KNO_3, and $AgNO_3$. Both the RbCl and the CsCl were "spectroscopically pure," and the cæsium salt was quite pure. The rubidium salt, however, contained sulphate, and it is struck out in Table VI. The detailed discussion of these results, as well as of those in Table IV, must stand over to the second part of this paper, but one or two facts may be pointed out. In the case of salts for which the values of $W(t-T)$ diminish as dilution increases, a minimum is reached when $(t-T)$ is something between $0°\cdot7$ and $1°\cdot0$ C., after which it increases in all so that for values of $(t-T)$ between $0°\cdot5$ and $1°\cdot2$ C. the values of $W(t-T)$ for all these salts are practically constant, and they follow Blagden's law. In the case of nitrates $W(t-T)$ generally increases with dilution, and a minimum, if it exists, would have to be sought in solutions saturated under high pressure. Strontium nitrate is an exception; it has a minimum at $1°\cdot0$ C.

In Table VII will be found the values of p, the vapour tension of pure water at the temperature (t) of the boiling

[1] See below in Note, p. 184.

brine, the relative reduction of this tension $\left(\dfrac{p-P}{p}\right)$ produced by the salt in solution, and the product of this ratio into the weight of steam condensed (W). The discussion of the contents of this table is deferred.

The discussion of the observations on mixtures is also deferred. It involves a very large amount of arithmetical work, which will take some time. The behaviour of mixtures is extremely interesting. In Table IV we have examples of two distinct types. Of the one type, mixtures of the chlorides of potassium and sodium may be taken as an example. For a mixture of equal molecules 0·2 $\dfrac{\text{K} + \text{Na}}{2}$ Cl, $(t - T) = 11°\!·87$ C. at sea-level, and $W_0 = 20·7$ grms. The concentration of its boiling mixture is, therefore, 50 per cent. greater than that of chloride of sodium alone. In other words, a boiling saturated solution of either KCl or NaCl has still plenty of room for the other.

Mixtures of the nitrates of strontium and lead give an example of another type. The amount of steam condensed is exactly the sum of the amounts required by the quantities of the respective salts separately. For $Sr(NO_3)_2$, $(t-T) = 6°\!·53$ C., and for $Pb(NO_3)_2$, $(t - T) = 3°\!·29$ C. In the mixture 0·2 $\dfrac{Sr_2 + Pb}{3}$ $(NO_3)_2$ a constant value of $(t - T) = 5°\!·98$ C. is observed for thirteen out of the twenty-five minutes that were required to dissolve all the salt. Therefore the saturation temperature of the mixture lies between those of the components. A third type is furnished by a mixture of the nitrates of strontium and barium. The elevation of boiling-point is not as great as with $Sr(NO_3)_2$ alone, and the maximum temperature does not remain constant for even one minute out of the fifty minutes which were required to dissolve the fifth of a molecule used. The water required to dissolve this mixture is about 25 per cent. more than that which is required to dissolve the salts separately.

The nitrates of strontium, barium, and lead are isomorphous salts, and no doubt are capable of forming mixed crystals. The chlorides of potassium and sodium, though both crystallising in the same form, do not form mixed crystals, and therefore do not eliminate each other from solution.

These few remarks will show the extent of the subject, and also its great interest.

The Device of the Elastic Tank of Uniform Depth.— In order to provide a standard of comparison between the effect of dissolved salt and that of increased pressure upon the boiling temperature of water, we imagine a quantity of water in a shallow tank kept boiling by steam. The tank is of uniform depth of 1 centimetre, but it can expand laterally to accommodate condensed steam, the depth of the enlarged tank remaining uniformly 1 centimetre. In these circumstances the number expressing the weight, in grammes, of the water, expresses also its volume and its surface in cubic and square centimetres respectively. Let the initial quantity of water be W grammes, and let it be at the boiling temperature corresponding to the atmospheric pressure, A, in kilogrammes per square centimetre (k./c.²). If we dissolve a quantity, say one-fifth, of a molecule in grammes of a salt in this quantity of water, the temperature can be raised above the boiling temperature, T, of pure water under the atmospheric pressure, A, and if the quantity, W, of water is exactly sufficient to dissolve the fifth of a molecule of salt at its boiling temperature, which is then the temperature of the boiling saturated solution of the salt under atmospheric pressure, A, which we represent by t_0, then the temperature of the boiling water has been raised from T to t_0. Now this effect could be produced without adding salt by increasing the load pressing upon the surface, W, of the water. It is already pressed by the weight of the atmosphere A kilogrs. per square centimetre, or a total weight of WA kilogrs. In order to confine the water so as to admit of its temperature being raised from T to t, we must add to WA a certain additional load, b kilogrammes, which is found by consulting Regnault's tables connecting the temperature and tension of saturated steam. In these tables we find directly the tension, p_0, of saturated steam of temperature t_0, in millimetres of mercury, which is converted into a_0 in k./c.². The excess of this above the atmospheric pressure multiplied by the surface, W, gives the extra load required:

$$b = W(a_0 - A),$$

and this quantity b kilogrammes is the *mechanical equivalent* of the fifth of a molecule of salt in so far as the raising of the boiling temperature of water, or the resistance to steam pressure, is concerned.

In the case of the boiling saturated solution of salt when steam is continued to be passed through and heat is lost, the solution is diluted by the addition of the condensed steam to the original quantity, W, of saturated water. This dilution or addition of pure water to the saturated water, is accompanied by a fall of the temperature of ebullition, which is very rapid at first, but becomes slower as the quantity of condensed steam increases, tending ultimately towards the boiling temperature of pure water at atmospheric pressure.

The more concentrated the solution is, the more accentuated are the specific properties of the dissolved salt, and they are most pronounced in the saturated solution which approximates to the condition of the liquefied salt, as the dilute solution approximates to that of pure water. The specific nature of the dissolved salt shows itself first in the maximum temperature to which the solvent water can be raised under a given pressure, and then in the rate of fall of boiling temperature with dilution. Different salts behave differently in these respects.

The uniformity observed in the physical properties of very dilute solutions is due in part to our limited powers of perception, and to arithmetical necessity. In proportion as the number expressing the dilution becomes very great it tends to occupy the whole field of view, and, consequently, to obscure or obliterate the specific properties of the substance dissolved.

Similarly, in considering the trigonometrical functions of angles, if we limit our contemplation to very small angles, we can perceive no difference between the sine, the arc, and the tangent, yet the difference is none the less real on that account.

In our mechanical experiment the excess of pressure of the water at any moment above that of the atmosphere is $\dfrac{b}{W_n}$, and when this is multiplied by the quantity of water present, W_n, the product is constantly b. Now b is the equivalent of the salt dissolved, therefore *our mechanical experiment represents the case where the increase of steam tension, neutralised by the salt, is proportional to the quantity of the salt.*

For convenience in reference, the values of the tension of saturated steam, at temperatures from 90° C. to 120° C., expressed in kilogrs. per sq. cm., are collected in Table VIII, page 224.

In Table IX we have the case of 100 grms. water in the elastic tank. The atmospheric pressure is 733·5 mm., or 1 kilogr. per sq. cm. The depth of the tank, which can be enlarged laterally, is uniformly 1 cm., therefore its volume at the beginning is 100 cub. cms., and its surface is 100 sq. cms. The boiling temperature of pure water at a pressure of 1 k./cm.^{2} is 99°·09 C. The tank is securely covered, and its temperature raised to 119°·57 C., at which temperature the tension of saturated steam is exactly 2 kilogrs. per sq. cm. Neglecting, or allowing for, any thermal expansion of the tank and its contents, the area of it has remained the same, namely, 100 cm.^{2}. Let the cover be loaded until its fastenings just become slack, then the surface of the water is pressed by the loaded cover and by the atmosphere; the latter on a surface of 100 cm.^{2} amounts to 100 kilogrs., and the former makes up the difference between this weight and 200 kilogrs., because the pressure of the steam at 119°·57 C. is 2 k./cm.^{2}. Therefore, our extra load, b_0, is 100 kilogrs., and it may be looked on as the weight of the cover, which, like the tank, is supposed to be capable of lateral extension of its area without alteration of weight, keeping pace with the lateral increase of volume and area of the tank, while its contents are being increased by the condensation of steam.

Let the temperature of the water in the tank be now reduced to $118°·03$ C., at which temperature its vapour tension is $1·9$ k./cm.^{2}, and let steam of this temperature be condensed in it. The volume of water in the tank increases while its area expands in the same ratio, until the weight pressing on its surface is at the rate of $1·9$ k./cm.^{2}, when the steam will lift the cover and escape. The weight of the cover has remained constant (= 100 kilogrs.), but the area exposed to the atmospheric pressure has increased. The resulting area and volume of water are given by the equation $1·9\,W = 100 + W$, whence $W = \dfrac{100}{0·9} = 111$.

Again, let the temperature of the water in the tank be reduced to $116°·29$ C., at which temperature its vapour tension is $1·8$ k./cm.^{2}, and let steam of this temperature be passed into it. It will be condensed until the volume and surface of the water have increased to such an extent that the total pressure on its surface is at the rate of $1·8$ k./cm.^{2}. As before, we find the value of W, when this point has been reached, to be $\dfrac{100}{0·8} = 125$, and so on.

In the principal table the pressure is reduced by $0·1$ k./cm.^{2} at a time, from $2·0$ to $1·1$ k./cm.^{2}, then by $0·01$ k./cm.^{2} down to $1·01$ k./cm.^{2} and by $0·001$ down to $1·001$ k./cm.^{2}. The value of W is inversely proportional to the difference of pressure, $a - A$, and becomes infinite when $a - A = 0$. On the other hand, it diminishes rapidly at high temperatures and pressures, and would become 0 when $a - A = \infty$.

The table is carried upwards to $a = 10$ k./cm.^{2}, and downwards to $a = 1·001$ k./cm.^{2}. In these limits the value of W varies from $11·1$ to $100,000$.

The temperature of saturated steam rises at a slower rate than its tension. Hence in our mechanical experiment with pure water, when the area is increased by a given amount, and the pressure per unit of area is correspondingly reduced, the consequent fall of temperature depends on the actual temperature, and is proportional to the reciprocals of the figures representing the mean difference of tension per degree

for the interval. In the following table we have the values of
W, each of which is the double of the one preceding it ; the
concentration is therefore halved in each case and corresponds
exactly with the values of $(a - A)$, which give the excess of
vapour tension above atmospheric pressure. Under $(t - T)$ we
have the corresponding excesses of temperature above that

W	$a - A$	$t - T$	Diff.
		° C.	° C.
125	0·8	17·20	...
250	0·4	9·62	7·58
500	0·2	5·14	4·48
1,000	0·1	2·66	2·48
2,000	0·05	1·35	1·31
4,000	0·025	0·68	0·67
8,000	0·0125	0·34	0·34
16,000	0·00625	0·17	0·17

of water boiling at atmospheric pressure (99°·09 C.). In the
last column we have the differences of successive values of
$(t - T)$. Did the value of $(t - T)$ fall in exact proportion to
the concentration, then each of these differences should be
identical with the value of $(t - T)$ opposite it, because the
temperature interval should be halved. When the value of
$(t - T)$ falls below 0°·68 C. the interval is apparently halved, and
at temperatures within half a degree of that of water boiling
at atmospheric pressure, the tension of saturated steam seems
to vary proportionately with its temperature; but the differences
which undoubtedly exist occur in places of decimals which
are excluded from the table.

NOTE (1916).

*Demonstration of the Identity of Blagden's Law with the
Thermal Law of Mixture.* The probability of this being the
case is stated on page 175.

The principle which finds expression in the thermal law of
mixture may be stated as follows :

When the mixture has been completed, the amount of heat lost by the hotter mass is equal to the amount of heat gained by the colder mass.

Taking the simplest case, namely that of the mixture of masses of the same liquid, for example, water :

Let W_0 be the weight of one of the masses of water, and let t_0 be its temperature.

Let w be the weight of the other mass of water, and let T be its temperature.

When the two masses have been mixed, let $W_1 = W_0 + w$ be the weight of the mixture, and let t_1 be its temperature.

Then, in terms of the above principle, we have

$$W_0(t_0 - t_1) = w(t_1 - T), \ldots\ldots\ldots\ldots\ldots(1)$$

whence $\qquad\qquad W_0 t_0 = W_1 t_1 - wT.$

Let $\qquad\qquad T = 0°,$

then $\qquad\qquad W_0 t_0 = W_1 t_1. \ldots\ldots\ldots\ldots\ldots\ldots(2)$

Equation (2) is the simplest expression of the law of thermal mixture, and it also expresses Blagden's law of the lowering of the freezing-point and that of the raising of the boiling-point of water by the dissolution of salt in it.

Therefore *the three processes are identical.*

They are cases of simple thermal mixture.

From equation (1) we obtain the temperature of the mixture :

$$t_1 = \frac{W_0 t_0 + wT}{w + W_0}.$$

In cases where masses of different liquids have been mixed, unless they have the same specific heat, the above equation will not give the observed temperature of the mixture.

Let $s_0 =$ the specific heat of the mass W_0 referred to that of w as unity.

Then the temperature of the mixture will be

$$t_1 = \frac{s_0 W_0 t_0 + wT}{s_0 W_0 + w},$$

whence we obtain

$$s_0 W_0 (t_1 - t_0) = w (T - t_1),$$

which expresses in such cases the principle of thermal mixture, namely, the amount of heat lost by the hotter mass is equal to that gained by the colder mass.

From this we obtain

$$s_0 = \frac{w (T - t_1)}{W_0 (t_1 - t_0)},$$

and s_0 is the *virtual specific heat* of the water W_0 under the *influence* of the salt, considered as an *imponderable*, to distinguish it from the specific heat of the solution in which the mass of the dissolved salt is taken into account.

The value of s_0 so obtained for the virtual specific heat of the water W_0 is affected by any thermal reactions which take place when two portions of the liquid are mixed at the same temperature. When no such thermal reaction takes place, then the values of s_0 obtained from the boiling-points of solutions of different degrees of dilution are identical. If the values so obtained for s_0 show a regular variation, then some such thermal reaction may be presumed and it will require at least one independent experiment for its determination.

In the following table three cases are chosen to illustrate the calculation of the *virtual specific heat* of the water W_0 of the solution of highest boiling-point by combination *seriatim* with the other values of W and the corresponding values of t, which is used to denote the elevation of the boiling temperature.

Potassium chloride has been taken as representative of salts under the influence of which the *virtual specific heat* of water is depressed below unity, and sodium nitrate has been chosen as representative of those which raise it above unity. The third case illustrates the influence of mechanical pressure under which the *virtual specific heat* of water is also raised above unity.

In all the three cases the values of the *virtual specific heat* are very concordant in each series. The slight rise exhibited by the values in the case of potassium chloride points to a thermal reaction due to dilution.

Linear Number	Influence under which the Temperature of Ebullition is elevated								
	Dissolution of						Mechanical Pressure		
	0·2 KCl			0·2 NaNO₃					
n	t	W	s_0	t	W	s_0	t	W	s_0
	° C.	grams		° C.	grams		° C.	grams	
0	8·39	26·30		17·93	8·90		17·20	125	
1	7·58	28·71	0·858	14·96	10·80	1·076	9·62	250	1·270
2	6·53	32·61	·842	12·99	12·54	1·078	5·14	500	1·278
3	5·54	37·31	·814	10·02	16·64	1·101	2·66	1000	1·280
4	4·56	44·31	·815	7·50	22·48	1·097	1·35	2000	1·278
5	3·56	54·91	·802	6·48	26·13	1·095	0·68	4000	1·276
6	2·58	73·31	·793	3·83	44·84	1·097	0·34	8000	1·271
7							0·17	16000	1·266

TABLE I.—*Particulars of Boiling Mixtures of Steam and NaCl at different Elevations.*

	Edinburgh	Edinburgh	Edinburgh	Zürich	Fiesch	Andermatt	Pontresina	Pontresina
Locality ...								
Date, 1897	21st Oct.	19th Oct.	15th Oct.	19th Aug.	16th Aug.	18th Aug.	12th Sept.	12th Sept.
Elevation in Metres	80	80	80	410	1054	1444	1820	1820
Temp. of boiling mixture, °C. t	108·98	108·56	107·65	107·20	105·05	103·82	102·57	102·57
,, water, °C. T	100·44	100·06	99·21	98·75	96·71	95·58	94·37	94·37
Difference, °C. ... $t-T$	8·54	8·50	8·44	8·45	8·34	8·24	8·20	8·20
Tension of saturated steam at t, mm. ... p	1039·0	1024·4	993·1	977·9	908·0	870·0	832·7	832·7
Atmospheric pressure, mm. ... P	772·0	761·6	738·8	726·7	674·9	647·6	619·4	619·4
$p-P$	268·0	262·8	254·3	251·2	233·1	222·4	213·3	213·3
$\dfrac{p-P}{p}$	0·2570	0·2565	0·2561	0·2569	0·2567	0·2556	0·2561	0·2561

	Pontresina	St Moritz	St Moritz	Eggischorn	Julier	Schafberg	Schafberg	Schafberg
Locality ...								
Date, 1897	17th Sept.	28th Aug.	23rd Aug.	15th Aug.	25th Aug.	5th Sept.	31st Aug.	10th Sept.
Elevation in Metres	1820	1860	1860	2193	2244	2733	2733	2733
Temp. of boiling mixture, °C. t	102·32	102·48	102·10	101·30	101·00	99·57	99·54	99·30
,, water, °C. T	94·13	94·28	94·00	93·12	92·85	91·47	91·46	91·20
Difference, °C. ... $t-T$	8·19	8·20	8·10	8·18	8·15	8·10	8·08	8·10
Tension of saturated steam at t, mm. ... p	825·4	829·9	821·9	796·0	787·6	748·4	747·6	741·2
Atmospheric pressure, mm. ... P	614·0	617·8	611·0	591·4	585·5	556·0	555·8	550·4
$p-P$	211·4	212·1	210·9	204·6	202·1	192·4	191·8	190·8
$\dfrac{p-P}{p}$	0·2561	0·2556	0·2566	0·2570	0·2566	0·2571	0·2566	0·2574

TABLE II.—*Temperature of a Boiling Mixture of Steam and Sodium Chloride at Barometric Pressures from 550 to 790 Millimetres.*

Atmospheric Pressure P	Temperature of Saturated Steam of Tension P T	Tension of Saturated Steam of Temperature t $(1\cdot345\,P)$ p	Temperature of Boiling Mixture t	Elevation of Boiling Point $t-T$
Millimetres	°C.	Millimetres	°C.	°C.
790	101·09	1062·5	109·64	8·55
780	100·75	1049·05	109·27	8·54
770	100·37	1035·6	108·88	8·51
760	100·00	1022·15	108·50	8·50
750	99·63	1008·7	108·11	8·48
740	99·26	995·25	107·72	8·46
730	98·88	981·8	107·32	8·44
720	98·49	968·35	106·92	8·43
710	98·11	954·9	106·51	8·40
700	97·71	941·45	106·10	8·39
690	97·32	928·0	105·68	8·36
680	96·92	914·55	105·29	8·37
670	96·51	901·1	104·83	8·32
660	96·10	887·65	104·39	8·29
650	95·68	874·2	103·96	8·28
640	95·26	860·75	103·51	8·25
630	94·83	847·3	103·07	8·24
620	94·40	833·85	102·61	8·21
610	93·96	820·4	102·15	8·19
600	93·51	806·95	101·68	8·17
590	93·06	793·5	101·21	8·15
580	92·60	780·05	100·73	8·13
570	92·13	766·6	100·24	8·11
560	91·66	753·15	99·75	8·09
550	91·18	739·7	99·24	8·06

KCl

t	$t-T$	$\dfrac{p-P}{p}$
°C.	°C.	
108·84	8·40	0·2534
108·45	8·35	0·2515
107·50	8·29	0·2522
108·26	8·35	0·2520
102·24	7·87	0·2474
102·23	7·89	0·2480
102·19	7·90	0·2482
102·00	7·88	0·2479
101·90	7·90	0·2489
102·11	7·89	0·2481
100·63	7·78	0·2467
99·12	7·65	0·2450
98·98	7·69	0·2463
98·89	7·69	0·2463
98·86	7·69	0·2464
98·96	7·68	0·2460

NaCl

t	$t-T$	$\dfrac{p-P}{p}$
°C.	°C.	
108·98	8·54	0·2570
108·56	8·50	0·2565
108·77	8·52	·2567
107·65	8·44	0·2561
107·20	8·45	0·2569
107·42	8·45	0·2565
105·05	8·34	0·2567
103·82	8·24	0·2556
104·44	8·29	0·2562
102·57	8·20	0·2561
102·57	8·20	0·2561
102·48	8·20	0·2556
102·32	8·19	0·2561
102·20	8·20	0·2566
102·43	8·20	0·2561
101·30	8·18	0·2570
101·00	8·15	0·2566
101·15	8·165	0·2568
99·57	8·10	0·2571
99·54	8·08	0·2566
99·30	8·10	0·2574

NH₄Cl

t	$t-T$	$\dfrac{p-P}{p}$
°C.	°C.	
113·76	14·43	0·3913
113·76	14·43	0·3913
107·92	13·52	0·3813
107·77	13·51	0·3813
107·77	13·51	0·3813
107·82	13·51	0·3813
107·79	13·49	0·3823
107·58	13·51	0·3817
107·68	13·50	0·3820
106·14	13·29	0·3792
106·14	13·29	0·3792
104·49	13·02	0·3757
104·28	12·98	0·3752
104·20	13·10	0·3757
104·32	13·03	0·3755

BaCl₂.2H₂O

t	$t-T$	$\dfrac{p-P}{p}$
°C.	°C.	
104·47	4·52	0·1476
104·47	4·52	0·1476
98·81	4·39	0·1478
98·46	4·37	0·1478
98·63	4·38	0·1478
95·70	4·23	0·1453
95·44	4·26	0·1465
95·57	4·24	0·1459

KClO₃

t	$t-T$	$\dfrac{p-P}{p}$
°C.	°C.	
104·08	3·80	0·1250
104·08	3·80	0·1250
97·73	3·30	0·1138
97·55	3·30	0·1138
97·64	3·39	0·1138
94·54	3·07	0·1080
94·54	3·06	0·1079
94·25	3·05	0·1075
94·44	3·06	0·1078

TABLE III (continued).—Particulars of Boiling Mixtures of Steam and various Salts.

$NaNO_3$			$Ba(NO_3)_2$			$Sr(NO_3)_2$			K_2SO_4			$(NH_4)_2SO_4$		
t	$t-T$	$\frac{p-P}{p}$	t	$t-T'$	$\frac{p-P}{p}$	t	$t-T$	$\frac{p-P}{p}$	t	$t-T$	$\frac{p-P}{p}$	t	$t-T$	$\frac{p-P}{p}$
°C.	°C.		°C.	°C.		°C.	°C.		°C.	°C.		°C.	°C.	
112·45	18·04	0·4686	101·19	1·28	0·0445	106·52	6·52	0·2045	101·70	1·42	0·0493	108·08	7·80	0·2383
112·22	17·80	0·4643	101·19	1·28	0·0445				101·70	1·42	0·0493	107·04	7·64	0·2350
112·33	17·92	0·4664	95·55	1·13	0·0404				95·72	1·30	0·0467	107·56	7·72	0·2367
108·39	17·09	0·4576	95·37	1·15	0·0415				95·50	1·28	0·0460	101·80	7·39	0·2345
108·29	17·11	0·4581	95·46	1·14	0·0409				95·61	1·29	0·0463	101·70	7·46	0·2365
108·25	16·95	0·4550	92·59	1·12	0·0411				92·74	1·25	0·0458	101·75	7·425	0·2355
108·31	17·05	0·4569	92·59	1·12	0·0411				92·54	1·24	0·0454	98·56	7·26	0·2346
									92·64	1·24	0·0456	98·42	7·28	0·2354
												98·49	7·27	0·2350

TABLE IV:—*Temperature of Condensation of Steam on Salts and in their Brines.*

No.	N	1			2		
Salt	rM	0·2 KCl			0·4 KCl		
Temp. of Saturated Steam	T	100°·44 C.			100°·10—100°·12 C.		
	n	$t-T$	W	$W(t-T)$	$t-T$	W	$W(t-T)$
	0	8·39	26·30	220·6	8·33	26·3	219·0
	1	8·28	25·91	214·5	8·22	26·14	214·9
	2	·7·58	28·71	217·6	7·37	29·09	214·4
	3	6·53	32·61	212·9	6·85	30·93	212·4
	4	6·02	34·91	210·1	6·34	33·04	209·2
	5	5·54	37·31	206·6	5·86	35·33	206·8
	6	5·05	40·61	205·0	5·37	38·29	205·7
	7	4·56	44·31	202·0	4·87	41·63	202·7
	8	4·06	48·91	198·5	...	...	...
	9	3·56	54·91	195·9	...	...	...
	10	3·06	62·81	192·2	...	...	...
	11	2·58	73·31	189·1	...	...	...

No.	N	3			4		
Salt	rM	0·2 KCl			0·2 KCl		
Temp. of Saturated Steam	T	99°·21 C.			94°·12—94°·13 C.		
	n	$t-T$	W	$W(t-T)$	$t-T$	W	$W(t-T)$
	0	8·30	26·56	220·5	7·88	27·21	214·4
	1	8·10	26·76	216·8	7·68	27·25	209·3
	2	7·08	30·26	214·2	5·94	34·35	204·0
	3	6·06	34·46	208·8	4·93	40·35	198·9
	4	5·05	40·36	203·8	4·43	44·25	196·0
	5	4·54	44·26	201·1	3·92	49·25	193·1
	6	4·04	49·16	198·8	3·42	55·65	190·3
	7	3·53	55·36	195·6	2·91	64·45	187·4
	8	3·02	63·66	192·4	2·41	76·85	185·2
	9	2·51	75·76	190·2	...	...	...
	10	...	...	...	...	...	...
	11	...	...	...	...	...	...

TABLE IV (*continued*).—*Temperature of Condensation of Steam on Salts and in their Brines.*

No.	N	5			6		
Salt	rM	0·2 KCl			0·2 KCl		
Temp. of Saturated Steam	T	94°·34—94°·35 C.			94°·37—94°·39 C.		
	n	$t-T$	W	$W(t-T)$	$t-T$	W	$W(t-T)$
	0	7·89	27·09	213·7	7·87	26·58	209·2
	1	7·66	27·76	212·6	6·82	30·04	204·9
	2	5·83	35·22	205·3	5·79	34·64	200·6
	3	4·81	41·58	200·0	4·78	41·00	196·0
	4	3·80	50·99	193·8	3·26	58·18	189·7
	5	2·79	67·63	188·8	2·75	68·43	188·2
	6	1·77	103·95	184·0	...	...	...

No.	N	7			8		
Salt	rM	0·2 KCl			0·2 KCl		
Temp. of Saturated Steam	T	91°·18 C.			91°·20 C.		
	n	$t-T$	W	$W(t-T)$	$t-T$	W	$W(t-T)$
	0	7·69	26·70	204·0	7·69	27·12	208·6
	1	7·43	27·42	203·7	7·50	27·08	203·1
	2	6·72	30·11	202·3	6·69	30·41	203·4
	3	5·71	34·74	198·4	5·67	34·96	198·2
	4	5·20	37·75	196·3	4·66	41·57	193·7
	5	4·69	41·19	193·2	3·65	51·78	189·0
	6	...	...	...	2·64	69·50	183·5

TABLE IV (*continued*).—*Temperature of Condensation of Steam on Salts and in their Brines.*

No.	N	9			10		
Salt	rM	0·2 NaCl			0·4 NaCl		
Temp. of Saturated Steam	T	100°·44 C.			100°·06 C.		
	n	$t - T$	W	$W(t - T)$	$t - T$	W	$W(t - T)$
	0	8·54	29·85	254·9	8·48	59·88	253·8
	1	8·36	30·41	254·2	8·38	60·52	253·6
	2	7·53	33·31	250·8	7·49	66·92	250·6
	3	6·54	37·21	243·3	6·90	70·92	244·6
	4	5·99	39·91	239·0	6·41	75·32	241·4
	5	5·55	42·51	235·9	5·92	80·22	237·4
	6	5·06	45·61	230·8	5·43	86·52	234·9
	7	4·57	49·51	226·3	4·93	93·02	229·3
	8	4·07	54·21	220·5	...	...	...
	9	3·58	60·41	216·3	...	...	...
	10	3·08	68·41	210·7	...	...	...
	11	2·59	79·81	206·6	...	...	...

No.	N	11			12		
Salt	rM	0·2 NaCl			0·2 NaCl		
Temp. of Saturated Steam	T	99°·21 C.			94°·13—94°·12 C.		
	n	$t - T$	W	$W(t - T)$	$t - T$	W	$W(t - T)$
	0	8·44	29·84	252·4	8·19	29·50	241·6
	1	8·31	30·3	251·8	8·07	29·85	240·9
	2	7·19	34·4	247·3	6·96	33·65	234·2
	3	6·18	38·9	240·4	5·94	38·25	227·2
	4	5·16	44·8	231·2	4·93	44·30	218·4
	5	4·65	48·6	226·0	4·44	48·35	214·7
	6	4·15	53·4	221·6	3·93	53·45	210·1
	7	3·64	59·5	216·6	3·43	60·25	206·7
	8	3·14	67·5	212·0	2·92	69·05	201·6
	9	2·63	78·5	206·5	2·42	81·35	196·9
	10	...	...	...	...	...	...
	11	...	...	...	...	...	...

TABLE IV (*continued*).—*Temperature of Condensation of Steam on Salts and in their Brines.*

No.	N	13			14		
Salt	$r.M$	0·2 NaCl			0·2 NaCl		
Temp. of Saturated Steam	T	94°·37 C.			94°·37—94°·34 C.		
	n	$t-T$	W	$W(t-T)$	$t-T$	W	$W(t-T)$
	0	8·21	29·21	239·8	8·21	28·65	235·2
	1	8·03	29·84	239·6	8·04	29·24	235·1
	2	7·13	33·44	238·4	7·14	32·83	234·4
	3	6·12	37·93	232·1	6·14	37·11	227·9
	4	5·10	43·72	223·0	5·12	43·04	220·4
	5	4·09	52·34	214·1	4·12	51·59	212·5
	6	3·08	66·13	203·7	...	...	...
	7	2·07	94·11	194·8	...	...	...

No.	N	15			16		
Salt	rM	0·2 NaCl			0·2 NaCl		
Temp. of Saturated Steam	T	91°·47 C.			91°·20 C.		
	n	$t-T$	W	$W(t-T)$	$t-T$	W	$W(t-T)$
	0	8·11	...	...	8·11	29·33	238·0
	1	8·06	28·63	230·8	7·90	30·02	237·2
	2	7·10	32·88	233·4	7·10	33·04	234·6
	3	6·09	38·47	234·3	6·09	37·66	229·4
	4	...	...	...	5·07	43·42	220·1
	5	...	...	...	4·06	52·20	211·9
	6	...	...	...	3·05	66·65	203·3
	7	...	...	...	...	...	...

On Steam and Brines

TABLE IV (*continued*).—*Temperature of Condensation of Steam on Salts and in their Brines.*

No.	N	17			18		
Salt	rM	0·2 NH$_4$Cl			0·4 NH$_4$Cl		
Temp. of Saturated Steam	T	99°·33—99°·38 C.			94°·26 C.		
	n	$t-T$	W	$W(t-T)$	$t-T$	W	$W(t-T)$
	0	14·46	13·62	197·0	13·48	26·50	178·6
	1	14·12	13·54	191·2	13·24	27·24	180·3
	2	11·75	16·64	195·5	12·08	30·15	182·1
	3	9·78	19·84	194·0	11·07	33·35	184·6
	4	7·80	24·84	193·7	10·06	37·03	186·2
	5	6·80	28·34	192·7	9·05	41·34	187·0
	6	5·81	32·94	191·4	8·03	45·37	182·1
	7	4·83	39·04	188·6	7·02	53·57	188·0
	8	4·33	43·24	187·2	...	...	...
	9	3·81	48·54	184·9	...	...	...
	10	3·33	54·94	182·9	...	...	...
	11	2·83	63·94	181·0	...	...	...
	12	2·32	76·84	178·2	...	...	...

No.	N	19			20		
Salt	rM	0·4 NH$_4$Cl			0·4 NH$_4$Cl		
Temp. of Saturated Steam	T	94°·26 C.			94°·40—94°·41 C.		
	n	$t-T$	W	$W(t-T)$	$t-T$	W	$W(t-T)$
	0	13·48	26·92	181·5	13·49	27·08	182·7
	1	13·24	27·47	181·8	12·44	29·66	184·4
	2	11·47	32·07	183·9	10·42	36·06	187·8
	3	9·96	37·30	185·7	9·41	40·06	188·5
	4	8·94	41·92	187·4	8·40	44·90	188·6
	5	7·93	47·15	186·9	7·39	49·62	183·3
	6	6·92	54·11	187·2	6·37	58·88	187·5
	7	5·91	63·04	186·3	5·35	69·30	185·3
	8	4·90	75·64	185·3	4·34	84·56	183·5
	9	...	...	...	3·33	109·10	181·6
	10	...	...	...	...	...	...
	11	...	...	...	...	...	...
	12	...	...	...	...	...	...

TABLE IV (*continued*).—*Temperature of Condensation of Steam on Salts and in their Brines.*

No.	N	21			22		
Salt	rM	0·4 NH$_4$Cl			0·1 BaCl$_2$. 2H$_2$O		
Temp. of Saturated Steam	T	91°·20 C.			99°·95 C.		
	n	$t-T$	W	$W(t-T)$	$t-T$	W	$W(t-T)$
	0	12·96	13·95	180·8	4·51	32·01	288·8
	1	12·77	28·5	181·9	4·48	32·07	287·4
	2	10·93	33·8	184·6	4·10	34·97	287·0
	3	9·92	37·6	186·5	3·59	39·67	284·8
	4	8·90	42·0	186·9	3·08	45·67	281·2
	5	7·89	47·3	186·6	2·58	53·27	274·4
	6	6·88	54·2	186·4	2·07	64·97	273·0
	7	5·87	63·1	185·2	1·57	83·47	262·2
	8	4·86	75·9	184·4	...	...	...
	9	...	...	...	...	...	...

No.	N	23			24		
Salt	rM	0·2 BaCl$_2$. 2H$_2$O			0·1 BaCl$_2$. 2H$_2$O		
Temp. of Saturated Steam	T	94°·42 C.			91°·18 C.		
	n	$t-T$	W	$W(t-T)$	$t-T$	W	$W(t-T)$
	0	4·38	73·7	322·8	4·25	36·6	310·9
	1	4·23	73·7	311·8	4·14	36·7	303·8
	2	3·62	85·6	309·9	3·66	41·4	303·0
	3	3·12	96·6	301·4	3·16	46·9	296·4
	4	2·71	108·5	294·0	2·65	54·3	287·8
	5	2·31	123·1	284·4	...	...	...
	6	2·00	138·7	277·4	...	...	...
	7	1·70	159·1	270·5	...	...	...
	8	1·40	186·8	261·5	...	...	...
	9	1·20	236·6	283·9	...	...	...

TABLE IV (*continued*).—*Temperature of Condensation of Steam on Salts and in their Brines.*

No.	N	25			26		
Salt	rM	0·2 KClO$_3$			0·2 KClO$_3$		
Temp. of Saturated Steam	T	100°·30 C.			98°·77—98°·79 C.		
	n	$t-T$	W	$W(t-T)$	$t-T$	W	$W(t-T)$
	0	3·78	40·10	151·6	3·61	42·2	152·3
	1	3·71	40·43	150·0	3·55	42·2	149·8
	2	3·45	44·63	154·0	3·28	46·5	152·5
	3	3·00	52·43	157·3	2·97	52·0	154·4
	4	2·50	63·73	159·3	2·67	58·6	156·5
	5	2·00	80·83	161·7	2·37	67·0	158·8
	6	1·51	111·13	167·8	2·06	77·8	160·3
	7	...	...	...	1·86	87·1	162·0
	8	...	...	...	1·65	99·3	163·8
	9	...	...	...	1·43	114·3	163·5

No.	N	27			28		
Salt	rM	0·2 KClO$_3$			0·2 KClO$_3$		
Temp. of Saturated Steam	T	94°·43 C.			91°·20 C.		
	n	$t-T$	W	$W(t-T)$	$t-T$	W	$W(t-T)$
	0	3·30	47·5	156·7	3·04	50·5	153·4
	1	3·22	47·0	151·3	2·93	50·55	148·6
	2	2·81	54·8	154·0	2·53	59·37	150·2
	3	2·41	64·8	156·2	2·03	75·50	153·3
	4	2·00	78·6	157·2	1·52	102·81	156·2
	5	1·70	93·8	159·5	...	...	...
	6	1·52	106·8	162·3	...	...	...
	7	1·29	125·8	162·3	...	...	...
	8	1·09	149·4	162·8	...	...	...
	9	...	...	...	...	...	...

TABLE IV (*continued*).—*Temperature of Condensation of Steam on Salts and in their Brines.*

No.	N	29			30		
Salt	rM	0·05 Ba(NO$_3$)$_2$			0·2 Ba(NO$_3$)$_2$		
Temp. of Saturated Steam	T	99°·91 C.			94°·42 C.		
	n	$t-T$	W	$W(t-T)$	$t-T$	W	$W(t-T)$
	0	1·28	37·5	192·1	1·12	163·5	183·2
	1	1·22	39·88	194·6	1·08	169·3	182·8
	2	1·10	44·88	197·5	0·98	190·5	186·7
	3	1·00	50·08	200·3	0·88	215·0	189·2
	4	0·80	65·28	208·9	0·78	247·3	192·9
	5	0·70	74·88	209·7	0·68	288·9	196·5
	6	0·60	92·48	224·4	0·58	348·4	202·1
	7	...	...	...	0·48	419·6	201·4
	8	...	...	...	0·38	584·1	221·9
	9	...	...	...	...	...	...

No.	N	31			32		
Salt	rM	0·2 Sr(NO$_3$)$_2$			0·2 Pb(NO$_3$)$_2$		
Temp. of Saturated Steam	T	100°·00 C.			99°·60 C.		
	n	$t-T$	W	$W(t-T)$	$t-T$	W	$W(t-T)$
	0	6·53	42·8	279·3	3·29	50·00	164·6
	1	6·32	42·3	267·3	3·15	51·7	162·8
	2	5·44	49·7	270·4	2·94	56·5	166·1
	3	4·94	53·9	266·3	2·64	62·9	166·1
	4	4·45	59·0	262·6	2·44	69·0	168·4
	5	3·95	66·0	260·7	2·24	75·2	168·6
	6	3·46	74·6	258·1	2·03	83·2	168·9
	7	2·96	86·0	254·5	1·82	92·9	169·1
	8	2·47	102·3	252·7	1·62	106·1	171·9
	9	...	...	...	1·42	126·8	180·0

TABLE IV (*continued*).—*Temperature of Condensation of Steam on Salts and in their Brines.*

No.	N	33			34		
Salt	rM	0·1 K$_2$SO$_4$			0·033 K$_2$SO$_4$		
Temp. of Saturated Steam	T	100°·28 C.			94°·42 C.		
	n	$t-T$	W	$W(t-T)$	$t-T$	W	$W(t-T)$
	0	1·38	73·0	201·5	1·28	...	...
	1	1·33	75·3	200·3	1·18	25·8	182·64
	2	1·21	84·1	203·5	0·98	30·5	179·3
	3	1·00	102·94	205·9	0·88	34·7	183·2
	4	...	...	...	0·78	38·4	179·7
	5	...	...	...	0·68	43·8	178·7
	6	...	...	...	0·58	51·6	179·6
	7	...	...	...	0·48	64·8	186·6
	8	...	...	...	...	...	...
	9	...	...	...	...	...	...
	10	...	...	•••	...	...	...
	11	...	...	...	...	...	...
	12	...	..	...	...	...	...

No.	N	35			36		
Salt	rM	0·05 K$_2$SO$_4$			0·1 (NH$_4$)$_2$SO$_4$		
Temp. of Saturated Steam	T	91°·30 C.			99°·40 C.		
	n	$t-T$	W	$W(t-T)$	$t-T$	W	$W(t-T)$
	0	1·23	...	...	7·63	12·64	192·88
	1	1·21	39·01	188·8	7·39	13·28	196·49
	2	0·81	61·56	199·5	5·97	16·18	191·14
	3	...	•••	...	5·06	18·79	190·00
	4	...	...	...	4·05	22·94	185·81
	5	...	...	...	3·54	25·95	183·69
	6	...	...	...	3·04	29·68	180·44
	7	...	...	...	2·53	35·28	178·51
	8	...	...	•••	2·02	43·78	176·87
	9	...	...	...	1·82	48·97	178·24
	10	...	...	•••	1·62	55·19	178·82
	11	...	...	•••	1·42	64·00	181·76
	12	...	...	...	1·21	75·62	182·99

TABLE IV (*continued*).—*Temperature of Condensation of Steam on Salts and in their Brines.*

No.	N	37			38		
Salt	rM	0·4 (NH₄)₂SO₄			0·4 (NH₄)₂SO₄		
Temp. of Saturated Steam	T	100°·28 C.			94°·24 C.		
	n	$t-T$	W	$W(t-T)$	$t-T$	W	$W(t-T)$
	0	7·81	50·7	197·9	7·46	51·9	193·6
	1	7·52	51·8	194·8	7·42	47·7	176·9
	2	6·93	56·2	194·7	6·94	55·2	191·5
	3	6·42	60·1	192·9	5·93	63·2	187·4
	4	5·93	64·8	192·1	...	...	...
	5	5·44	69·9	190·1	...	...	...
	6	4·94	76·2	188·2	...	...	...
	7	4·45	83·7	186·2	...	...	...
	8	3·95	93·5	184·7	...	...	...
	9	3·46	106·2	183·7	...	...	...

No.	N	39			40		
Salt	rM	0·2 (NH₄)₂SO₄			0·2 (NH₄)₂SO₄		
Temp. of Saturated Steam	T	94°·41 C.			91°·15—91°·17 C.		
	n	$t-T$	W	$W(t-T)$	$t-T$	W	$W(t-T)$
	0	7·38	25·8	190·1	7·25	26·2	190·1
	1	7·17	24·8	177·8	7·16	22·7	162·5
	2	6·36	29·5	187·6	5·91	31·5	186·2
	3	5·35	34·6	185·1	4·90	37·4	183·3
	4	4·34	41·7	181·0	3·88	46·4	180·0
	5	3·33	53·7	178·8	...	...	...
	6	2·32	75·4	174·9	...	...	...
	7	...	...	...	...	...	...
	8	...	...	...	...	...	...
	9	...	...	...	...	...	...

TABLE IV (*continued*).—*Temperature of Condensation of Steam on Salts and in their Brines.*

No.	N	41			42		
Salt	rM	$0.2\,(NH_4)_2SO_4$			$0.2\,\dfrac{K+Na}{2}Cl$		
Temp. of Saturated Steam	T	$91°\!\cdot\!30$ C.			$100°\!\cdot\!44$ C.		
	n	$t-T$	W	$W(t-T)$	$t-T$	W	$W(t-T)$
	0	7·24	26·4	191·1	11·87	20·7	245·7
	1	6·98	26·2	182·7	11·53	20·95	241·6
	2	6·17	30·4	187·5	10·44	23·35	243·8
	3	5·16	35·7	184·1	9·01	26·35	237·4
	4	4·15	43·5	180·4	8·01	29·15	233·5
	5	3·04	57·9	176·0	7·04	32·25	227·0
	6	2·13	82·1	174·8	6·04	36·75	222·0
	7	...	...	...	5·55	39·45	218·9
	8	...	...	...	5·06	42·65	215·8
	9	...	...	...	4·57	46·45	212·3
	10	...	...	...	4·07	51·15	208·2
	11	...	...	...	3·58	57·25	205·0
	12	...	...	...	3·08	64·95	200·0
	13	...	...	...	2·59	75·85	196·5

No.	N	43			44		
Salt	rM	$0.4\,\dfrac{K+Na}{2}Cl$			$0.4\,\dfrac{K+Na}{2}Cl$		
Temp. of Saturated Steam	T	$100°\!\cdot\!13$—$100°\!\cdot\!15$ C.			$94°\!\cdot\!26$—$94°\!\cdot\!25$ C.		
	n	$t-T$	W	$W(t-T)$	$t-T$	W	$W(t-T)$
	0	11·81	40·06	241·3	11·19	...	...
	1	11·46	41·98	240·5	10·89	41·9	228·2
	2	10·80	44·68	241·2	9·87	46·1	227·5
	3	9·86	48·38	238·0	9·17	49·2	225·6
	4	8·88	52·98	235·2	8·17	54·4	222·3
	5	8·38	55·68	233·4	7·65	57·6	220·3
	6	7·88	58·58	230·9	7·15	61·0	218·1
	7	7·39	61·78	228·3	6·64	65·1	216·1
	8	6·89	65·78	226·7	6·14	69·7	214·0
	9	6·39	70·08	223·9	5·62	75·2	211·3
	10	5·90	74·88	220·9	...	...	...
	11	5·41	80·88	218·8	...	...	...
	12	4·90	87·48	214·3	...	...	...
	13	...	...	...	...	...	...

TABLE IV (*continued*).—*Temperature of Condensation of Steam on Salts and in their Brines.*

No.	N	45			46		
Salt	rM	$0 \cdot 2 \dfrac{K_2 + Na_3}{5} Cl$			$0 \cdot 2 \dfrac{K_3 + Na_2}{5} Cl$		
Temp. of Saturated Steam	T	$100^\circ \cdot 18$ C.			$100^\circ \cdot 18$ C.		
	n	$t-T$	W	$W(t-T)$	$t-T$	W	$W(t-T)$
	0	11·81	...	...	11·84	...	...
	1	10·58	23·0	243·3	10·80	22·15	239·2
	2	9·38	25·9	243·0	9·38	25·35	236·9
	3	8·39	28·6	239·0	8·39	27·85	233·7
	4	7·40	31·5	233·1	7·40	30·85	228·3
	5	6·41	35·5	227·6	6·41	34·85	223·4
	6	5·92	38·0	225·0	5·92	37·25	220·5
	7	5·43	40·8	221·5	5·43	40·25	218·6
	8	4·93	44·2	217·9	4·93	43·55	214·7
	9	4·44	48·2	214·0	4·44	47·55	211·1
	10	3·94	53·2	209·6	3·94	52·65	207·4
	11	3·45	59·7	206·0	3·45	59·15	204·1
	12	2·95	68·3	203·5	2·95	67·85	200·2
	13	2·46	80·5	198·0	2·46	80·35	197·7

No.	N	47			48		
Salt	rM	$0 \cdot 2 \dfrac{K + Na_2}{3} Cl$			$0 \cdot 2 \dfrac{K_2 + Na}{3} Cl$		
Temp. of Saturated Steam	T	$100^\circ \cdot 23$ C.			$100^\circ \cdot 23$ C.		
	n	$t-T$	W	$W(t-T)$	$t-T$	W	$W(t-T)$
	0	11·86	...	...	11·80	...	...
	1	9·98	24·59	245·4	10·32	22·43	231·5
	2	8·89	27·49	244·4	9·39	25·83	242·5
	3	7·91	30·29	239·6	7·91	28·63	226·5
	4	6·92	33·79	233·9	6·92	32·13	222·3
	5	6·36	36·29	230·9	5·93	36·63	217·2
	6	5·93	38·29	227·1	5·44	39·33	213·9
	7	5·38	41·69	224·3	4·94	42·63	210·4
	8	4·94	45·69	225·8	4·45	46·63	207·4
	9	4·45	48·49	215·8	3·95	51·43	203·0
	10	3·95	53·59	211·7	3·46	58·03	200·7
	11	3·46	60·19	208·3	2·96	66·43	196·5
	12	2·96	68·69	203·4	2·45	78·63	192·6
	13	2·47	80·59	199·1	...	...	...

TABLE IV (*continued*).—*Temperature of Condensation of Steam on Salts and in their Brines.*

No. ... ... N	49			50		
Salt ... ... rM	$0\cdot2\,\dfrac{K+Na_3}{4}Cl$			$0\cdot2\,\dfrac{K_3+Na}{4}Cl$		
Temp. of Saturated Steam T	$100°\cdot22$ C.			$100°\cdot17$ C.		
n	$t-T$	W	$W(t-T)$	$t-T$	W	$W(t-T)$
0	11·82	...	...	11·80	...	...
1	9·56	25·7	245·7	9·66	23·56	227·6
2	8·40	29·4	247·0	8·40	27·06	227·3
3	7·41	32·1	237·9	7·50	30·16	226·2
4	6·87	34·3	235·6	6·42	33·86	217·4
5	6·42	36·4	233·7	5·93	36·46	216·2
6	5·93	38·6	228·9	5·44	39·06	212·4
7	5·44	41·3	224·7	4·94	42·46	209·8
8	4·94	44·9	221·8	4·45	46·36	206·3
9	4·45	48·8	217·2	3·95	51·36	202·8
10	3·95	53·9	212·9	3·46	57·56	200·2
11	3·46	59·3	205·8	2·96	66·16	195·8
12	2·96	68·6	203·0	2·47	77·86	192·3
13	2·47	80·5	198·9	...	...	...

No. ... ... N	51			52		
Salt ... ... rM	$0\cdot2\,\dfrac{K+Na_4}{5}Cl$			$0\cdot2\,\dfrac{K_4+Na}{5}Cl$		
Temp. of Saturated Steam T	$100°\cdot18$ C.			$100°\cdot18$ C.		
n	$t-T$	W	$W(t-T)$	$t-T$	W	$W(t-T)$
0	11·80	...	...	11·78	...	...
1	9·21	26·86	247·4	9·16	25·0	229·0
2	8·38	29·36	246·0	8·39	26·75	224·4
3	7·40	32·46	240·2	7·40	29·95	221·6
4	6·41	36·46	233·7	6·41	33·95	217·6
5	5·92	38·86	230·0	5·92	36·25	214·6
6	5·43	41·86	227·2	5·43	39·15	212·5
7	4·93	44·96	221·6	4·93	42·45	209·2
8	4·44	50·26	223·2	4·44	46·25	205·3
9	3·94	54·26	213·7	3·94	51·25	201·9
10·	3·45	60·66	209·3	3·45	57·45	198·0
11	2·95	69·66	205·5	2·95	66·15	195·2
12	2·46	82·16	202·1	2·46	78·25	192·5
13	...	...	...	...	...	...

TABLE IV (*continued*).—*Temperature of Condensation of Steam on Salts and in their Brines.*

No.	N	53			54		
Salt	rM	$0 \cdot 2 \dfrac{K + Na_2}{3} Cl$			$0 \cdot 2 \dfrac{K_2 + Na}{3} Cl$		
Temp. of Saturated Steam	T	94°·2c C.			94°·20 C.		
	n	$t - T$	W	$W(t-T)$	$t - T$	W	$W(t-T)$
	0	11·19	...	...	11·18	...	...
	1	9·63	24·0	231·1	9·43	23·26	219·3
	2	8·62	26·2	225·8	8·42	25·74	216·7
	3	7·60	29·3	222·7	7·40	28·86	212·6
	4	6·59	33·1	218·1	6·39	32·66	208·7
	5	6·08	35·3	214·6	5·87	35·00	205·5
	6	5·57	37·9	211·1	5·37	37·94	203·7
	7	5·07	41·2	208·9	4·86	41·14	200·0
	8	4·56	44·9	204·7	4·36	45·34	193·1
	9	4·06	49·9	202·6	3·85	50·60	194·8
	10	...	...	...	...	...	...

No.	N	55			56		
Salt	rM	$0 \cdot 2 \dfrac{K + Na_3}{4} Cl$			$0 \cdot 2 \dfrac{K_3 + Na}{4} Cl$		
Temp. of Saturated Steam	T	94°·20 C.			94°·17 C.		
	n	$t - T$	W	$W(t-T)$	$t - T$	W	$W(t-T)$
	0	11·17	...	...	11·19	...	...
	1	9·13	25·35	231·4	8·95	24·33	217·7
	2	8·11	27·75	225·1	7·93	27·18	215·5
	3	7·10	31·05	220·5	6·92	30·63	212·0
	4	6·08	35·25	214·3	6·42	32·52	208·8
	5	5·57	37·89	211·0	5·90	35·07	206·9
	6	5·07	41·01	207·9	5·40	37·83	204·3
	7	4·57	44·85	205·0	4·89	41·22	201·6
	8	4·06	49·35	200·4	4·39	45·27	198·7
	9	3·55	55·35	196·5	3·88	50·28	195·1
	10	3·05	63·30	193·1	3·38	57·03	192·7

TABLE IV (*continued*).—*Temperature of Condensation of Steam on Salts and in their Brines.*

No.	N	57			58		
Salt	rM	$0.2\dfrac{K+Na_4}{5}Cl$			$0.2\dfrac{K_4+Na}{5}Cl$		
Temp. of Saturated Steam	T	94°·16—94°·14 C.			94°·14—94°·13 C.		
	n	$t-T$	W	$W(t-T)$	$t-T$	W	$W(t-T)$
	0	11·18	...	...	11·01	24·00	...
	1	8·87	25·90	229·7	8·27	24·00	198·5
	2	7·95	28·44	226·1	6·95	28·16	195·7
	3	6·94	31·16	216·2	5·94	32·16	191·0
	4	5·92	36·36	215·2	5·43	34·72	188·5
	5	5·42	39·16	212·2	4·93	37·76	186·1
	6	4·91	42·52	208·8	4·43	41·36	183·2
	7	4·41	46·36	204·4	3·92	46·32	181·6
	8	3·91	51·56	201·6	3·42	52·56	179·7
	9	3·41	57·96	196·7	2·91	60·80	176·9

No.	N	59			60		
Salt	rM	$0.4\dfrac{NaNH_4}{2}Cl$			$0.4\dfrac{NaNH_4}{2}Cl$		
Temp. of Saturated Steam	T	94°·30 C.			94°·30 C.		
	n	$t-T$	W	$W(t-T)$	$t-T$	W	$W(t-T)$
	0	15·52	...	...	15·52	...	...
	1	10·60	41·6	220·5	10·57	40·7	215·1
	2	9·62	45·7	219·8	9·59	44·9	215·3
	3	8·63	50·3	217·0	8·60	49·3	212·0
	4	7·64	56·1	214·3	7·61	55·1	209·6
	5	6·65	63·4	210·8	6·62	62·0	205·2
	6	6·16	67·7	208·5	6·13	66·2	202·9
	7	5·66	72·7	205·7	5·63	71·2	200·4
	8	5·17	78·6	203·2	5·14	77·4	198·9
	9	...	...	...	...	...	...

TABLE IV (*continued*).—*Temperature of Condensation of Steam on Salts and in their Brines.*

No.	N	61			62		
Salt	rM	$0.2\,\dfrac{2KCl + KClO_3}{3}$			$0.4\,\dfrac{3KCl + KClO_3}{4}$		
Temp. of Saturated Steam	T	100°.28 C.			100°.28 C.		
	n	$t-T$	W	$W(t-T)$	$t-T$	W	$W(t-T)$
	0	9.82	...	...	9.87	...	...
	1	7.09	26.83	190.2	9.34	43.15	201.5
	2	6.37	29.93	190.6	8.39	47.65	199.9
	3	5.82	32.63	190.0	7.41	53.65	198.8
	4	5.22	36.23	189.1	6.92	57.25	198.1
	5	4.45	41.93	186.6	6.42	61.35	197.0
	6	3.95	46.93	185.4	5.93	66.15	196.1
	7	3.46	53.03	183.5	5.43	71.65	194.5
	8	2.96	61.43	181.8	4.94	78.55	194.0
	9	2.47	72.93	180.1	4.45	86.25	191.9
	10	...	...	...	3.95	96.85	191.3
	11	...	...	...	...	...	...

No.	N	63			64		
Salt	rM	$0.2\,(KCl + KClO_3)$			$0.2\left(\dfrac{(NH_4)_2SO_4}{2} + NH_4Cl\right)$		
Temp. of Saturated Steam	T	94°.10 C.			93°.86—93°.85 C.		
	n	$t-T$	W	$W(t-T)$	$t-T$	W	$W(t-T)$
	0	8.98	...	...	15.49	...	...
	1	5.15	33.8	174.1	12.64	21.25	179.1
	2	4.66	37.3	173.8	10.98	25.45	186.3
	3	4.16	41.8	173.9	9.99	27.95	186.1
	4	3.65	47.3	172.6	9.01	30.95	185.9
	5	3.15	54.6	172.0	8.02	34.75	185.8
	6	2.64	65.1	171.8	7.04	39.55	185.6
	7	2.23	77.4	172.6	6.56	41.35	181.2
	8	...	...	...	6.06	45.55	184.0
	9	...	...	...	5.08	53.85	182.4
	10	...	...	...	4.59	58.90	180.2
	11	...	...	...	4.09	65.80	179.4

TABLE IV (*continued*).—*Temperature of Condensation of Steam on Salts and in their Brines.*

No.	N	65			66		
Salt	rM	$0\cdot1\left(\dfrac{K_2SO_4}{2}+KCl\right)$			$0\cdot1\left(\dfrac{K_2SO_4}{2}+NaCl\right)$		
Temp. of Saturated Steam	T	94°·12 C.			93°·83—93°·80 C.		
	n	$t-T$	W	$W(t-T)$	$t-T$	W	$W(t-T)$
	0	7·93	...	...	10·32	...	...
	1	2·11	66·35	186·6	3·44	42·95	196·9
	2	2·01	70·35	188·5	3·03	48·05	194·1
	3	1·91	73·65	187·6	2·63	54·65	191·6
	4	1·80	77·05	184·9	2·43	58·95	190·6
	5	1·70	81·65	185·1	2·23	63·85	189·9
	6	...	...	...	2·04	69·95	190·3

No.	N	67			68		
Salt	rM	$0\cdot05\,Ba(NO_3)_2 +0\cdot2\,NaCl$			$0\cdot05\,Ba(NO_3)_2 +0\cdot1\,NaCl$		
Temp. of Saturated Steam	T	94°·10 C.			93°·97—93°·95 C.		
	n	$t-T$	W	$W(t-T)$	$t-T$	W	$W(t-T)$
	0	9·75	...	...	9·72	...	...
	1	6·69	41·7	223·2	3·59	42·25	202·3
	2	6·18	44·7	221·0	3·09	48·65	199·9
	3	5·67	47·9	217·3	2·59	57·95	200·1
	4	5·17	52·6	217·5	2·09	70·85	197·5
	5	4·66	57·5	214·3	...	...	...
	6	4·16	64·0	213·0	...	...	...

TABLE IV (*continued*).— *Temperature of Condensation of Steam on Salts and in their Brines.*

No.	N	69			70		
Salt	rM	$0 \cdot 1 \dfrac{4BaCl_2 . 2H_2O + Ba(NO_3)_2}{5}$			$0 \cdot 1 \dfrac{5BaCl_2 . 2H_2O + 2Ba(NO_3)_2}{7}$		
Temp. of Saturated Steam	T	99°·90 C.			99°·98 C.		
	n	$t-T$	W	$W(t-T)$	$t-T$	W	$W(t-T)$
	0	5·56	...	...	5·59	...	...
	1	4·95	25·2	259·4	5·28	22·9	241·8
	2	4·15	31·1	258·2	4·57	27·0	246·8
	3	3·64	35·7	259·8	4·07	30·6	249·3
	4	3·13	41·1	257·2	3·56	34·8	247·8
	5	2·63	48·5	255·2	3·06	40·7	249·1
	6	2·12	59·4	251·8	2·55	48·4	246·8
	7	1·62	76·5	247·8	2·04	59·7	243·6
	8	1·11	110·3	244·8	1·54	77·9	239·7

No.	N	71			72		
Salt	rM	$0 \cdot 05 (BaCl_2 . 2H_2O + Ba(NO_3)_2)$			$0 \cdot 4 \dfrac{6 (NH_4)_2 SO_4 + K_2 SO_4}{7}$		
Temp. of Saturated Steam	T	93°·94—93°·91 C.			100°·28 C.		
	n	$t-T$	W	$W(t-T)$	$t-T$	W	$W(t-T)$
	0	5·40	...	...	7·24	...	...
	1	2·32	47·7	221·4	5·99	66·15	198·1
	2	2·12	52·7	223·4	5·45	70·65	192·5
	3	1·91	57·7	220·4	4·96	77·25	191·6
	4	1·72	66·2	227·8	4·46	84·75	189·0
	5	1·61	69·8	226·8	3·97	94·55	192·7
	6	...	...	...	...	...	...
	7	...	...	...	...	...	...
	8	...	...	...	...	...	...

TABLE IV (*continued*).—*Temperature of Condensation of Steam on Salts and in their Brines.*

No.	N	73			74		
Salt	rM	$0 \cdot 1 \left.\begin{array}{l}K\\NH_4\end{array}\right\} SO_4$			$0 \cdot 16\, Sr\,(NO_3)_2$ $+\, 0 \cdot 423\, Ba\,(NO_3)_2$		
Temp. of Saturated Steam	T	$94° \cdot 13$ C.			$100° \cdot 00$ C.		
	n	$t-T$	W	$W(t-T)$	$t-T$	W	$W(t-T)$
	0	5·36	...	...	6·31	...	...
	1	2·31	39·25	181·4	3·02	79·7	237·9
	2	2·11	43·35	183·0	2·74	88·4	239·8
	3	1·90	47·65	181·0	2·47	99·2	242·5
	4	1·70	53·55	182·0	...	...	...
	5	1·60	56·95	182·2	...	...	...
	6	1·49	60·45	180·2	...	...	...
	7	1·39	65·15	181·0	...	...	...
	8	1·29	70·35	181·4	...	...	...
	9	1·19	77·35	184·0	...	...	...

No.	N	75			76		
Salt	rM	$0 \cdot 1\, \dfrac{Sr_2+Ba}{3}\,(NO_3)_2$			$0 \cdot 2\, \dfrac{3Pb+Ba}{4}\,(NO_3)_2$		
Temp. of Saturated Steam	T	$99° \cdot 60$ C.			$99° \cdot 60$ C.		
	n	$t-T$	W	$W(t-T)$	$t-T$	W	$W(t-T)$
	0	5·43	...	...	2·90	...	...
	1	2·23	50·1	222·4	2·42	68·6	166·0
	2	2·02	56·0	226·2	2·22	75·6	167·8
	3	1·72	66·1	227·4	2·01	84·8	170·4
	4	1·52	75·1	228·4	1·81	94·9	171·8
	5	1·32	88·2	232·8	1·61	110·6	178·1
	6	1·11	106·3	236·0	1·41	132·6	186·9
	7	1·01	119·8	242·0	...	...	...
	8	...	...	...	...	...	...
	9	...	...	...	...	...	...

TABLE IV (*continued*).—*Temperature of Condensation of Steam on Salts and in their Brines.*

No.	N	77			78		
Salt	rM	$0.1\dfrac{2Pb+Ba}{3}(NO_3)_2$			$0.2\dfrac{5Sr+4Pb}{9}(NO_3)_2$		
Temp. of Saturated Steam	T	99°.62—99°.63 C.			100°.29 C.		
	n	$t-T$	W	$W(t-T)$	$t-T$	W	$W(t-T)$
	0	2·89	...	...	5·93	...	...
	1	2·00	42·2	168·8	4·77	45·7	218·0
	2	1·70	51·3	174·4	4·16	52·3	221·7
	3	1·50	59·7	179·2	3·55	60·1	213·4
	4	1·30	71·4	185·6	3·05	69·5	212·0
	5	1·08	88·2	190·6	2·54	83·8	212·9
	6	0·98	100·3	196·6	2·33	90·7	211·3
	7	0·88	113·6	200·0	2·13	100·1	213·2
	8	...	...	...	1·93	111·0	214·2
	9	...	...	...	1·73	126·7	219·2
	10	...	...	...	...	...	...
	11	...	...	...	...	...	...

No.	N	79			80		
Salt	rM	$0.2\dfrac{2Sr+Pb}{3}(NO_3)_2$			$0.2\ NaNO_3$		
Temp. of Saturated Steam	T	100°.29 C.			94°.41 C.		
	n	$t-T$	W	$W(t-T)$	$t-T$	W	$W(t-T)$
	0	5·98	...	...	17·93	8·90	159·5
	1	5·27	44·0	231·9	17·60	8·84	155·6
	2	4·46	51·4	229·2	14·96	10·80	161·6
	3	4·06	56·1	227·8	12·99	12·54	163·0
	4	3·55	63·7	226·1	11·01	14·97	164·8
	5	3·05	73·6	224·5	10·02	16·64	166·7
	6	2·54	88·2	224·0	8·77	19·02	166·8
	7	2·34	95·6	223·7	7·50	22·48	168·6
	8	2·13	104·7	223·0	6·48	26·13	169·3
	9	1·93	116·5	224·8	5·49	30·86	169·4
	10	1·73	132·4	229·0	4·47	38·27	171·0
	11	...	...	...	3·83	44·84	171·7

TABLE IV (*continued*).—*Temperature of Condensation of Steam on Salts and in their Brines.*

No.	N	81			82		
Salt	rM	0·5 $NaNO_3$			0·5 $NaNO_3$		
Temp. of Saturated Steam	T	91°·18 C.			91°·30 C.		
	n	$t-T$	W	$W(t-T)$	$t-T$	W	$W(t-T)$
	0	16·91	21·9	148·6	16·89	24·5	164·7
	1	16·30	22·00	143·4	16·29	25·00	162·8
	2	14·21	26·7	151·8	13·92	30·8	171·5
	3	12·56	30·5	153·2	11·39	38·9	177·2
	4	10·91	35·4	154·5	10·25	41·9	171·7
	5	...	...	...	9·42	45·8	172·8
	6	...	...	...	8·43	51·0	172·0
	7	...	...	...	7·50	57·5	172·5

TABLE V.—*Temperature of Condensation of Steam in diluted Brines.*

No.	N	83			84		
Salt	rM	0·05 LiCl			0·05 KCl		
Temp. of Saturated Steam	T	99°·88—99°·87 C.			99°·75 C.		
	n	$t-T$	W	$W(t-T)$	$t-T$	W	$W(t-T)$
	0	3·97	16·1	63·9	3·59	14·2	51·2
	1	2·95	19·8	58·4	2·58	19·0	49·0
	2	2·45	23·0	56·3	2·07	23·0	47·6
	3	1·94	27·4	53·2	1·57	29·9	47·0
	4	1·74	30·3	52·7	1·37	34·0	46·6
	5	1·54	33·6	51·7	1·16	39·9	46·3
	6	1·34	37·9	50·8	0·96	47·9	46·0
	7	1·13	43·8	49·5	0·76	60·3	45·8
	8	0·93	52·8	49·1	0·66	69·7	46·0
	9	0·84	59·2	49·7	0·56	83·0	46·5
	10	0·74	66·9	49·5	0·46	106·6	49·0
	11	0·64	77·2	49·4	...	...	...
	12	0·54	90·8	49·0	...	...	...
	13	0·44	112·6	49·5	...	...	...

TABLE V (*continued*).—*Temperature of Condensation of Steam in diluted Brines.*

No.	N	85			86		
Salt	$r.M$	0·05 NaCl			0·05 $\dfrac{Na + K}{2}$ Cl		
Temp. of Saturated Steam	T	99°·75 C.			99°·42—99°·45 C.		
	n	$t-T$	W	$W(t-T)$	$t-T$	W	$W(t-T)$
	0	5·82	10·7	62·3	5·54	10·5	58·2
	1	3·79	15·0	56·8	4·03	13·4	54·0
	2	2·58	20·7	53·4	3·02	17·2	52·0
	3	2·07	24·7	51·1	2·51	20·1	50·5
	4	1·57	31·7	50·0	2·00	24·2	48·4
	5	1·37	36·1	49·5	1·70	28·2	48·0
	6	1·16	42·2	49·0	1·49	31·7	47·2
	7	0·96	51·4	49·3	1·29	36·3	46·8
	8	0·76	65·6	49·3	1·08	42·7	46·1
	9	0·66	77·9	51·4	0·98	46·8	45·9
	10	0·56	92·2	51·6	0·88	52·7	46·4
	11	0·46	113·7	52·3	0·78	59·3	46·3
	12	...	...	...	0·68	68·6	46·7
	13	...	...	...	0·58	81·8	47·5
	14	...	...	...	0·47	101·9	47·9
	15	...	...	...	0·36	141·1	50·8

No.	N	87			88		
Salt	rM	0·05 NH$_4$Cl			0·05 RbCl		
Temp. of Saturated Steam	T	99°·46—99°·47 C.			99°·51—99°·52 C.		
	n	$t-T$	W	$W(t-T)$	$t-T$	W	$W(t-T)$
	0	4·49	9·1	40·9	5·05	10·1	51·0
	1	4·09	11·8	48·3	4·04	12·4	50·1
	2	3·08	15·3	47·2	3·03	15·8	47·9
	3	2·57	17·9	46·0	2·52	18·6	46·9
	4	2·06	21·8	44·9	2·01	22·7	45·6
	5	1·66	26·9	44·6	1·81	25·1	45·4
	6	1·46	30·3	44·2	1·61	28·0	45·1
	7	1·25	35·3	44·1	1·41	32·0	45·1
	8	1·05	42·2	44·3	1·20	37·3	44·8
	9	0·85	51·4	43·7	1·00	45·7	45·7
	10	0·74	64·8	...	0·90	49·9	44·9
	11	0·64	69·6	44·6	0·80	56·1	44·9
	12	0·54	82·0	44·3	0·70	64·4	45·1
	13	0·44	101·4	44·6	0·60	77·9	46·7
	14	...	...	...	0·49	95·1	46·6
	15	...	...	...	0·39	124·1	48·4

TABLE V (*continued*).—*Temperature of Condensation of Steam in diluted Brines.*

No.	N	89			90		
Salt	rM	0·05 CsCl			0·05 BaCl$_2$. 2H$_2$O		
Temp. of Saturated Steam	T	99°·51 C.			99°·49—99°·50 C.		
	n	$t-T$	W	$W(t-T)$	$t-T$	W	$W(t-T)$
	0	5·55	9·5	52·7	2·54	27·1	68·8
	1	4·04	12·5	50·5	2·03	32·6	66·2
	2	3·03	15·8	47·9	1·83	35·7	65·3
	3	2·02	22·7	45·9	1·63	39·7	64·7
	4	1·61	28·2	45·4	1·43	44·7	63·9
	5	1·41	32·2	45·4	1·23	51·4	63·2
	6	1·20	37·2	44·6	1·02	60·7	61·9
	7	1·00	45·0	45·0	0·92	68·2	62·7
	8	0·90	49·8	44·8	0·82	75·7	62·1
	9	0·80	56·5	45·2	0·72	86·6	62·4
	10	0·70	63·9	44·7	0·62	101·7	63·6
	11	0·60	78·1	46·9	0·51	126·0	64·2
	12	0·50	96·8	48·4	...	...	...
	13	0·40	120·3	48·1	...	...	...
	14	...	...	...	...	...	...

No.	N	91			92		
Salt	rM	0·05 KI			0·05 KBr		
Temp. of Saturated Steam	T	99°·59 C.			99°·59 C.		
	n	$t-T$	W	$W(t-T)$	$t-T$	W	$W(t-T)$
	0	6·08	10·2	62·0	5·07	11·0	55·8
	1	4·56	12·8	54·4	3·55	14·8	52·5
	2	3·55	15·7	55·7	2·54	19·7	50·0
	3	2·54	20·7	52·6	2·03	24·2	49·1
	4	2·03	25·1	51·0	1·63	29·2	47·6
	5	1·53	32·4	49·6	1·43	33·0	47·2
	6	1·43	34·9	50·0	1·22	38·4	46·8
	7	1·22	39·7	48·4	1·02	45·6	46·5
	8	1·02	46·9	47·8	0·92	50·7	46·6
	9	0·92	51·6	47·5	0·82	56·5	46·3
	10	0·82	57·7	47·3	0·72	64·7	46·6
	11	0·72	65·2	46·4	0·62	75·1	46·6
	12	0·62	75·8	47·0	0·52	89·4	46·5
	13	0·52	90·8	47·2	0·42	116·1	48·8
	14	0·42	116·0	48·7	...	...	...

TABLE V (*continued*).—*Temperature of Condensation of Steam in diluted Brines.*

No.	N	93			94		
Salt	rM	0·05 KClO₃			0·05 KNO₃		
Temp. of Saturated Steam	T	99°·31—99°·29 C.			99°·57—99°·55 C.		
	n	$t-T$	W	$W(t-T)$	$t-T$	W	$W(t-T)$
	0	3·02	13·2	39·9	3·27	6·8	22·2
	1	2·52	16·2	40·8	2·56	14·8	37·9
	2	2·01	20·3	40·8	2·04	18·5	37·7
	3	1·71	24·4	41·7	1·54	25·2	37·8
	4	1·51	27·6	41·7	1·34	29·4	39·4
	5	1·31	32·3	42·3	1·13	35·0	39·6
	6	1·10	38·1	41·9	1·02	38·8	39·6
	7	1·00	41·7	41·7	0·92	42·6	39·2
	8	0·90	45·9	41·3	0·82	48·3	39·6
	9	0·80	51·6	41·3	0·72	55·6	40·0
	10	0·70	59·8	41·9	0·62	63·9	39·6
	11	0·61	69·6	42·5	0·53	78·4	41·6
	12	0·52	91·5	47·6	0·43	95·2	40·9
	13	0·42	108·9	45·7	0·35	137·4	48·1
	14	...	...	...	...	...	...

No.	N	95			96		
Salt	rM	0·05 NaNO₃			0·05 Sr(NO₃)₂		
Temp. of Saturated Steam	T	99°·87 C.			99°·30—99°·31—99°·30 C.		
	n	$t-T$	W	$W(t-T)$	$t-T$	W	$W(t-T)$
	0	4·08	11·4	46·4	5·06	14·7	74·4
	1	3·06	14·9	45·6	3·54	18·4	65·1
	2	2·56	17·9	45·8	3·04	21·1	64·1
	3	2·05	22·1	45·3	2·53	24·7	62·5
	4	1·85	24·5	45·3	2·02	30·5	61·6
	5	1·65	27·3	45·0	1·82	34·0	61·9
	6	1·45	31·0	44·9	1·61	37·9	61·0
	7	1·24	36·4	45·1	1·41	42·9	60·5
	8	1·04	43·4	45·1	1·20	49·9	59·9
	9	0·94	47·3	44·5	1·00	59·5	59·5
	10	0·84	54·1	45·4	0·90	66·0	59·4
	11	0·74	61·0	45·1	0·80	75·0	60·0
	12	0·64	70·4	45·0	0·71	84·4	59·9
	13	0·54	83·5	45·7	0·61	99·8	60·9
	14	0·44	106·1	46·7	...	...	...

TABLE V (*continued*).—*Temperature of Condensation of Steam in diluted Brines.*

No.	N	97			98		
Salt	rM	$0.05\ Pb(NO_3)_2$			$0.05\ AgNO_3$		
Temp. of Saturated Steam	T	$99°.29$ C.			$99°.87$ C.		
	n	$t-T$	W	$W(t-T)$	$t-T$	W	$W(t-T)$
	0	2·34	18·1	42·4	3·06	10·1	30·9
	1	2·03	20·8	42·2	2·56	12·7	32·5
	2	1·83	23·5	43·0	2·05	16·6	34·0
	3	1·63	26·5	43·2	1·85	19·0	35·2
	4	1·43	30·6	43·7	1·65	21·6	35·6
	5	1·22	36·3	44·3	1·45	24·7	35·8
	6	1·02	44·2	45·1	1·24	29·7	36·8
	7	0·92	49·4	45·4	1·04	36·3	37·8
	8	0·82	57·0	46·7	0·94	40·5	38·1
	9	0·72	65·6	47·2	0·84	46·3	38·9
	10	0·62	79·6	49·4	0·74	52·7	39·0
	11	0·52	96·5	50·2	0·64	62·6	40·1
	12	...	...	...	0·54	76·1	41·1
	13	...	...	...	0·44	98·2	43·2

TABLE VI.—Values of $W(t-T)$ for even Temperatures taken from Curves.

$t-T$	Chlorides						Potassium Salts					Nitrates					
	LiCl	NaCl	NaK Cl/2	KCl	CsCl*	NH₄Cl	KI	KBr	KClO₃	KCl	KNO₃	NaNO₃	Ba(NO₃)₂	Sr(NO₃)₂	Pb(NO₃)₂	AgNO₃	KClO₃
°C.																	
3·5	...	...	...	...	...	...	...	...	...	...	...	...	...	...	...	...	...
3·0	...	54·8	52·0	50·0	47·8	47·0	54·1	51·3	40·0	49·9	...	...	...	65·0	...	...	...
2·5	56·4	52·9	50·3	48·8	46·9	45·9	52·6	50·0	40·8	48·8	37·9	45·5	...	64·0	...	...	40·5
2·0	53·8	50·9	48·6	47·5	45·8	44·8	50·9	48·5	41·2	47·7	37·8	45·5	...	62·5	42·4	39·4	40·9
1·8	52·8	50·4	48·1	47·3	45·6	44·6	50·3	48·1	41·6	47·2	37·9	45·3	...	61·8	42·9	40·3	41·2
1·6	52·0	50·0	47·6	47·1	45·4	44·4	49·7	47·6	41·9	46·9	38·1	45·1	...	61·4	43·3	40·7	41·7
1·4	50·9	49·5	47·0	46·7	45·3	44·3	49·0	47·1	42·0	46·6	39·0	45·0	...	61·0	43·7	41·0	42·2
1·2	49·9	49·0	46·3	46·3	44·9	44·2	48·4	46·8	42·0	46·3	39·5	45·0	48·7	60·5	44·3	41·7	42·2
1·0	49·5	48·8	46·0	46·0	44·8	43·9	47·7	46·5	41·4	45·9	39·5	45·0	50·2	59·9	45·0	42·8	41·7
0·9	49·3	48·9	46·1	45·9	44·9	43·9	47·4	46·4	41·3	45·8	39·5	45·0	51·2	59·4	45·7	43·3	41·5
0·8	49·2	49·2	46·3	45·8	45·0	44·0	47·1	46·3	41·4	45·8	39·6	45·1	52·0	59·6	46·7	43·9	41·4
0·7	49·1	50·0	46·7	45·9	45·5	44·2	46·7	46·3	41·9	46·3	39·8	45·1	52·5	59·8	47·9	44·6	42·0
0·6	49·3	51·1	47·2	46·2	46·8	44·4	46·8	46·5	42·9	48·2	40·3	45·2	...	60·3	49·4	45·8	42·7
0·5	49·4	52·4	48·0	48·1	48·3	...	47·4	47·0	44·4	...	41·4	45·9	...	60·7	50·5	47·2	44·2
0·4	49·7	...	...	...	...	...	48·5	...	...	...	44·9	...	...	...	...	48·8	45·7

* Reason for excluding RbCl, see Contents, p. xiv.

TABLE VII.—*Relations between relative Reduction of Vapour Tension and Dilution.*

No. ... N	1			2		
Salt ... rM	0·2 KCl			0·4 KCl		
P	772·0 mm.			762·7—763·3 mm.		
n	p mm.	$\frac{p-P}{p}$	$W\frac{p-P}{p}$	p mm.	$\frac{p-P}{p}$	$W\frac{p-P}{p}$
0	1037·6	0·2560	6·733	1019·9	0·2522	6·633
1	...	...	...	...	...	...
2	1005·6	0·2323	6·669	987·3	0·2272	6·609
3	970·2	0·2043	6·662	969·8	0·2132	6·592
4	953·4	0·1903	6·643	953·4	0·1994	6·588
5	937·7	0·1767	6·593	937·7	0·1860	6·571
6	922·0	0·1628	6·611	925·2	0·1750	6·701
7	906·4	0·1483	6·571	906·1	0·1576	6·561
8	890·9	0·1334	6·525	...	...	...
9	875·4	0·1181	6·485	...	...	...
10	860·4	0·1027	6·451	...	...	...
11	845·9	0·0874	6·407	...	...	...

No. ... N	3			4		
Salt ... rM	0·2 KCl			0·2 KCl		
P	738·8 mm.			613·7—614·0 mm.		
n	p mm.	$\frac{p-P}{p}$	$W\frac{p-P}{p}$	p mm.	$\frac{p-P}{p}$	$W\frac{p-P}{p}$
0	988·4	0·2525	6·756	816·0	0·2479	6·732
1	...	...	...	...	...	...
2	947·6	0·2205	6·672	761·9	0·1941	6·666
3	915·0	0·1926	6·635	734·8	0·1644	6·633
4	883·5	0·1637	6·606	721·7	0·1492	6·601
5	867·9	0·1488	6·584	708·6	0·1335	6·573
6	852·8	0·1337	6·571	695·9	0·1177	6·548
7	837·6	0·1180	6·529	683·1	0·1011	6·515
8	822·7	0·1020	6·493	670·8	0·0846	6·501
9	808·0	0·0857	6·493	...	...	...
10	...	...	...	...	...	...
11	...	...	...	...	...	...

TABLE VII (*continued*).—*Relations between relative Reduction of Vapour Tension and Dilution.*

No. ... N	5			7		
Salt ... rM	0·2 KCl			0·2 KCl		
P	618·7 mm.			549·8 mm.		
n	p mm.	$\frac{p-P}{p}$	$W\frac{p-P}{p}$	p mm.	$\frac{p-P}{p}$	$W\frac{p-P}{p}$
0	822·5	0·2478	6·772	729·6	0·2464	6·611
1	...	...	...	...	...	...
2	764·6	0·1909	6·722	704·5	0·2196	6·611
3	737·5	0·1611	6·697	679·0	0·1903	6·610
4	711·1	0·1299	6·630	666·6	0·1752	6·612
5	685·6	0·0976	6·601	654·3	0·1597	6·576
6	660·6	0·0635	6·600	...	...	...
7	...	...	...	...	...	...
8	...	...	...	...	...	...
9	...	...	...	...	...	...
10	...	...	...	...	...	...
11	...	...	...	...	...	...

No. ... N	8			9		
Salt ... rM	0·2 KCl			0·2 NaCl		
P	550·4 mm.			772·0 mm.		
n	p mm.	$\frac{p-P}{p}$	$W\frac{p-P}{p}$	p mm.	$\frac{p-P}{p}$	$W\frac{p-P}{p}$
0	730·3	0·2463	6·680	1038·9	0·2569	7·668
1	...	...	...	...	...	...
2	704·5	0·2187	6·650	1003·9	0·2309	7·725
3	678·9	0·1894	6·619	970·4	0·2045	7·609
4	654·3	0·1588	6·599	952·4	0·1894	7·565
5	630·5	0·1270	6·574	938·0	0·1769	7·520
6	607·4	0·0938	6·518	922·4	0·1631	7·439
7	...	...	...	906·7	0·1485	7·352
8	...	...	...	891·2	0·1338	7·253
9	...	...	...	876·0	0·1187	7·171
10	...	...	...	860·9	0·1033	7·067
11	...	...	...	846·3	0·0878	7·007

TABLE VII (*continued*).—*Relations between relative Reduction of Vapour Tension and Dilution.*

No.	N	12			16		
Salt	rM	0·2 NaCl			0·2 NaCl		
	P	614·0—613·7 mm.			550·4 mm.		
	n	p mm.	$\dfrac{p-P}{p}$	$W\dfrac{p-P}{p}$	p mm.	$\dfrac{p-P}{p}$	$W\dfrac{p-P}{p}$
	0	825·4	0·2561	7·555	741·5	0·2577	7·558
	1	...	...		...	...	
	2	790·1	0·2229	7·500	715·0	0·2302	7·606
	3	761·9	0·1941	7·424	689·3	0·2015	7·588
	4	734·8	0·1644	7·183	664·2	0·1713	7·438
	5	721·7	0·1496	7·233	640·1	0·1401	7·313
	6	708·6	0·1339	7·157	616·7	0·1075	7·165
	7	695·9	0·1181	7·116	...	...	...
	8	683·1	0·1016	7·015	...	...	...
	9	670·8	0·0851	6·923	...	...	...
	10	...	...	...	...	...	...

No.	N	17			19		
Salt	rM	0·2 NH$_4$Cl			0·4 NH$_4$Cl		
	P	742·0—743·3 mm.			619·9 mm.		
	n	p mm.	$\dfrac{p-P}{p}$	$W\dfrac{p-P}{p}$	p mm.	$\dfrac{p-P}{p}$	$W\dfrac{p-P}{p}$
	0	1220·1	0·3918	...	996·13	0·3807	...
	1	...	...	...	...	...	...
	2	1115·1	0·3346	5·558	929·70	0·3364	5·394
	3	1043·6	0·2890	5·734	882·23	0·3007	5·608
	4	975·9	0·2394	5·937	851·31	0·2753	5·770
	5	942·9	0·2127	6·028	821·57	0·2491	5·872
	6	911·2	0·1853	6·114	792·64	0·2217	5·998
	7	880·7	0·1571	6·133	764·63	0·1932	6·089
	8	865·8	0·1423	6·153	737·47	0·1635	6·183
	9	850·4	0·1264	6·135	...	...	...
	10	836·2	0·1116	6·131	...	...	...
	11	821·6	0·0955	6·106	...	...	...
	12	807·4	0·0794	6·101	...	...	...

TABLE VII (*continued*).—*Relations between relative Reduction of Vapour Tension and Dilution.*

No.	N	21			22		
Salt	rM	0·4 NH$_4$Cl			0·1 BaCl$_2$. 2H$_2$O		
	P	550·35 mm.			758·64 mm.		
	n	p mm.	$\frac{p-P}{p}$	$W\frac{p-P}{p}$	p mm.	$\frac{p-P}{p}$	$W\frac{p-P}{p}$
	0	880·4	0·3782	...	889·7	0·1473	...
	1	...	...	...	...	...	...
	2	819·8	0·3287	5·555	876·9	0·1349	9·435
	3	790·9	0·3042	5·719	861·5	0·1194	9·473
	4	762·7	0·2784	5·846	846·2	0·1035	9·454
	5	735·6	0·2519	5·957	831·5	0·0877	9·343
	6	709·3	0·2241	6·073	816·6	0·0710	9·226
	7	683·8	0·1966	6·203	802·3	0·0545	9·103
	8	659·1	0·1650	6·261	...	...	...
	9	...	...	...	...	...	...

No.	N	23			24		
Salt	rM	0·2 BaCl$_2$. 2H$_2$O			0·1 BaCl$_2$. 2H$_2$O		
	P	620·56 mm.			549·94 mm.		
	n	p mm.	$\frac{p-P}{p}$	$W\frac{p-P}{p}$	p mm.	$\frac{p-P}{p}$	$W\frac{p-P}{p}$
	0	728·0	0·1475	...	644·07	...	...
	1	...	...	...	...	...	...
	2	708·3	0·1238	10·597	630·25	0·1274	10·549
	3	695·6	0·1078	10·413	618·73	0·1112	10·431
	4	685·3	0·0944	10·242	607·17	0·0942	10·230
	5	675·4	0·0811	9·983	...	...	...
	6	667·8	0·0708	9·820	...	...	...
	7	660·6	0·0605	9·626	...	...	...
	8	653·3	0·0502	9·377	...	...	...
	9	646·2	0·0396	9·369	...	...	...

TABLE VII (*continued*).—*Relations between relative Reduction of Vapour Tension and Dilution.*

No. ... N	25			27		
Salt ... rM	0·2 KClO$_3$			0·2 KClO$_3$		
P	768·2 mm.			620·8 mm.		
n	p mm.	$\dfrac{p-P}{p}$	$W\dfrac{p-P}{p}$	p mm.	$\dfrac{p-P}{p}$	$W\dfrac{p-P}{p}$
0	879·4	0·1264	...	700·4	0·1136	...
1	...	...	...	...	...	...
2	867·9	0·1149	5·128	688·1	0·0979	5·365
3	854·3	0·1008	5·285	678·1	0·0845	5·476
4	839·4	0·0848	5·404	668·1	0·0708	5·565
5	824·8	0·0686	5·545	660·8	0·0605	5·675
6	811·1	0·0529	5·879	656·5	0·0544	5·810
7	...	...	...	650·9	0·0463	5·824
8	...	...	...	646·2	0·0393	5·871

No. ... N	28			29		
Salt ... rM	0·2 KClO$_3$			0·05 Ba(NO$_3$)$_2$		
P	550·35 mm.			757·56 mm.		
n	p mm.	$\dfrac{p-P}{p}$	$W\dfrac{p-P}{p}$	p mm.	$\dfrac{p-P}{p}$	$W\dfrac{p-P}{p}$
0	616·4	0·1072	...	792·9	0·0445	...
1	...	...	...	...	...	...
2	604·9	0·0902	5·355	787·9	0·0385	6·911
3	593·8	0·0732	5·527	785·1	0·0350	7·011
4	582·6	0·0584	5·696	779·5	0·0281	7·337
5	...	...	...	776·7	0·0246	7·368
6	...	...	...	774·0	0·0212	7·927
7	...	...	...	...	...	...
8	...	...	...	...	...	...

TABLE VII (*continued*).—*Relations between relative Reduction of Vapour Tension and Dilution.*

No. ... N		30			31		
Salt ... rM		0·2 Ba(NO₃)₂			0·2 Sr(NO₃)₂		
P		620·6 mm.			760·00 mm.		
	n	p mm.	$\frac{p-P}{p}$	$W\frac{p-P}{p}$	p mm.	$\frac{p-P}{p}$	$W\frac{p-P}{p}$
	0	646·7	0·0403	...	955·7	0·2048	8·376
	1	...	...	...	...	...	...
	2	643·4	0·0355	6·763	920·5	0·1743	8·663
	3	641·0	0·0318	6·837	904·6	0·1598	8·613
	4	638·6	0·0283	6·998	889·4	0·1455	8·584
	5	636·3	0·0247	7·136	873·9	0·1303	8·600
	6	634·0	0·0211	7·351	859·1	0·1154	8·610
	7	631·6	0·0175	7·343	844·1	0·0996	8·514
	8	629·3	0·0138	8·061	823·7	0·0773	7·910

No. ... N		32			36		
Salt ... rM		0·2 Pb(NO₃)₂			0·1 (NH₄)₂SO₄		
P		749·2 mm.			743·9 mm.		
	n	p mm.	$\frac{p-P}{p}$	$W\frac{p-P}{p}$	p mm.	$\frac{p-P}{p}$	$W\frac{p-P}{p}$
	0	842·0	0·1102	5·510	972·2	0·2348	5·941
	1	...	...	...	...	...	...
	2	831·8	0·0993	5·610	918·2	0·1898	6·149
	3	822·8	0·0895	5·629	889·7	0·1638	6·155
	4	817·2	0·0832	5·741	858·8	0·1338	6·139
	5	811·4	0·0766	5·760	843·5	0·1181	6·130
	6	805·4	0·0698	5·807	828·9	0·1025	6·085
	7	799·4	0·0628	5·834	814·0	0·0861	6·074
	8	793·8	0·0562	5·963	799·4	0·0694	6·077
	9	788·1	0·0493	6·251	793·8	0·0628	6·149
	10	...	...	...	788·1	0·0561	6·191
	11	...	...	...	782·6	0·0494	6·323
	12	...	...	...	776·7	0·0422	6·381

TABLE VII (*continued*).—*Relations between relative Reduction of Vapour Tension and Dilution.*

No. ... N	39			41		
Salt ... rM	$0{\cdot}2\ (NH_4)_2SO_4$			$0{\cdot}2\ (NH_4)_2SO_4$		
P	620·33 mm.			552·43 mm.		
n	p mm.	$\dfrac{p-P}{p}$	$W\dfrac{p-P}{p}$	p mm.	$\dfrac{p-P}{p}$	$W\dfrac{p-P}{p}$
0	810·0	0·2342	...	721·2	0·2345	...
1	...	...	...	...	...	...
2	781·2	0·2059	6·074	693·8	0·2038	6·196
3	753·5	0·1769	6·121	668·8	0·1740	6·212
4	726·7	0·1464	6·105	644·5	0·1429	6·316
5	700·6	0·1146	6·154	618·7	0·1071	6·201
6	675·4	0·0816	6·153	598·2	0·0765	6·281

No. ... N	42			44		
Salt ... rM	$0{\cdot}2\ \dfrac{K+Na}{2}Cl$			$0{\cdot}4\ \dfrac{K+Na}{2}Cl$		
P	772·0 mm.			616·9—616·68 mm.		
n	p mm.	$\dfrac{p-P}{p}$	$W\dfrac{p-P}{p}$	p mm.	$\dfrac{p-P}{p}$	$W\dfrac{p-P}{p}$
0	1161·9	0·3356	6·947	921·1	0·3302	...
1	...	...		...	...	...
2	1107·7	0·3031	7·077	879·4	0·2985	6·880
3	1055·7	0·2686	7·038	858·2	0·2812	6·917
4	1020·6	0·2436	7·101	828·3	0·2554	6·947
5	987·3	0·2181	7·034	813·2	0·2416	6·958
6	954·0	0·1908	7·012	798·9	0·2281	6·957
7	938·0	0·1770	6·983	784·5	0·2139	6·962
8	922·4	0·1631	6·956	770·7	0·1998	6·963
9	906·7	0·1485	6·898	756·5	0·1848	6·948
10	891·2	0·1337	6·839	...	...	...
11	876·0	0·1187	6·795	...	...	...
12	860·9	0·1032	6·702	...	...	...
13	846·2	0·0877	6·652	...	...	...

TABLE VII (*continued*).—*Relations between relative Reduction of Vapour Tension and Dilution.*

No. ...	N	80			82		
Salt ...	rM	0·2 NaNO$_3$			0·5 NaNO$_3$		
	P	620·3 mm.			552·43 mm.		
	n	p mm.	$\dfrac{p-P}{p}$	$W\dfrac{p-P}{p}$	p mm.	$\dfrac{p-P}{p}$	$W\dfrac{p-P}{p}$
	0	1163·0	0·4666	...	1011·5	0·4539	...
	1	...	...	...	...	...	...
	2	1052·9	0·4109	4·438	913·4	0·3952	4·869
	3	984·6	0·3700	4·640	836·2	0·3394	5·281
	4	919·8	0·3256	4·874	803·1	0·3122	5·232
	5	888·7	0·3020	5·025	779·8	0·2916	5·342
	6	850·7	0·2708	5·151	752·7	0·2661	5·428
	7	813·4	0·2374	5·337	728·0	0·2412	5·548
	8	784·8	0·2096	5·477	...	...	...
	9	757·3	0·1809	5·582	...	...	...
	10	730·1	0·1504	5·756	...	...	...
	11	713·4	0·1305	5·872	...	...	...

TABLE VIII.—*Vapour Tension of Water in Kilogrammes per Square Centimetre for whole Degrees Centigrade.*

° C.	Kilogrammes per Square Centimetre	Difference	2nd Difference
120	2·0275	...	...
119	1·9639	0·0636	...
118	1·9020	0·0619	0·0017
117	1·8417	0·0603	0·0016
116	1·7830	0·0587	0·0016
115	1·7258	0·0572	0·0015
114	1·6702	0·0556	0·0016
113	1·6160	0·0542	0·0014
112	1·5633	0·0527	0·0015
111	1·5120	0·0513	0·0014
110	1·4620	0·0500	0·0013
109	1·4135	0·0485	0·0015
108	1·3662	0·0473	0·0012
107	1·3203	0·0459	0·0014
106	1·2757	0·0446	0·0013
105	1·2323	0·0434	0·0012
104	1·1902	0·0421	0·0013
103	1·1492	0·0410	0·0011
102	1·1094	0·0398	0·0012
101	1·0708	0·0386	0·0012
100	1·0333	0·0375	0·0011
99	0·9970	0·0363	0·0012
98	0·9616	0·0354	0·0009
97	0·9273	0·0343	0·0011
96	0·8940	0·0333	0·0010
95	0·8617	0·0323	0·0010
94	0·8303	0·0314	0·0009
93	0·8000	0·0303	0·0011
92	0·7705	0·0295	0·0008
91	0·7420	0·0285	0·0010
90	0·7144	0·0276	0·0009

TABLE IX.—*Example of Use of Elastic Tank.*

$T = 99°{\cdot}09$ C. $A = 1{\cdot}000$ k./cm.^{2}. $W_0 = 100 \begin{Bmatrix} \text{gr.} \\ \text{c.c.} \\ \text{cm.}^2. \end{Bmatrix}$. $a_0 = 2{\cdot}000$ k./cm.^{2}.

$t_0 = 119°{\cdot}57$ C. $b_0 = 100$ Kilogrammes. $\dfrac{b_0}{W} = a - A$.

$t - T$ °C.	a k./cm.^{2}	t °C.	$(a - A)$ $\dfrac{b_0}{W}$	W	$W(t - T)$
79·79	10·0	178·88	9·0	11·1	886
75·29	9·0	174·38	8·0	12·5	941
70·37	8·0	169·46	7·0	14·3	1006
64·94	7·0	164·03	6·0	16·7	1084
58·85	6·0	157·94	5·0	20·0	1177
51·91	5·0	151·00	4·0	25·0	1298
43·73	4·0	142·82	3·0	33·3	1456
33·71	3·0	132·80	2·0	50·0	1686
27·63	2·5	126·72	1·5	66·7	1843
20·48	2·0	119·57	1·0	100·0	2048
18·94	1·9	118·03	0·9	111·0	2102
17·20	1·8	116·29	0·8	125	2150
15·44	1·7	114·53	0·7	143	2208
13·61	1·6	112·70	0·6	167	2272
11·67	1·5	110·76	0·5	200	2334
9·62	1·4	108·71	0·4	250	2405
7·45	1·3	106·54	0·3	333	2481
5·14	1·2	104·23	0·2	500	2570
2·66	1·1	101·75	0·1	1000	2660
2·40	1·09	101·49	0·09	1110	2664
2·14	1·08	101·23	0·08	1250	2675
1·88	1·07	100·97	0·07	1430	2688
1·62	1·06	100·71	0·06	1670	2705
1·35	1·05	100·44	0·05	2000	2700
1·08	1·04	100·17	0·04	2500	2700
0·82	1·03	99·91	0·03	3330	2730
0·54	1·02	99·63	0·02	5000	2700
0·27	1·01	99·36	0·01	10000	2700

No. 5. [*From Nature, August* 22, 1901, *Vol. LXIV*, *pp.* 399–400.]

THE SIZE OF THE ICE-GRAIN IN GLACIERS

IN referring to the size of the grain of the glacier in the chapter on chemistry and physics in the *Antarctic Manual*, I have given 700 grammes as the maximum weight which I have observed. In August, 1895, I made an extended study of the structure of glacier-ice, ·principally from the Aletsch Glacier. The fragments of this glacier, which float as icebergs in the Mergelin See, are exposed to the powerful weathering influence of the summer sun, and are comparatively easily dissected into their constituent grains. A number of blocks were so dissected in order to ascertain the weight and size of the largest grains. The following weights of single grains were determined:—700, 590, 450, 270, 255, 170, 155 and 100 grammes. It was observed that blocks of ice contained grains of all sizes, which fitted each other so exactly that, in the fresh unweathered block, the whole volume was filled with ice.

It was not then thought necessary to determine the weight of the smaller grains. On revisiting the Mergelin See in the latter part of July of this year (1901), I dissected several blocks of ice more or less completely and weighed their constituent grains. In order to effect the dissection a powerful sun is requisite, and a powerful sun means a high atmospheric temperature, under the influence of which the small grains melt and disappear very quickly. All the grains in the block are melting at the same time, but the smaller the grain the greater is the ratio of its external surface to its mass. Therefore the weights of the large grains are reduced to a less degree than those of the small ones. Hence it is impossible to furnish an exact statistical account of any block of ice, but the figures in the following tables give a very fair idea of the structural composition of the ice examined. The analyses of blocks E and F are the most complete.

The first block, A, is from the lower end of the Glacier des Bossons in the Chamonix valley, and it was examined on July 17, 1901, which was one of the hottest days of that very hot week. The other blocks are all from the Aletsch Glacier, as they are found floating in the Mergelin See, the waters of which are retained at one end by the ice of that glacier. The Aletsch Glacier is the largest in Switzerland and it contains the largest ice-grains that I have met with. Different parts of the glacier, even in the immediate vicinity of the lake, are of different grain, and the fragments are easily distinguished as they float in the water. Thus block F is a block of large-grained ice, while E is of comparatively small-grained ice, though it is by no means of the smallest grain.

List of blocks dissected.

Block A.—Chamonix, July 17, 1901. From the end of the Glacier des Bossons.

Blocks B, C and D were taken from the Mergelin See on July 21, 1901, and exposed to the sun on a rock for some hours. B and C were then dissected, though not completely; that is, a certain comparatively small portion of each of them remained undissected. D was dissected only in so far as to enable a prominent and very large grain (570 grms.) to be removed and weighed. The remainder was left till the next day. Owing to the high temperature of the air both by night and by day, its size was very much reduced. It is called Block d, and it was dissected on July 22, 1901.

Block E from the Mergelin See was collected and dissected on July 22, 1901.

Block F had suffered far-reaching sun weathering. It was not removed from the lake but was dissected in the water on July 24, 1901.

The results are embodied in the two following tables. All the weights are given in grammes. They were determined on a Salter's spring-balance which carried 500 grammes, and its scale was divided into intervals of 10 grammes each. Ice-grains which weighed more than 500 grammes were divided in two.

TABLE I.

Weight, in grammes, of single ice-grains			
A	*C*	*d*	*F*
160	230	125	590
110	210	70	550
90	200	65	460
85	150	60	360
80	130	30	325
80	75	25	250
75	60	25	240
75	60	—	240
60	50	400	190
40	50		180
40	50		155
35	25	*E*	150
30	25	120	150
30	25	115	140
30	25	105	130
25	—	85	130
—	1365	70	125
1045		60	120
		60	120
		50	120
		50	110
B		50	110
315		45	100
220		35	100
150		35	90
90	*D*	30	90
75	570	30	80
60	225	30	75
50	95	30	60
50	75	25	60
40	75	25	45
25	50	25	45
20		20	45
			30
1095	1090	1125	5765

TABLE II.

Number of grains weighed	Aggregate weight of grains	Average weight of one grain	Number of grains weighed	Aggregate weight of grains	Average weight of one grain
	A			*d*	
16	1045	65·3	7	400	57·1
10	110	11·0	10	30	3·0
4	25	6·25			
10	25	2·5	17	430	25·3
40	1205	30·1			
				E	
	B		22	1125	51·1
			10	150	15·0
11	1095	99·5	10	135	13·5
10	95	9·5	10	120	12·0
23	60	2·6	10	60	6·0
			10	50	5·0
44	1250	28·5	10	40	4·0
			10	20	2·0
			10	15	1·5
	C				
15	1365	91·0	102	1715	16·8
5	50	10·0			
10	25	2·5		*F*	
30	1440	48·0			
			34	5765	169·6
			10	190	19·0
	D		10	190	19·0
			10	140	14·0
6	1090	182·0	6	60	10·0
5	30	6·0	40	70	1·75
11	1120	102·0	110	6415	58·3

As has been already pointed out, the figures in the tables do not give an exact statistical account of the blocks of ice. The smallest grains have most frequently escaped being weighed, therefore the average size of the grain comes out higher than the truth. The figures in the tables give a general idea of the constitution or anatomy of a block of ice taken from the lower part of a large glacier. They are particularly interesting when we reflect that every grain, even the largest, has grown, according to the rigid laws of crystal-lomorphic development, from a single snow-crystal which probably weighed no more than one or two centigrammes.

In the Mergelin See, glacier-ice can be studied in a way that is possible in no other place.   The fragments of the Aletsch Glacier which float in it are veritable icebergs, and behave in the same way as their relatives in the Arctic or Antarctic Ocean.   In the middle of summer, however, they are exposed to a much more powerful sun than either the northern or the southern bergs.   Consequently, the weathering and disintegration, as well as the melting, proceed at a much more rapid rate.

The action of the sun's rays on glacier-ice is twofold; it disarticulates the ice into its constituent grains, and it splits the individual grain up into laminæ perpendicular to the principal axis of the crystal and bounded by the planes of fusion discovered and described by Tyndall.   These planes are the distinguishing characteristic of the individual ice-grain.

Under the influence of radiant heat an ice-crystal begins to melt at the surface which separates these laminæ, and the process of disintegration and decay is directed by their plane. On the other hand, an ice-crystal, floating in water and losing heat, generates ice laminæ which are directed by the same planes, which form the continuation of the corresponding laminæ of the parent crystal.   This was well observed at the end of August 1895.   Every night a thin skin of ice was formed at the shallow end of the lake, where the ice-blocks are collected.   As the grains in a block of glacier-ice are distributed quite irregularly, the water line of a floating block necessarily cuts a great number of grains, all of which are oriented differently.   The ice which was formed during the night along this line was oriented crystallographically by the grain with which it was in contact and from which it appeared to spring in continuation of its crystalline laminæ.   This produces a remarkable pattern of lines on the surface of the lake-ice contiguous to a block of glacier-ice.

Tyndall has described and figured the minute features of

the disintegration of the crystal under the absorption of radiant heat. Similar and complementary features are observed when ice is generated from an existing crystal under the dissipation of heat. To do justice to them, however, would require the services of a skilful, patient and resourceful artist.

The disarticulating and analysing action of the sun's rays is not accomplished without the selection and expenditure of energy. Accordingly we observe that one grain protects another. The disarticulation into separate grains, although very thorough near the surface of a glacier, does not penetrate far. A stroke or two with an ice-axe reveals the fresh blue ice. The analysis of the individual grain into crystallographically oriented laminæ can be particularly well studied in the Mergelin See. It is only the grains that are exposed to the sky, and above water, that are so analysed; and prolonged exposure of this kind reduces a grain to the last stage of dilapidation. The grains beneath the surface, *whether of ice or water*, are almost completely unattacked.

The importance, or rather the necessity, of direct skylight for the disarticulation of glacier-ice into its constituent grains is very well seen in the artificial grottos which are maintained at easily accessible parts of most popular glaciers. The thickness of the layer of completely disarticulated ice is so small that it is hardly noticed, and the whole grotto appears to be cut out of pure blue ice. If the observer, on penetrating for a few paces, turns round and looks outwards, he sees the surface of the ice-walls of the grotto etched with strange line-figures. These are most strongly marked near the opening, and they cease exactly at the spot where the last ray of direct sky-light strikes the ice. The lines so developed are formed by the intersection of the surface of the ice-wall of the cave with the separating surfaces of contiguous ice-grains. The photographic picture thus presented is one of very great interest.

It is only perfectly pure water, received directly as it flows from the still, that can be frozen into homogeneous glass-like ice. All natural ice proceeds from impure water.

In lake-ice of moderate thickness the crystalline axis is perpendicular to the surface of the lake. Consequently, Tyndall's planes of fusion are parallel to this surface. When exposed to a powerful sun, and with an air temperature even much below 0° C., the ice weathers into horizontal laminæ separated by Tyndall's planes of fusion, and into vertical columns. The column in lake-ice and the grain in glacier-ice are homologous features. They express the form which the individual crystal takes in these different varieties of natural ice.

Were it not for the fact that a glacier is made up of distinct grains of ice, and that this substance has the property of melting and freezing at different temperatures, according to the composition of the water with which it comes in contact and to the pressure to which it is subjected, there is little doubt that a glacier would be as motionless as any other mass of crystalline rock.

No. 6. [*From the Proceedings of the Royal Institution of Great Britain*, 1909, *Vol. XIX, p.* 243.]

ICE AND ITS NATURAL HISTORY

In a single lecture it will be impossible for me to do more than deal with those points in the natural history of ice which have come under my own notice, and which I have made the subject of special investigation.

The Nature of the Ice formed by Freezing Saline Solutions.— During the Antarctic cruise of the "Challenger," the old question arose as to whether the salt, which is always found in the water produced by melting sea-ice, was present in the solid state as part of the crystalline ice, or in the liquid state as part of the adhering brine. It was one of considerable economical importance to whalers and other mariners who frequent Polar seas. It had been found that freshly frozen genuine sea-water ice was not drinkable. Genuine land-ice, which is easily recognised, gave, of course, good water. The great mass, however, of the ice found floating in Polar seas is of mixed origin. Thus, in its first stage it may be the ice formed by the primary freezing of sea-water. On this falls snow. When there is wind the sea breaks on the edge of the floating ice and throws salt spray over it, which freezes in due time. Hence a piece of sea-ice, picked up at random, is likely to be a very heterogeneous substance; and no two pieces can be expected to behave exactly alike. It is, therefore, not surprising that the reports of different navigators regarding the potability of the water formed by melting sea-ice differed. Some maintained that it was undrinkable; others held that it could be drunk if the first portions melted were rejected, and only the last fraction used.

In February 1874, when the "Challenger" reached her furthest south, much ice of all kinds was met with, and I took

the opportunity to make a study of the floating sea-ice.
I collected many samples, observed their melting temperatures, and determined the percentage of chlorine in the water
produced by their fusion.   The results obtained showed that
their melting temperature was very variable, and always below
0° C.   It was further observed that this temperature was the
lower the greater the percentage of chlorine found in the
melted ice.   The advantage of fractional melting for the
purpose of obtaining water of less salinity was confirmed ;
but it was found impossible, by any means, to produce pure
water by melting the ice.   This, combined with the fact that
its melting-point was considerably lower than that of pure
ice, was for me convincing evidence, at that date, that the salt
was present in it in the solid state, and that, consequently, the
crystalline body formed by freezing sea-water and similar
saline solutions was not pure ice.

About ten years later Dr Otto Pettersson's treatise[1] on
the properties of water and ice came before me, and I studied
it with great interest.   In it he refers to my "Challenger"
work, and rejects the view that "sea-ice is itself wholly destitute of salts, and only mechanically encloses a certain quantity
of unfrozen and concentrated sea-water."   I was much gratified
to find that he had, quite independently, arrived at the same
conclusion as I had.

In the careful study which I made of his work, the following passage, p. 318, arrested my attention : "A thermometer
immersed in a mixture of snow and sea-water, which is constantly stirred, indicates $-1°\cdot8$ C."   If this statement was
exact, it was clear that the evidence furnished by the melting
temperature of the sea-ice was not entitled to the weight
which I had attached to it, and that the conclusion, at which
we had both independently arrived, was open to doubt.   On
making the experiment, I was able to confirm his statement.
I thereupon decided to proceed without delay to investigate
the subject experimentally.

The question at issue concerned saline solutions generally,

---

[1] *On the Properties of Water and Ice*, by Dr Otto Pettersson: Publications of
the "Vega" expedition, 1883.

and, in the research, which I undertook in 1886, solutions of single salts were considered in the first line, and sea-water was included as a particular case of a composite saline solution[1].

The principle which guided the research was the following: if the crystalline body which is formed when a non-saturated saline solution is partially frozen is pure ice, then pure ice of independent origin, such as snow, must, when mixed with the same saline solution, and heat is supplied, melt at the same temperature, when the concentration is the same.

The result of the research was definitively to establish the validity of this principle on experimental evidence.

It was found that when a non-saturated saline solution is gradually frozen, certain crystals, which we call ice-crystals, separate out; and, during this process, the temperature of the mixture gradually falls, while its concentration increases. When this mixture was warmed, the crystals gradually melted, the temperature rose, and the concentration of the solution diminished. When, in the process of cooling and freezing, the temperature had fallen to a certain point, say t, and the salinity to s, it was found, when this process was reversed, and the same temperature t was reached during the process of warming and melting, that the solution had regained the same salinity s. Therefore, the substance which forms the ice-crystals which separate out at a temperature t during cooling, melts again at the same temperature during warming; and the concentration of the solution at that temperature is the same whether the temperature is arrived at by cooling or warming.

When a solution of the same salt, having a higher con-

---

[1] The results of the research which I began in the year 1886 were communicated to the Royal Society of Edinburgh in a paper on 'Ice and Brines,' which was read on March 21, 1887, and was published in the *Proceedings* of the Society, vol. xiv. pp. 129–149. A full account of it was also published in *Nature*, 1887, vol. xxxv. p. 608, and vol. xxxvi. p. 9. (See above, p. 130.) The whole subject of the influence of dissolved salt on the state of aggregation of the substance water at temperatures below its normal freezing and melting point, and above its normal boiling and condensing point was passed in review in my 'Chemical and Physical Notes' in the *Antarctic Manual*, 1901, pp. 73–108 (above, pp. 44–68).

centration than that which was used in the previous experiment, was cooled to nearly 0° C., and was then mixed with sufficient freshly fallen snow to form a sludge, and heat was supplied, there arrived a moment in the course of the melting when the temperature of the mixture had returned to t, the concentration of the brine was then found to be represented by the same salinity s, which had been found to correspond to the temperature t in the previous operation. Therefore, the snow melted in exactly the same way as did the ice-crystals which had been formed by the freezing of the solution itself, and at the same temperature for the same concentration.

But it was only a question whether the salt was in the ice or in the brine. There is no salt in snow; yet it behaved in the saline solution in the same way as the crystals formed by freezing that solution; **therefore, the crystals formed by freezing the saline solution must be equally free from salt with the snow.**

It was thus proved that the crystals formed in freezing a non-saturated saline solution are pure ice, and that the salt from which they cannot be freed does belong to the adhering brine.

After the main principle had been established, the determination of the temperature at which snow or comminuted ice melts in saline solutions was substituted for the determination of their freezing-points; and in the case of many of the common salts and acids, even at high concentrations, yielded results which agreed with those previously determined by others by the more laborious freezing method.

The evidence on which Dr Pettersson arrived at the conclusion that the ice, formed on freezing saline solutions, contains salt otherwise than as mechanically enclosed brine, was furnished by his analyses of sea-water and of melted sea-ice, brought home from the Arctic regions. In these samples he found that the ratio $Cl : SO_3$ in the samples of sea-water was, as was to be expected, almost constant, while, in those of the sea-ice, it varied within very wide limits; and he justly observes that, if it is claimed that the ice of sea-water is salt

only in virtue of adhering sea-water of slightly greater concentration, the saline contents of the water adhering to the
ice-crystals must have the same composition as that of the
water before freezing. Having found that this was not the
case, he was apparently justified in concluding that the act of
freezing in sea-water has a selective or distributive effect on
its saline ingredients. But, in freezing, the solution is separated only into two parts, the crystals and the mother-liquor.
If there is found to be a deficiency of a saline ingredient in
the mother-liquor, this can be due only to its transference
to the crystals. Therefore, the crystals must contain salt
solidly.

This reasoning is perfectly valid if we admit what is
tacitly assumed, that the samples of ice analysed were produced by the *primary* freezing of sea-water of the composition
of the samples of unaltered sea-water, which were analysed
at the same time. But it was perfectly evident to me that
this was not the case, especially as regarded the samples
brought home by the "Vega" which were characteristic
samples of the ice encountered by a ship having her Arctic
experience. From the records of this expedition, which agree
with those of all other Arctic voyages, primary sea-ice is to
be seen only at the beginning of winter, or the commence-
ment of freezing. Almost before the ice is a day old secondary
changes take place and become more and more pronounced
as the season advances and the ice thickens. An excellent
idea of the nature of these changes is furnished by Dr Pettersson's investigations and by those of Dr Karl Weyprecht[1].

But the question of the nature of the ice formed by
freezing non-saturated saline solutions can be solved only by
the phenomena attending their primary congelation, where
all secondary complications are excluded. I therefore made
a number of experiments in which different samples of sea-
water, in sufficient quantity (300 grms.), were frozen in a bath
having a temperature between two and three degrees lower
than the probable freezing-point of the sea-water. Freezing
was continued until about one-third of the solution had passed

---

[1] *Die Metamorphosen des Polareises*, Wien, 1879.

into the crystalline state. The mother-liquor was drawn off, as perfectly as possible, before removing the beaker, in which the freezing had been effected, from the bath. The crystals were transferred quickly to a funnel, and drained by means of the jet pump. The chlorine and sulphuric acid were then determined in each fraction, and the ratio Cl : SO$_3$ calculated (Cl = 100). As an example, the following values of this ratio were obtained for water from the Firth of Forth:—In the original water 11·83, in the mother-liquor 11·67, and in the crystals 11·62. In water from the Firth of Clyde the ratios were: in the original water 11·58, in the mother-liquor 11·57, and in the crystals 11·67. These ratios agree with each other as closely as the analytical possibilities permit.

Inasmuch as the chlorides and sulphates together make up more than 99 per cent. of the solid contents of sea-water, the constancy of these ratios makes it so probable as to be certain that the ice of sea-water is salt only in virtue of adhering sea-water of slightly increased concentration.

It follows, therefore, that the evidence furnished by the quantitative freezing of a composite saline solution confirms the conclusion arrived at on the basis of the identity of the temperature at which a saline solution freezes, with that at which ice melts in it; namely, that the crystalline body formed by freezing a non-saturated saline solution is pure ice.

It was not until after this had been established, in 1887, that it became legitimate to say: " The freezing-point of *water* is lowered by the presence of salt dissolved in it," instead of saying : " The freezing-point of a *saline solution* is so much lower than that of pure water." The former of these statements expresses the fundamental principle of cryometric chemistry.

Distinction between the Melting-Point of a Substance and the Temperature at which it melts under given conditions.—An important consequence of this research was that the melting temperature of a body is not necessarily identical with the temperature at which it melts under particular circumstances.

I define the freezing-point of a substance to be :—The temperature at which it, as a liquid, passes into itself as a solid; and its melting-point to be the temperature at which it, as a solid, passes into itself as a liquid.

Under "substance" I understand a single substance, completely defined as a chemical individual. A good example of what I mean is afforded by the substance, of which eighteen parts, by weight, consist of two parts of hydrogen and sixteen parts of oxygen, which have combined with the liberation of a quantity of heat sufficient to raise the temperature of the product by some 2000° C. In chemistry this substance is expressed by the symbol H_2O. In the solid state it is called ice, in the liquid state water, and in the gaseous state steam.

If we have a quantity of this substance, partly in the solid and partly in the liquid state, and in such conditions that, if ever so little heat be removed from the mixture, the quantity of ice is increased, and, if ever so little heat be added to the mixture, the quantity of ice is diminished and that of the water correspondingly increased, the temperature of the mixture is the freezing and melting temperature of the substance H_2O.

When ice is melting in a mixture of ice and water, immersed in a melting-bath, the temperature of the water must be a little higher than that of the ice, else there would be no inducement for heat to pass from the water to the ice; similarly, when ice is being formed in a mixture of ice and water, immersed in a freezing-bath, the temperature of the water must be a little lower than that of the ice produced. Therefore, the observed freezing temperature of water, and melting temperature of ice must always be different, if quite exactly determined. But it is found that, by supplying heat at as low a temperature as possible in the melting experiment, and removing it at as high a temperature as possible in the freezing experiment, the observed melting and freezing temperatures of the substance H_2O approach each other more and more closely. Therefore, as a matter of experiment alone, it is legitimate to conclude that the limiting values of these temperatures are identical. It is called 0° on the thermometric scales of Réaumur and Celsius, and 32° on that of Fahrenheit.

This temperature was for long held to be invariable; indeed, it is little more than half a century since it was established that it is lowered by increase of pressure and raised by relief of the same, the quantitative proportion being that of $1°$ C. to 135 atmospheres.

In the above definition of the freezing and melting temperature of H_2O, that substance is specified as being present partly in the solid and partly in the liquid state. **The temperature at which ice begins to take form in water which is cooled when in contact only with itself, or with a solid other than ice, has not been determined, and is in fact uncertain.** The moment the smallest particle of ice is present, the water has the opportunity of **passing, as a liquid, into itself as a solid**: but not till then.

Evidence of the uncertainty which exists regarding the temperature at which ice begins to form in water, when it is cooled in contact only with a solid other than ice, is furnished by the wet-bulb thermometer when it is being prepared for use at temperatures below $0°$ C., by freezing on it the quantity of water which is supported, against gravity, by the **perfectly clean** bulb. When this is rotated in air of $-10°$ to $-20°$ C. ice never begins to form until the temperature of the bulb of the thermometer has fallen to $-2°$ or $-3°$ C., and rarely before it has fallen to $-4°$ C. In many cases I have observed it fall to temperatures as low as $-8°$ or $-9°$ C.; and in such cases, when freezing begins, the whole drop of water is frozen without its being able, by the liberation of latent heat alone, to raise the temperature of the bulb of the thermometer to $0°$ C.

In our definition of the freezing and melting temperature of H_2O, no substance is specified except H_2O. But H_2O, that is, absolutely pure water, rarely, if ever, occurs in nature. Therefore our definition is not directly applicable to water as it occurs in nature. It has been found by experiment that when, in a mixture of ice and water, the water contains ever so little foreign, especially saline, matter in solution, the temperature at which it freezes, and that at which pure ice melts in it, is lowered. It is therefore probable that **in Nature,**

ice never melts and water never freezes exactly at 0° C. In fact the temperature at which ice melts in nature depends on the medium in which it melts as well as on the pressure to which it is subjected. **If the pressure is constant, it varies with the nature of the medium; and, if the nature of the medium is constant, it varies with the pressure.**

The effect produced by both these agencies is the same in kind: each, by its presence, induces the melting of ice and the freezing of water at a lower temperature than would be the case in its absence. In the case of dissolved salt this inducing power is active only at temperatures lying between 0° C. and the cryohydric temperature of the salt. Between these temperatures the solid salt, when exposed to an atmosphere which is not perfectly dry, is deliquescent (*Antarctic Manual*, p. 81). Below the cryohydric temperature ice and salt are indifferent to each other. Of the two agencies the one which has the more potent influence on the natural history of ice is the nature of the medium in which it freezes or melts.

The Influence of Salt in inducing the Melting of Ice at Temperatures between 0° *C. and its Cryohydric Point furnishes a quantitative explanation of observed Anomalies in its Physical Constants.*—This influence furnishes a simple and natural explanation of the anomalies so often noticed in the physical behaviour of ice. Thus the belief that ice, at temperatures near 0° C., does not contract but expands on being cooled, has been maintained by such experienced observers as Hugi and Petzold besides Pettersson. It is impossible to arrive at this conclusion without close and accurate observation. The observations in each case were perfectly exact, but the interpretation of them was faulty.

When due weight is given to the influence of the medium, the anomaly disappears, and it is found that ice does not behave in the capricious way supposed, but conforms to the usual custom by expanding when warmed and contracting when cooled. The disturbing agency is the impurity which is present in even the purest water. This is excluded from the ice in the process of freezing, and remains in solution in the residual water, which becomes more and more concentrated as the freezing proceeds. The amount of such solution, which remains liquid at any time, depends on the temperature of the ice and liquid, and the concentration of the liquid or solution. When these are given, and the nature of the dissolved impurity is known, the amount of liquid present and the consequent contraction follow necessarily.

In order to illustrate this, it is necessary to select some substance as representative impurity. Chloride of sodium has been chosen, because it is the most widely disseminated and the best studied ingredient of natural waters.

The following extract from my paper on 'Ice and Brines' explains this in detail.

"All natural waters, including rain-water, contain some foreign and usually saline ingredients. If we take chloride of sodium as the type of such ingredients, and suppose a water to contain a quantity of this salt equivalent to one part by weight of chlorine in a million parts of water, then we should have a solution containing 0·0001 per cent. of chlorine, and it would begin to freeze and to deposit pure ice at a temperature of −0°·0001 C.; and it would continue to do so until, say, 999,000 parts of water had been deposited as ice. There would then remain 1000 parts of residual water, which would retain the salt, and would contain, therefore, 0·1 per cent. of chlorine, and would not freeze until the temperature had fallen to −0°·1 C. This water would then deposit ice at temperatures becoming progressively lower, until, when 900 more parts of ice had been deposited, we should have 100 parts residual water, or brine as it might now be called, containing 1 per cent. of chlorine, and remaining liquid at temperatures above −1°·0 C. When 90 more parts of ice had been deposited, we should have 10 parts of concentrated brine containing 10 per cent. chlorine and remaining liquid as low as −13° C. In the case imagined, we assume the saline contents to consist of NaCl only, and with further concentration the cryohydrate would no doubt separate out and the mass become really solid. On reversing the operation, that is, warming the ice just formed, we should, when the temperature had risen to about −13° C., have 999,990 parts ice and 10 brine containing 10 per cent. chlorine. Now, owing to the remarkable fact that pure ice, in contact with a saline solution, melts at a temperature which depends on the nature and the amount of the salt in the solution, and is identical with the temperature at which ice separates from a solution of the same composition on cooling, the brine liquefies more and more ice at progressively rising temperatures, until, as before, when the temperature of the mass has risen to −0°·1 C. it consists of 999,000 parts of ice and 1000 parts of liquid water containing 1 part of chlorine. The remainder of the ice will melt at a temperature gradually rising from $-0°·1$ to $0°·0$ C.

"The consideration of this example furnishes an easy explanation of the anomalous behaviour of ice formed from anything but the very purest distilled water, in the neighbourhood of its melting-point. This subject has been studied with great care and thoroughness by Pettersson. The apparent expansion of all but the very purest ice, when cooled below $0°$ C., is ascribed by him in part to solid saline contents of the ice, which exercise a disturbing and unexplained influence on its physical properties. Viewed in the light of the fact that the presence of even the smallest quantity of saline matter in solution prevents the formation of ice at $0°$ C., and promotes its liquefaction at temperatures below $0°$ C., we see that this apparent expansion of the ice on cooling is probably due to the fact that we are dealing, not with homogeneous solid ice, but with a mixture of ice and saline solutions. As the temperature falls this solution deposits more and more ice, and its volume increases. But the increase of volume is due to the formation of ice out of water, and not to the expansion of a crystalline solid already formed.

TABLE IX.—*Water containing 7 parts Cl in* 1,000,000.

Temp. °C.	Water Frozen c.c.	Ice Formed c.c.	Brine remaining c.c.	Ice and Brine c.c.	Pettersson III. Vol. of Ice at $T°$ c.c.	Difference c.c.
T	V_1	v_1	V_2	v_2	P	$P - v_2$
$-0·07$	99,000	107,979	1000	108,979	108,980	1
$-0·10$	99,300	108,306	700	109,006	109,007	1
$-0·15$	99,533	108,561	467	109,028	109,038	10
$-0·20$	99,650	108,687	350	109,037	109,048	11
$-0·40$	99,825	108,879	175	109,054	109,057	3

"In Table IX are given the volumes occupied by the ice (with enclosed brine) formed by freezing 100,000 c.c. (at $0°$ C.) of a water containing chloride of sodium equivalent to 7 grams chlorine in 1,000,000 c.c. (at $0°$ C.).

"The volume (v_2) of the ice and brine formed on freezing this water is compared with (P) that observed by Pettersson in freezing a sample of the distilled water in ordinary use in the laboratory. It will be seen that the volumes observed agree very closely with those calculated for a water containing 7 parts of chlorine in a million, on the assumption that the saline matter is contained entirely in adhering liquid brine."

16—2

In the same paper (above, p. 148), the very low values found by Dr Pettersson for the latent heat of fusion of different samples of ice prepared by freezing sea-water, were quantitatively explained on the basis of the above law; and the conclusion was experimentally confirmed by the thermal exchanges which were observed to take place in the freezing of a sample of sea-water, as the result of which chemical analysis showed that the ratio $Cl : SO_3$ was the same in the original water, in the ice formed, and in the brine remaining.

I give these quotations from my paper of 1887 at length because, though published so many years ago, they appear to have been but little read. I know of no manual of physical chemistry in which the above demonstration of the fundamental fact of cryoscopic chemistry is given, nor am I acquainted with any recent treatise on natural ice in which the influence of the nature of the medium, in which it melts, on the melting temperature of ice at constant pressure is even mentioned, still less taken into account.

For these reasons, in preparing the lecture, I have given the greatest space to this all-important subject, and I proceed with its development.

In Table I we have, under v, the volume, in cubic centimetres, of ice, the melting of which is induced at the temperature t by the presence of 1·5105 gram chloride of sodium. The values of v are derived from determinations of the freezing-point of solutions of chloride of sodium. Under $w\,(= 0.9167v)$, we have the volume of water so produced, and under $c\,(= v - w)$, the contraction due to the melting. The cryohydric temperature of solution of chloride of sodium is taken as $-21°·72$, and its concentration as 29·97 grams salt in 100 grams water.

The coefficient of cubic dilatation by heat of pure ice is taken as 0·00016, and it is assumed to be constant at the temperatures under consideration. The specific gravity of pure ice at t referred to that of water at the same temperature, is taken as 0·9167. The volume of the salt diffused through the ice is disregarded.

Using these constants, and those of Table I, we will apply

the principle to the discussion of the apparent variations of volume of a block of ice, the volume of which at 0° C. is 1000 c.c. It contains diffused through it 1·5105 gram NaCl, which we assume to be provisionally in the *inert* state, in which it is deprived of the power to induce the melting of ice at temperatures between 0° C. and − 21°·72 C. Let the temperature of the block containing the inert NaCl be reduced to − 23° C.; its volume will be reduced to 996·320 c.c., and as the temperature is below the cryohydric temperature, the salt is by nature inert; at such temperatures ice and common salt are indifferent to each other. Let the temperature of the block of ice be now raised to − 22°; the salt remains inert, and the volume of the ice increases to 996·48 c.c. If the temperature is further increased to − 21°·721, the NaCl will still remain inert, and the volume of the ice will become 996·525 c.c.

If the heating is continued the temperature rises exactly to the cryohydric point, − 21°·72, at which temperature the indifference of chloride of sodium to ice ceases, and induced melting at that temperature takes place. It will then be observed that the temperature remains constant for a time, while the volume of the block diminishes. When the temperature begins to rise, the volume of ice melted will be 5·498 c.c. As this produces 5·040 c.c. water, the diminution of volume is 0·458 c.c., and the apparent volume of the block is 996·067 c.c.

Let us now go back to the initial state, in which we have the block of 1000 c.c. ice, containing 1·5105 gram inert NaCl diffused through it, at the temperature 0° C. Let the temperature be reduced to − 21° C., the ice remaining inert. The volume of the ice will then be 996·64 c.c. Let the NaCl recover its activity, it will melt 5·629 c.c. ice, producing 5·160 c.c. water under a contraction of 0·469 c.c., so that the apparent volume of the ice at − 21° C. is

$$996·64 − 0·469 = 996·171 \text{ c.c.}$$

By the aid of Table I and the other constants we can calculate the composition of a block of ice of any weight or volume, which contains, diffused through it, the constant quantity 1·5105 gram NaCl. The results of such a calculation are given in Table II for a block of ice having the volume 1000 c.c. at 0°, the salt being supposed *inert*.

TABLE I.

t	v	w	$c = v - w$	t	v	w	$c = v - w$
° C.	c.c.	c.c.	c.c.	° C.	c.c.	c.c.	c.c.
− 21·72	5·498	5·040	0·458	− 10	10·437	9·567	0·870
− 21	5·629	5·160	·469	− 9	11·327	10·373	0·954
− 20	5·825	5·340	·485	− 8	12·532	11·488	1·044
− 19	6·035	5·532	·503	− 7	14·195	13·012	1·183
− 18	6·284	5·760	·524	− 6	16·406	15·039	1·367
− 17	6·598	6·048	·550	− 5	19·538	17·910	1·628
− 16	6·978	6·396	·582	− 4	23·946	21·951	1·995
− 15	7·409	6·792	·617	− 3	31·685	29·045	2·640
− 14	7·858	7·203	·655	− 2	48·352	44·323	4·029
− 13	8·372	7·674	·698	− 1	98·352	90·156	8·196
− 12	8·964	8·217	·747	− 0·5	198·350	181·821	16·529
− 11	9·648	8·844	·804				

TABLE II.—1000 *c.c.* Ice + 1·5105 *gram NaCl.*

t	V	v	w	c	U	H
° C.	c.c.	c.c.	c.c.	c.c.	c.c.	c.c.
− 25	996·000	...	...	...	996·000	996·000
− 22	996·480	...	...	...	996·480	996·480
− 21·721	996·525	...	...	...	996·520	996·520
− 21·72	996·520	5·498	5·040	0·458	996·067	991·022
− 21	996·640	5·629	5·162	0·467	996·171	991·011
− 20	996·800	5·825	5·340	0·485	996·315	990·975
− 15	997·600	7·409	6·792	0·617	996·983	990·191
− 10	998·400	10·437	9·567	0·870	997·530	987·963
− 8	998·720	12·532	11·488	1·044	997·676	986·188
− 7·2	998·848	13·884	12·727	1·157	997·691	984·964
− 7·0	998·880	14·195	13·012	1·183	997·697	984·685
− 6·8	998·912	14·628	13·409	1·219	997·693	984·284
− 6·0	999·040	16·406	15·039	1·367	997·673	982·634
− 4	999·360	23·946	21·951	1·995	997·365	975·414
− 2	999·680	48·352	44·323	4·029	995·651	951·328
− 1	999·840	98·352	90·156	8·196	991·644	901·488
− 0·1	999·984	998·350	915·154	83·196	916·788	1·634
0	1000	1000	916·7	83·3	916·7	0

Proceeding by steps in this way we obtain the numbers to be found under U in Table II. In it $V[= V_0(1 - 0·00016t)]$

represents the volume of pure ice at t, the salt being inert;
v, w and c have the same signification as in Table I.
$H(=V-v)$ represents the volume of pure ice included in U.
It will be seen that H, the volume of real ice present at t,
suffers a sudden diminution at $-21°\cdot72$, and then continues

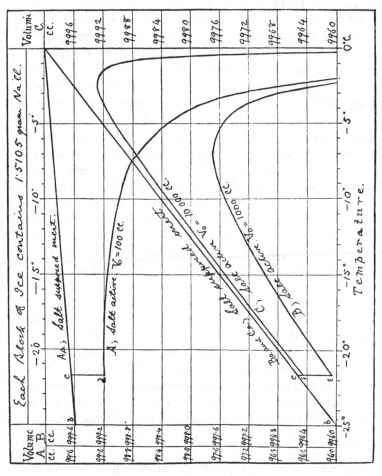

Fig. 1. The Thermal Expansion of Ice containing 1·5105 gram NaCl.

to diminish, at first slowly, and then at an increasing rate,
until liquefaction is complete.

In Fig. 1, curve B represents the variation of apparent
volume U with temperature t, the salt being active; while

the straight line Ba represents the variation of V with t, the salt being inert. The line Ba represents the dilatation of pure ice, on the basis of a constant coefficient of dilatation 0·00016. As chloride of sodium is, by nature, inert at temperatures below the cryohydric point, the straight line bc is part of the curve B, as well as of Ba. Between $-25°$ and $-21°·72$ the ice expands uniformly at the same rate, whether it is pure or contains salt. We have in c the first singular point of the curve, and it occurs in all the curves of expansion of ice contaminated by chloride of sodium. When the temperature of the ice rises to $-21°·72$, the inertness of the salt is exchanged for activity; and, if the requisite supply of heat is available, 5·498 c.c. of ice are melted, producing 5·040 c.c. water, under a contraction of 0·458 c.c. While this amount of ice is melting, the temperature remains constant, but the volume U contracts. Graphically, this is represented by a straight line (ce) parallel to the axis of volume. Therefore the curve of volume of the ice between $-25°$ and the point where the temperature begins to rise above $-21°·72$ is represented by two straight lines bc and ce which meet each other in an acute angle at c. When the temperature rises above $-21°·72$, another, generally acute, angle is formed at e, so that this portion of the curve of volumes takes the form of the letter **Z**.

Between $-25°$ and $-21°·721$ the coefficient of dilatation is 0·00016. At the point c, corresponding in Table II to the temperature $-21°·721$, the cryohydric point has been reached on a rising gradient, but no melting has taken place. Melting is supposed to begin only when the temperature has reached $-21°·72$ exactly. Then there is a contraction of 0·458 c.c. with no change of temperature, so that at this stage the coefficient of dilatation is $-\dfrac{0·458}{0} = -\infty$. When the temperature rises above $-21°·72$, the apparent volume U increases with the temperature, but at a gradually diminishing rate until, at $-7°·0$, the increase of volume due to simple expansion of the ice is exactly balanced by the contraction due to induced melting. At this temperature the coefficient of expansion changes sign, and between $-7°·0$ and $-0°·1$, at which the ice has practically all melted, the coefficient of expansion is negative. **Therefore the coefficient of thermal expansion of this ice changes sign three times when it is warmed from a temperature below the cryohydric point of solution of chloride of sodium to that at which liquefaction is complete.**

Curves C and Ca refer to ice of which 9167 grams or 10,000 c.c. at $0°$ C. contain 1·5105 gram NaCl. Above the Z-like portion the curve shows a maximum volume of 9992·834 c.c. at $-2°·3$. At temperatures higher than this the coefficient of apparent expansion is negative.

In the same figure curves A and Aa represent the volumes of a block of ice weighing 91·67 grams and containing 1·5105 gram NaCl diffused through it. The line Aa represents its volume V at temperature t, the salt being inert; at $0°$ the volume is 100 c.c. The curve A, after the Z-like portion, shows a maximum volume at $-20°·5$, above which its apparent volume diminishes, as the temperature rises, and at $-1°·0$ the ice is practically all melted.

When 1·5105 gram NaCl is diffused in a block of ice weighing 88 grams, then the volume of the ice, the salt being inert, is 96 c.c. at $0°$; and when the salt is active, the maximum volume is at the cryohydric temperature, so that from $-21°·72$, the apparent volume U diminishes as the temperature rises. **Therefore for blocks of ice which contain, per 100 parts by weight of ice, less than 29·97, and more than 1·7164 parts of NaCl, the coefficient of apparent expansion is negative at all temperatures above $-21°·72$.**

In Fig. 2 we have curves D and Da, and E and Ea. The same constant quantity of chloride of sodium, 1·5105 gram, is diffused, in the case of D, in 1 cubic metre, and, in that of E, in 10 cubic metres of ice. When melted these blocks of ice would furnish waters containing chlorine in the proportion of 1 gram to 1 ton of water in D and to 10 tons of water in E. The critical temperature at which the coefficient of expansion changes sign is $-0°·2275$ in D and $-0°·0725$ in E.

In Table III we have the upper critical temperature (τ) at which the coefficient of apparent dilatation changes sign for blocks of ice having volumes ranging from 100 cubic

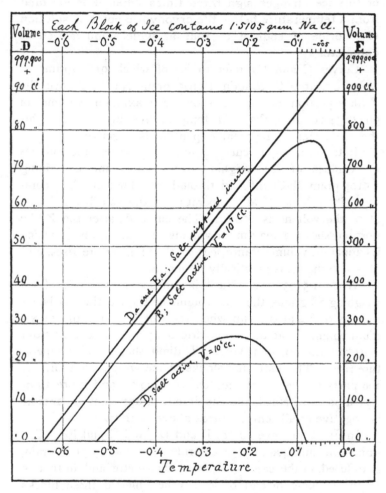

Fig. 2. The Thermal Expansion of Ice containing 1·5105 gram NaCl.

centimetres to 100 cubic metres, each containing 1·5105 gram NaCl. Under V_0 we have the initial volume of the block of ice supposed pure and solid at 0° C., and under v the volume

of ice which can be melted under the inducing influence of
1·5105 gram of chloride of sodium at the critical temperature τ, at which the apparent coefficient of cubic expansion of
the ice is equal to o.

A block of 100 c.c. of ice, which contains 1·5105 gram of
NaCl diffused through it, furnishes on being melted 91·67 c.c.
of water, which contain 0·9167 gram of chlorine, dissolved in
it as chloride of sodium. This water contains chlorine in the
proportion 1 gram to 100 grams of water, and represents a
concentration about one-half that of average sea-water. When
the volume of ice, V_0, is 1 cubic metre, the water produced by
its melting contains chlorine in the proportion of one part to
one million parts of water by weight.

TABLE III.

V_0	v	τ	V_0	v	τ	V_0	v	τ
c.c.	c.c.	°C.	c.c.	c.c.	°C.	c.m.	c.c.	°C.
100	5·73	− 20·5	1000	14·20	− 7·0	0·01	41·83	− 2·3
200	6·74	− 16·6	2000	20·00	− 4·9	0·1	136·3	− 0·725
400	9·85	− 10·75	4000	27·80	− 3·5	1·0	438	− 0·2275
600	11·75	− 8·65	6000	32·24	− 2·95	10·0	1377	− 0·0725
800	12·85	− 7·8	8000	37·57	− 2·55	100·0	4306	− 0·02275

Waters which contain dissolved matter equivalent to no
more than 1 gram of chlorine in 10,000 grams of water are
in the category of ordinary fresh waters, and we see that the
critical temperature of ice which furnishes such water lies as
low as − 2°·3. When the dissolved matter is equivalent only
to 1 gram of chlorine in 100,000 grams of water, the critical
temperature is − 0°·725. The other waters are in the category of distilled waters, and it is doubtful if, by any chemical
means whatever, we could determine as little dissolved matter
as 1 gram chlorine in 1 ton of water; yet the critical temperature of such ice lies nearly a quarter of a degree below
the melting temperature of pure ice. The critical temperature
of expansion of ice affords a means of detecting impurity
equivalent to quantities of chlorine as small as 1 gram in
10 tons, and even 1 gram in 100 tons of water.

In his work on *The Properties of Water and Ice*, Pettersson gives in Plate XXI the curves of the change of volume with temperature, of three samples of ice frozen from samples of melted sea-ice. They are numbered in descending order of concentration, VI, V and IV. For the details, which are interesting, the reader must consult the original work. The nature of these samples is roughly indicated by their specific gravity at 0° C., referred to that of distilled water at 4° C., and by the percentage of chlorine in them.

No.	VI	V	IV
Specific gravity	1·00940	1·00539	1·00030
Per cent. chlorine	0·649	0·273	0·014

I calculated by the method which I have above developed, the volume of ice at 0°, containing 1·5105 gram chloride of sodium, which would show the same volumetric behaviour as these samples. The results are given in Table IV.

Pettersson's volumes (P) for certain whole degrees of temperature (t) are taken. The water which agreed with the volumetric behaviour of each of Pettersson's samples was found empirically, and the volume of the ice at each temperature, t, for each sample calculated as in Table II. The temperature, t, at which Pettersson observed the volume to be a maximum was the chief guide. The volumes obtained by my method are found in line U. I then took my volume and Pettersson's volume for the temperature at or near the maximum of volume, and obtained a factor by which I reduced my volumes at the other temperatures to the same denomination as those of Pettersson. These are given in line B, those of Pettersson being given in line P. It will be seen that the agreement between calculation and observation is very good.

Cryoscopic Equivalence between Pressure and Salinity.— We have seen that the freezing-point of water and melting-point of ice is lowered by increase of pressure and by addition of salt in solution. The effect produced by these agencies is the same in kind.

When water, which holds salt in solution, freezes at a temperature below 0° C., the salt forms no part of the ice produced. But it is the cause of the lowering of the freezing-point, because, if it is removed, the freezing-point reverts to the normal. The depressed melting-point of the ice depends for its persistence on its continued contact with the saline solution from which it sprang.

When water, which is subjected to any pressure, freezes

TABLE IV.

t	$-18°$	$-14°$	$-12°$	$-10°$	$-8°$	$-6°$	$-3°$
			Sea-ice VI. $V_0 = 145$ c.c.				
U	144·058	144·020	143·975	143·898	143·770	143·494	...
B	...	108,363	108,330	108,272	108,175	107,968	...
P	108,383	108,358	108,330	108,261	108,148	107,923	...
$B-P$	...	5	0	11	27	45	...
			Sea-ice V. $V_0 = 250$ c.c.				
U	...	248·785	248·773	248·730	248·636	248·393	...
B	...	...	108,634	108,615	108,574	108,468	...
P	...	108,640	108,630	108,615	108,572	108,485	...
$B-P$	...	...	4	0	2	-17	...
			Sea-ice IV. $V_0 = 8000$ c.c.				
U	...	7981·435	7983·893	7986·330	7988·716	7990·953	7993·520
B	...	108,683	108,717	108,750	108,782	108,813	...
P	...	108,695	108,729	108,761	108,790	108,815	108,847
$B-P$	...	-12	-12	-11	-12	-2	...

at a lower temperature than it does in a vacuum, the pressure, being immaterial, cannot form part of the ice. But it is the cause of the lowering of the freezing-point; because, on its removal, the freezing-point reverts to the normal. The depression of the freezing-point depends for its persistence on the continuance of the pressure to which the ice is exposed. We can explain its action only by describing it. It acts by **influence**. This influence, however, is subject to quantitative law which has been well investigated.

On the same grounds it seems to be legitimate to say that the salt in solution acts also by influence; and its influence is subject to quantitative law which has been well investigated.

As both influences are subject to known laws, we are able to ascertain the equivalence which exists between them.

Further, the pressure, being immaterial, is of one kind, while the salt, which is material, is of as many kinds as there are chemically individual substances soluble in water.

In considering the two agencies, we must compare a certain absolute pressure with a certain weight of a particular salt. In this discourse we have generally considered the constant quantity 1·5105 gram chloride of sodium. The generally accepted pressure which lowers the freezing-point of water by 1° C. is 135 atmospheres, which is here taken as equivalent to 140 kilograms per square centimetre (kg./cm.²). If we consider the surface of the water which is exposed to the pressure as 100 square centimetres, then the total pressure required is 14,000 kg. We therefore take as our constant absolute pressure 14,000 kg. When the distilled water exposes to it a surface of 100 cm.² its freezing temperature is lowered by 1° C. If the lowering of the freezing-point of water by pressure were simply proportional to the pressure, then, when the surface exposed to the pressure of 14,000 kg. is s cm.², the lowering of the freezing-point would be given by t in Table V. It results, however, from the experiments of Tammann[1] that proportionality of the lowering of the freezing-point of water to the pressure to which it is exposed, is of the same order as that of the lowering of the freezing-point of water to the amount of salt dissolved in it. In Table V, s' gives the surface (cm.²) of water which must be exposed to a pressure of 14,000 kg. in order that, according to Tammann, its freezing-point may be lowered by t degrees. The corresponding volumes of ice melted, at ordinary pressure, under the influence of 1·5105 gram NaCl, are given by v. To be strictly comparable with s and s', the numbers v should be

---

[1] 'Ueber die Grenzen des festen Zustandes,' von G. Tammann, *Annalen der Physik*, [4], ii. 1.

increased in the ratio 98·352 : 100; but this has been considered unnecessary.

It will be seen that the *surface* of the water exposed to the constant pressure, 14,000 kilograms, is roughly proportional to the *volume* of the ice exposed to the influence of the constant quantity of 1·5105 gram, or 0·0258 gram-molecule of chloride of sodium. The proportionality is the closer the greater the surface of the water compressed and the greater the volume of ice which contains the salt.

In comparing the power of inducing the freezing of water at temperatures below 0° C., possessed by a given absolute pressure and by a given mass of a particular salt, we see that the following law holds : **The surface of the water exposed**

TABLE V.

Lowering of freez- ing-point	t	1°	2°	5°	10°	15°	20°	21·72°
Surface of water exposed to pressure of 14,000 kilograms	s'	cm.² 100	cm.² 50·3	cm.² 20·05	cm.² 11·7	cm.² 8·3	cm.² 6·64	cm.² 6·13
	s	100	50	20	10	6·67	5·0	4·55
Volume of ice containing 1·5105 gram NaCl	v	c.c. 98·35	c.c. 48·35	c.c. 19·54	c.c. 10·44	c.c. 7·41	c.c. 5·82	c.c. 5·50

to the pressure and the volume of the water which holds the salt in solution are approximately proportional.

It was shown in my paper[1] on 'Steam and Brines,' that the elevation of the boiling-point of water by pressure, and by dissolved salt, follows the same law : **there is the same approximate proportionality between the surface which supports the pressure and the volume which holds the salt.**

Influence of Impurity on the Apparent Latent Heat of Ice.— This is illustrated by the numbers in Table I. Thus, at − 1° C., the apparent volume of the block of ice is 991·644 c.c.,

[1] 'On Steam and Brines,' by J. Y. Buchanan, F.R.S., *Transactions of the Royal Society of Edinburgh* (1899), xxxix. p. 549. (Above, p. 151.)

and it is made up of 901·49 c.c. ice and 90·154 c.c. water. When this is warmed to − 0°·1, we may take it that the whole of the ice is melted. Taking the latent heat per unit volume of ice as 66·5 at − 0°·1, and its specific heat per unit volume as 0·45, the heat required to raise the ice from − 1° to − 0°·1 is 365·1 gram-degrees (g.°); that required to raise the temperature of the water by the same amount is 81·14 g.°, and the heat required to melt the ice at − 0°·1 is 59,949 g.°, the total heat used being 60,395·2 g.°. If we ignore the possibility of partial melting, and assume that we have 999·84 c.c. solid ice at − 1°, and that its temperature is raised to 0°, at which temperature it melts, we have the following expenditure of heat: for rise of temperature 449·9 g.°, and for melting 66,489·3 g.°, making together 66,939·2 g.°, as against 60,391·5 g.°. If from 60,395·2 g.° we deduct the heat calculated for warming the ice in the second case, 449·9 g.°, we obtain 59.945·3 g.° as the heat required to melt 1000 c.c., or 916·7 grams, of ice at 0°, whence the latent heat would be, per unit volume, 59·94, and per unit weight 65·39.

This example illustrates also the effect of impurity on the apparent specific heat of ice.

It will thus be seen how powerful is the influence of medium on the behaviour of ice in the laboratory; it will now be shown that, in nature, this influence is equally powerful and much more far-reaching.

The nature of the medium is responsible, in the case of sea-ice, for depressions of freezing and melting temperature of 30, 40, and even more degrees of Celsius' thermometer, while the greatest pressure to which fresh-water ice is exposed in nature cannot well produce an alteration of freezing and melting-point amounting to as many hundredths of a degree.

The ice which surrounded the "Vega" during her winter's imprisonment in the Arctic Ocean had a pasty semi-liquid consistence, although the temperature of the air was at or below − 30° C. It remained stationary only because it was on a level surface. Had it been shovelled up on an inclined plane it would have quickly flowed down it until it reached the lowest level again. If we pick up a piece of ice floating in the Polar Sea, we know that it will prove to be very far from homogeneous. It may have a foundation of genuine primary sea-ice; but the ice forming the superstructure is sure to consist of snow, frozen spray, and very likely fragments of land-ice, all cemented together into a species of conglomerate. We have seen that when this is exposed to warmth it begins to melt at a temperature which may be one or two degrees below the melting-point of pure ice; and the liquid furnished by the melting is salt water. The further melting takes place in the ascending order of temperature; the salt ice of low melting-point disappearing first, and the purer ice melting later. We thus see how ice can be cemented by ice, just as metallic objects may be united by solder. In both cases the binding material differs from the objects united, chiefly in being more easily fusible.

If we have a number of cubes of pure ice, which fit each other exactly, and, if after being moistened with salt water they are exposed to frost, they will solidify to a single block. If this be exposed to the sun, the cementing salt ice will melt first, and, when it ceases to bind, the constituent cubes of pure ice will fall asunder, having themselves suffered practically no diminution due to melting.

The Glacier Grains.—Now this is precisely what happens when a block of sound glacier-ice is exposed to the rays of the sun for a short time; and it is one of the most striking and instructive experiments that can be made. Under the influence of the sun's rays, the binding material melts first, the continuity of the block is destroyed, the individual grains become loose and rattle if the block be shaken, and finally they fall into a heap. A block of glacier-ice is a geometrical curiosity. It consists of a number of solid bodies of different sizes and of quite irregular shapes, yet they fit into each other as exactly and fill space as completely as could the cubes above referred to.

The granular constitution of glacier-ice can nowhere be better studied than at the Mergelin See, the well-known lake which exists so long as the Aletsch glacier maintains a water-tight dam across the little side valley, which its waters occupy.

It is roughly triangular in shape. As the ice of the glacier is subject to more or less disintegration, there are always icebergs and small fragments of ice floating on its waters. The portions projecting above the water are exposed in summer often to a very powerful sun, and are loosened by the radiant heat into their constituent grains. A lump lifted out is found to consist of these disarticulated grains which rattle when shaken, and in a strong sun fall to pieces.

Size of Glacier Grains.—In August 1895 I made an extended study of the structure of glacier-ice, principally from the Aletsch glacier. The fragments of this glacier, which float as icebergs in the Mergelin See, are exposed to the powerful weathering influence of the summer sun, and are comparatively easily dissected into their constituent grains. A number of blocks were so dissected, in order to ascertain the weight and size of the largest grains. The following weights of single grains were determined: 700, 590, 450, 270, 255, 170, 155, and 100 grams. It was observed that blocks of ice contained grains of all sizes, which fitted each other so exactly that, in the fresh unweathered block, the whole volume was filled with ice.

The following table gives a summary of the results of dissecting seven blocks and weighing the grains.

The figures in the table do not give an exact statistical account of the blocks of ice. The smallest grains have most frequently escaped being weighed, therefore the average size of the grain comes out higher than the truth. On the other hand all were melting rapidly, so that the grains when on the balance weighed less than they did one hour before, and much less than they did the day before. The figures in the table give a general idea of the constitution or anatomy of a block of ice taken from the lower part of a large glacier. They are particularly interesting when we reflect that every grain, even the largest, has grown, according to the rigid laws of crystallomorphic development, from a single snow-crystal which probably weighed no more than one or two centigrammes.

The action of the sun's rays on glacier-ice is twofold; it

Number of Grains Weighed	Aggregate Weight of Grains	Average Weight of one grain	Number of Grains Weighed	Aggregate Weight of Grains	Average Weight of one grain
	A			*d*	
16	1045	65·3	7	400	57·1
10	110	11·0	10	30	3·0
4	25	6·25			
10	25	2·5	17	430	25·3
40	1205	30·1			
				E	
	B		22	1125	51·1
			10	150	15·0
11	1095	99·5	10	135	13·5
10	95	9·5	10	120	12·0
23	60	2·6	10	60	6·0
			10	50	5·0
44	1250	28·5	10	40	4·0
			10	20	2·0
	C		10	15	1·5
15	1365	91·0	102	1715	16·8
5	50	10·0			
10	25	2·5			
				F	
30	1440	48·0	34	5765	169·6
			10	190	19·0
	D		10	190	19·0
			10	140	14·0
6	1090	182·0	6	60	10·0
5	30	6·0	40	70	1·75
11	1120	102·0	110	6415	58·3

disarticulates the ice into its constituent grains, and it splits the individual grain into laminæ perpendicular to the principal axis of the crystal and bounded by the planes of fusion described by Tyndall. These planes are the distinguishing characteristic of the individual ice-grain.

Under the influence of radiant heat an ice-crystal begins to melt at the contiguous surfaces of these laminæ, and the process of disintegration and decay is directed by their plane. On the other hand, an ice-crystal, floating in water and losing heat, generates ice laminæ which are directed by the same planes, which form the continuation of the corresponding laminæ of the parent crystal. This was well observed at the end of August, 1895. Every night a thin skin of ice was formed at the shallow end of the lake, where the ice-blocks were collected. As the grains in a block of glacier-ice are distributed quite irregularly, the water line of a floating block necessarily cuts a great number of grains, all of which are oriented differently. The ice which was formed during the night along this line was oriented crystallographically by the grain with which it was in contact and from which it appeared to spring in continuation of its crystalline laminæ. This produces a remarkable pattern of lines on the surface of the lake-ice contiguous to a block of glacier-ice.

Tyndall has described and figured the minute features of the disintegration of the crystal under the absorption of radiant heat. Similar and complementary features are observed when ice is generated from an existing crystal under the dissipation of heat. To do justice to them, however, would require the services of a skilful, patient and resourceful artist.

The disarticulating and analysing action of the sun's rays is not accomplished without selection and the expenditure of energy. Accordingly we observe that one grain protects another. The disarticulation into separate grains, although very thorough near the surface of a glacier, does not penetrate far. A stroke or two with an ice-axe reveals the fresh blue ice. The analysis of the individual grain into crystallographically oriented laminæ can be particularly well studied in the Mergelin See. It is only the grains that are exposed to the sky, and above water, that are so analysed; and prolonged exposure of this kind reduces a grain to the last stage of dilapidation. The grains beneath the surface, *whether of ice or water*, are almost completely unattacked.

The importance of direct sky-light for the disarticulation of glacier-ice into its constituent grains is very well seen in the artificial grottos which are maintained at easily accessible parts of most popular glaciers. One of the galleries in the grotto of the Morteratsch glacier is represented in Fig. 3. It

Plate II

Fig. 3. A Gallery in the Grotto of the Morteratsch Glacier
in January 1907.

Plate III

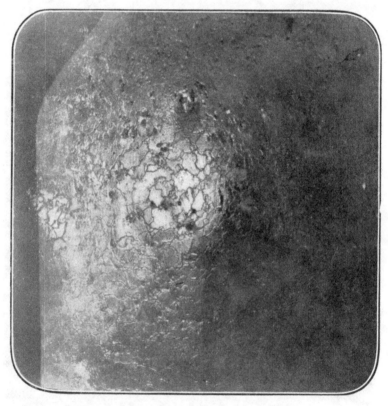

Fig. 4. Solar Etching of Ice in the Morteratsch Grotto
in September 1907.

is from a photograph which I took in January 1907. The hoar-frost on the roof and the sharp line where it ceases on the walls are well shown. The white object on the left is a stone working its way out through the ice. Many of these were visible in the body of the ice. The thickness of the layer of completely disarticulated ice is so small that it is hardly noticed, and the whole grotto appears to be cut out of pure blue ice. If the observer, on penetrating for a few paces, turns round and looks outwards, he sees the surface of the ice-walls of the grotto etched with strange linear figures. These are most strongly marked near the opening, and they extend as far as direct sky-light strikes the ice. The lines so developed are formed by the intersection of the surface of the ice-wall of the cave with the separating surfaces of contiguous ice-grains. The photographic picture thus presented is one of very great interest. The illustration, Fig. 4, shows this etching on a buttress in the grotto of the Morteratsch glacier, taken in September 1907.

After the autumnal equinox very little melting of ice takes place, and by the end of October it has, as a rule, ceased entirely. The etched figures on the walls of the entrance of the grotto, which were developed during summer, disappear quickly with the arrival of winter. But the winter brings with it another means of delineation of the grain which does not depend on solar radiation. Even at the lowest of winter temperatures, the atmosphere contains vapour of water, which it is prepared to relinquish under the same conditions as those under which dew is formed in summer. In the Alpine winter, however, it is deposited not as dew but as rime, that is, not as water but as ice. It is well known that very fine etching on a polished surface, which can with difficulty be seen without assistance, at once becomes visible if the surface be breathed on. In winter, the walls and roof of the grotto are cold, dry, smooth and polished like glass. The winter air entering from without and circulating in the grotto *breathes* on the polished surface of ice and develops the figure of the ice by the rime which is deposited on it. As rime always settles by preference on sharp edges, it seeks out the lines of separation between the grains and settles on them, showing the whole granular structure. In January 1907, there was a wonderful exhibition of this natural dama-scening on the roof of the cave of the Morteratsch glacier; in January 1908, however, it was quite inferior and would not have struck the eye. The illustration, Fig. 5, represents a portion of the roof of the cave, which I photographed in January 1907. As the roof is not flat, but made up of shell-like cavities, worn by the hot air in summer, the delineation of the grain is sharp in some parts of the photograph and faint in others.

A precisely similar phenomenon was observed in 1886 by Professor Forel, in the remarkable natural grotto of the Arollo glacier, of which he has given so fascinating a de-scription in the *Archives des Sciences Physiques et Naturelles*, Genève, 1887, xvii. p. 498. The delineation of the etched figures by rime was observed by him in the month of July in a remote and secluded chamber nearly 250 metres from the entrance of the grotto. In artificial grottos, like that of the Morteratsch glacier, in which the air circulates freely, the hoar-frost disappears very quickly with the end of winter.

Sun-Weathering of Granular Ice Produces White Surface of Glacier.—The surface features of glaciers cannot be better studied anywhere than on the Aletsch glacier, which is the largest in Switzerland. From the Mergelin See up to the Concordia hut, the surface is without danger, and is easily travelled. If the glacier be here crossed, the beaten track of the mountaineers is left, and, near the north side, the ice, though perfectly smooth and almost level, is quite untrodden. I often made expeditions alone over this part of the glacier in the summer of 1895, and frequently met with the remains of dead animals of different kinds, chiefly birds. At one place I fell in with what was evidently a family of chamois which had perished on the ice. There were the two parents and the kid. One of the parents and the kid were lying on a ridge of ice, and, having remained dry, were in the condition of mummies, with their skins drawn tightly over them. The other parent had been lying in a furrow, and had been

Plate IV

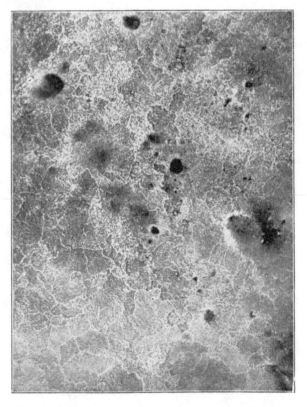

Fig. 5. Etching of Ice by Hoar-Frost: Morteratsch Grotto,
January 1907.

completely macerated, leaving its skeleton in a broken-up condition. I was unable to arrive at a satisfactory solution of how these animals had met their death. It was evident, however, that beasts or birds of prey must be rare, else the remains would not have been so preserved. As a souvenir, I collected a number of the vertebræ of the macerated individual, and took them home. In picking them up, I was much struck with their resemblance to the disarticulated grains of a block of glacier-ice; or, rather, it struck me that they were the only objects which I had seen, with which I could compare the grains in respect of their outward form. Just as the vertebræ of the chamois fit exactly to each other, to form the vertebral column of the animal, so do the grains of the glacier fit exactly into each other to form a compact block. The vertebræ are united and held together by ligaments, the grains of the glacier are united by an aqueous cement, which has a slightly lower melting temperature than their own.

When we walk on the glacier we crush under foot nothing but the grains of the glacier, which have been loosened for our benefit by the radiant energy of the sun. If the white surface layer of the disintegrated ice be chipped away with an ice-axe, so as to expose a smooth surface of blue ice, in the course of a single summer's day this smooth blue surface will become as white and crumbling as any other part of the surface of the glacier. If it were not for this interaction between the solar rays and the granular ice, traversing a glacier in summer would be almost an impossibility.

Snow Névé, and Glacier.—In the lowlands, snow falls and melts again, and we have no opportunity of witnessing the metamorphoses which it may experience when lying for a long time on the ground. It is otherwise with the snow which falls in the high mountains. There the temperature of the air is nearly always below the melting-point, and the snow may remain for years without reaching that temperature. It is often assumed that the higher we climb amongst the mountains, the greater is the quantity of snow which falls in the year. But this is a mistake. In the Alps the greatest amount of snow falls at a height of from 2000 to 2500 metres above the sea. The crystalline snow of the mountains takes the granular form much more easily than does the flaky snow of the lowlands. The snow that falls on glaciers in the winter melts and disappears during summer like that on the neighbouring lands. Snow in the higher regions which has persisted through a summer passes into firn or névé. This is always in clearly granular form.

It is to Hugi that we owe most of our exact knowledge and detailed description of the névé or firn, of its genesis and of its metamorphoses. He built a hut on the Finsteraarfirn at an elevation of 3300 metres, and inhabited it for a considerable time for the sole purpose of studying the firn or névé and its natural history. He traces the development of the névé from the fine crystalline snow of the highest levels, and observes it as it passes into glacier. At a height of 3000 metres the transformation has taken place at a depth of 7 metres below the surface of the névé; at an elevation of 2700 metres it is met with at a depth of a few feet, and at a height of 2400 metres the névé has passed into glacier at the surface. In experimenting on the névé, he found that when a hard compact mass of it was exposed to the influence of rising temperature, the binding material of the grains soon dissolved to water without the grains themselves being apparently attacked at all. A lower temperature then reunites the grains so that the whole appears as a uniform compact mass. This shows the lower melting-point of the less pure cementing mass of ice. He sums up the whole history of the development of the glacier in a remarkable passage (p. 73) of his work, *Ueber das Wesen der Gletscher.* All the changes which we witness taking place on the incline between the most elevated névé and the lowest extremity of the glacier in the valley, are repeated on the vertical, between the upper and the under surfaces of the névé and the glacier. In both directions we observe greater age and more definite development of the mass. Further, what we observe in both these directions we observe also in the individual grain. The older kernel of the névé is compact and blue like the lower glacier, while the white spongy rind on the outside is more of the nature of snow, like the highest névé, and passes by layers into the compact central grain. Also in the case of the individual grain, the nucleus or kernel is the first and oldest, and only by continued development does the rind shape itself and gradually pass into the mass of the nucleus and so become a glacier-grain, which then continues its development as the glacier itself continues its own development. In these relations lies the foundation of the whole natural history of the glacier. It will be observed that eighty years ago Hugi held very modern ideas on the subject of development.

The Grain of Lake-Ice.—It is not glacier-ice alone which suffers disintegration when exposed to a powerful sun. Lake-ice behaves in a similar way. Beautiful examples of this can be seen in Alpine seas every winter. During the harvesting of the ice from the lake, the blocks often lie for a day or more before they are carted away to the ice houses. Occasionally some of them get overlooked and remain for many days exposed to the powerful sun of February, while maintaining the low temperature of the air usual in that month. No melting takes place; but, after even a few hours' exposure to the sun, the block shows the figure of its grain in development. It is being etched by the sun's radiation.

The grain of lake-ice has a very different appearance from that of glacier-ice, but both are individual crystals. The difference in their appearance is to be traced to the difference of treatment which they have received during their existence. The glacier grains have been practically rolling over each other during their descent, while those of the lake have established themselves at right angles to the surface of the water and have remained there. So long as the ice is increasing in thickness, the temperature of its upper surface is very low. It is perfectly transparent, and its surface is smooth, dry and polished like glass, and it shows no trace of crystalline figure. When the ice is undisturbed, this develops itself only at the end of the season when the thaw sets in. Then the whole ice-sheet rises to its melting temperature and is at the same time exposed to the direct radiation of the sun. This produces disarticulation of the ice into groups of vertical prisms, which are then floating independently: they are kept together only by crowding. Ice in this state is said to be rotten; and it will be recognised that, however thick the ice-sheet may be, when it gets into this condition it is dangerous. In the neighbourhood of the outflow the crowding is relieved. The disarticulated groups become disengaged, the smaller groups and individual prisms are able to assume their attitude of stability and to float on their sides. All then drift towards the outlet. The ice breaks up, and the lake is cleared in an astonishingly short time.

If it were not for the law that even impure water in freezing always forms pure ice, the impurity remaining in the liquid and generally entangled in the interstices of the grains, and that the pure ice which is in contact with this impure liquid melts at a lower temperature than that which is in contact with nothing but the water formed by its own melting, the ice covering a lake would be a continuous sheet offering no points of weakness, and it would have to melt as a whole. It is doubtful if lakes such as those met with in the upper Engadine would get rid of their ice-covering at all. On the Silser See the ice is usually over 60 centimetres thick when the thaw sets in, but when once the ice begins to break up, the lake is cleared in a day. Sixty centimetres of ice would take a long time to disappear on the basis of surface melting alone.

While the winter lasts the ice on the lake shows no crystalline structure. This develops only after removal from the water and exposure to the sun. The ice then splits up into prisms in a vertical plane. These are at first of irregular section, and as sun-weathering proceeds the thicker prisms split up into thinner. When a block has lain exposed to the February sun and cold, it may fall to pieces, each piece being a long, thin triangular prism, with some resemblance to a razor blade. When the ice is cold and dry, the outlines of the grains are lines; when the ice has a temperature of 0° C., it melts perfectly round the grain, forming troughs in which the water collects, and the aspect is that of a dark polygon, surrounded by light-coloured canals. In one piece, which was much weathered, I counted 24 such grains in an area of 9 square centimetres. In a slab which had not been lying long, I counted 23 grains in an area of 150 square centimetres, giving an average area of 6·5 square centimetres per grain; the largest had an area of 12 square centimetres. In another slab there was a very large grain which measured 7 centimetres in one direction and 4 centimetres at right angles to it. In a slab in which the sun-weathering had proceeded very far, I counted 113 grains in a disc of 5 centimetres radius, which gives 0·69 square centimetre as the average area per grain.

In the absence of actual experience, one is apt to expect a slab of lake-ice, when subjected to sun-weathering, to be disarticulated into hexagonal columns; but this expectation is quite gratuitous. Ice may crystallise in a form bounded by plane faces, according to the laws of its crystallographic system if it has the freedom which it possesses when crystallising out of an independent medium such as a saline solution or air. But the foreign matter dissolved in fresh water is present in so small quantity that what we have before us is the solidification rather than the crystallisation of ice, and each column as it tries to develop itself is interfered with by its neighbour, and the resulting slab of ice is made up of elementary prisms crowded together, but preserving parallelism of crystallographic axis.

Characteristics of an Advancing Glacier.—In the month of August 1895 I visited the valley of Chamonix, and had the good fortune to find that the Glacier des Bossons was advancing. This was particularly noticeable when crossing the moraine on its eastern flank. The ice was everywhere undermining it, and keeping it generally on the move. On the western side, where the glacier terminated in several points, these acted like ploughshares, in turning over the detritus in front of them. The glacier stream escaped from an *Antrum*, having a remarkably small entrance, about 2 metres high and 3 metres wide.

In pursuance of certain experiments which I was making

regarding the melting of glaciers, I entered the antrum and bored a nearly vertical hole, about 12 centimetres deep and 1½ centimetres wide, in the ice of the eastern wall. It was within a metre of the entrance, but so situated that the water produced by the melting of the surface ice, which poured over the entrance, could not reach it.

Although everywhere there appeared to be melting, the hole remained empty and dry; even the snow-turnings which remained in the hole were quite dry. It was still perfectly dry when I left the glacier at 6 p.m. When I visited it again the next day at 3.45 p.m., it was still perfectly dry. The stream flowing out of the antrum was so voluminous that it was impossible to explore it that afternoon. The next day, however, the 15th, at 11 a.m., the volume of water was small, and it was possible to penetrate into the antrum. It ran into the ice in a straight line for 21 metres, and terminated in a rock 2 metres high by 3 metres broad, which filled the whole area of the cavern.

Grooving of Ice by Rock.—While making these measurements, my attention was arrested by an appearance on the roof of the cave, close to where the top of the rock bore against the ice. From the line of contact, and for a distance of about half a metre towards the entrance, the ice was deeply scored and grooved. At first I could hardly believe it; but there was no doubt about it. Although I had often heard of glacial action, the idea of glacial *reaction* had never occurred to me, still less had I ever expected to witness it. The experience that in an auger-hole bored in the ice within a metre of the entrance, and in very hot weather, the delicate ice-turnings which were left in it remained unmelted for at least twenty-four hours, made it no longer wonderful that the surface of the ice at a distance of 20 metres further from the entrance should remain unmelted long enough to show the effect on it of the rock over which, on evidence external to the glacier, it was passing with quite sensible velocity. In fact, the antrum of the Glacier des Bossons, in the summer of 1895, offered an example on a large scale of a cold-pressed tube, of which the material was ice, which yielded perfectly to continued pressure, while the rock was the resisting button.

It will be readily recognised that the scoring and grooving of the ice surface by the opposing rock can be witnessed only when a glacier is advancing, and, in summer, only in localities adequately protected from melting agency.

For years I imagined that I was the only person who had seen glacial action thus reversed, but later, when I acquired Hugi's works and after I had studied them carefully, I found that he had observed this, as well as almost everything else that it is possible to see in a glacier. Before starting on his memorable winter expedition to the Eismeer, he had the position of the lower extremities of both the upper and the lower glacier of Grindelwald exactly marked out, so as to be able, on his return, to ascertain, by direct measurement, any advance or retreat which had taken place. His way to the Eismeer lay along the flank of the upper glacier which at that date, January 1832, was advancing in great volume into the plain where it spread itself out like a fan, and he observed that on its western side the motion of the glacier was opposed by a mass of rock. It pushed itself over this rock with great energy, and for a distance of 41 feet down the valley the surface of the ice, which had passed over the rock, was deeply scored and grooved. It was observed that the glacier shoved itself over the rock at the rate of $5\frac{1}{2}$ or 6 inches per day; it was also noticed that the ice on the eastern and opposite side of the glacier hardly moved at all, not more than a few inches in three weeks. It was, however, reported in the following spring by the man whom Hugi employed to observe the glacier daily, that in February the western flank of the glacier became stationary, while the eastern flank pushed forward, digging up great masses of stone and earth.

This passage, which I have reproduced at length, not only gives valuable objective information about the glacier, it also enables us to form a subjective appreciation of the man who, with nothing but the stipend of a schoolmaster, was able to undertake an enterprise on so large a scale and so difficult as a *Winterreise in das Eismeer*, and who carried it out with such attention to detail as to have the two principal glaciers which were fed by the Eismeer on which he was going to sojourn for a fortnight kept under observation during the whole winter.

In comparing my observation of the striation of the glacier by the rock with that of Hugi, it will be noticed that his was made in January, the coldest month of winter, and in the year 1832, when the glaciers were in the plenitude of their power, while mine was made in August, the hottest month of summer, in the year 1895, when almost all the glaciers of Switzerland were, and had for long been in an advanced state of decay, the Glacier des Bossons forming, as it turned out, a temporary exception. Owing to the high external temperature, I was obliged to creep into the innermost part of the glacier in order to observe the relatively slight effect of the agency, which, in Hugi's case, produced such a striking effect in the broad light of day. But, notwithstanding the fact that the advance of the Glacier des Bossons which I witnessed in 1895 was slight and temporary, it sufficed to show me the fundamental difference between the effects produced by a glacier on its surroundings, according as it is on the increase or on the decrease. When it is growing, the glacier is a living bearer of energy like a river; when it is shrinking, it is inert as a salt lake.

It has been noticed by those who have observed the occasional advance of European glaciers in recent years, that there is little that is gradual about their start. They get under way at once at a fair speed, and proceed without delay in the work of handling the débris that has accumulated in front of them during their repose. When a glacier is really advancing, there is no doubt about it; where there is doubt, it may be taken that the advance has not begun.

External Work of a Glacier.—Owing to the enormous diminution in the amount of land-ice in the course of the last half-century, we are accustomed to talk of the retreat of the glaciers. But a glacier never retreats; it stops advancing, and melts where it stands. Even in a stationary glacier, however, the flow of the ice in all its parts continues; but its effect outside of the glacier is almost if not quite *nil*. In the stationary state its function in nature is conservation. In the advancing state it adds the function of distribution. Its destructive effect is very small. It protects the rock beneath it from weathering, which is a chemical process, by the constant maintenance of a low temperature and the practical exclusion of the atmosphere. Any destructive action which it exerts must therefore be mechanical.

When two substances meet each other in mechanical strife it is the harder that wears the softer, and in the strife between rock and ice Nature makes no exception to this law. We have seen from Hugi's description of a strife between rock and ice which went on under his own eyes, what the effect was on the ice, which was at a temperature much below that of melting. In a week or two, however, the evidence of the effect on the ice would be obliterated, great though it was, while the reactive effect on the rock would remain, but it is doubtful if it would be perceptible. In the battle between ice and rock, the ice suffers much; the rock comes out with a scar or two. The scars abide, but the destruction disappears and leaves no record; hence the neglect of the major effect and the exaggerated importance generally attached to the minor.

It has been said above that when the glacier begins to advance, it performs a distributive function by digging up and pushing before it the rock débris which it finds in front of it in the valley. This débris consists partly of matter which has been brought down by the glacier at some previous time when its dimensions were greater, and partly from the general wasting of rock on the faces of the mountains, which form the sides of the valley. This is the source of the detritus which gravitates surely into the valley, whether it be occupied by glacier or river. The function of the river also is mainly distributive: it has very little share in the duty of procuring or shaping the pebbles in its bed. The stones in the river-bed are furnished by the chemical process of weathering on the rocks above. This loosens and separates, and, in many cases, rounds the fragments *in situ*. Gravity does the rest. It often helps the disintegration by fracturing blocks, the support of which has been sapped by weathering along the joints, and in all cases it brings the fragments down into the valley so soon as they have lost their support. The river cleans the stones that arrive in its bed. When a flood comes it pushes on a certain quantity of them, and during each flood the fragments suffer some little attrition. But the greater the flood the more rapidly are the pebbles hurried on towards the region where the stream becomes a depositing rather than a moving agency. No single pebble can be exposed to the wearing action of the flood over a greater distance than that from the spot where it fell from the mountain into the valley to the mouth or beginning of the delta of the stream; the motion is always downwards, always in the same direction, and takes place only in floods. The low water of the stream rather protects than wastes the stones.

The real region of mechanical erosion and attrition is the sea-shore. In comparison with it, every other is insignificant. The power available and spent over it is enormous. It is provided by the energy of the winds which blow over the opposing ocean, and are accumulated in the form of undulations by the waves which they generate. This energy is carried without sensible dissipation by the waves until they meet the shallow water of the coast, and there it is discharged in the form of breakers. These are the symbol of the conversion of the potential energy of waves into the kinetic energy of currents. They sweep the pebbles up the beach, and both return together by their own weight. The work of that wave has been done; but with the extinction of one wave another follows, and so on for ever.

The great potential energy residing in rocks which occupy an elevated position, the imminence of its conversion into kinetic energy, by any, even the least decay of the material, and the far-reaching effects which the conversion can produce, have not received adequate appreciation.

A rock precipice is a seat of weathering, with gravity always at its foot. When weathering has produced decay, and decay has removed support, gravity claims the fragment as its own, and the result is a *talus*.

Whether the precipice sheds a fragment in a year, or a thousand of them in a day, the primary shape of the talus is the same. If there is no lower level for it to descend to, it weathers *in situ*, and its final state is a *pampa*.

Advantage of the Study of Tropical Lands.—The study of all matters relating to surface geology in temperate latitudes is made difficult by the variability of the meteorological conditions. Had the study of physical geography and of meteorology taken its rise in countries within the tropics, advance would have been much quicker. The surface geology of a place depends mainly on the meteorological conditions. If these are simple, the dependence on them of the conditions of the surface geology is easily traced. The determining primary features of tropical meteorology are heat and moisture, and these together produce a secondary agency which is all-important, namely intensity of chemical action. This secondary agency is by far the most effective in altering the relief of the surface of the land. Its importance cannot even be guessed by those who have not visited tropical or equatorial regions, and studied the soil and the rocks from which it has been formed. An important difference between the climate of tropical and of temperate regions is that, in the former, the wet weather is concentrated into a· few months of rainy season, and the dry weather into a few months of dry season, while in the latter there is no such separation; wet and dry weather are distributed indiscriminately throughout the year. One secondary effect of the temperate climate is that the local streams have some water in them all the year round, which may be swelled to floods at any time of the year. In tropical, or rather sub-tropical regions, the stream beds are dry during the greater part of the year; and, in some cases, for whole years together. During the long, hot summer the rocks on the mountain sides, which always, even in the driest season, retain moisture below the surface, are eaten into along every discontinuity and loosened up into fragments, which, so long as they remain *in situ*, tend more and more to lose their edges and corners by the chemical action of weathering. When gravity dislodges and brings them down into the valley, they only require to be hurried along by one flood to come out clean and round. In these places no one can fail to recognise that the function of the river or stream is mainly distributive. Now, whether the river flows only during one or two days in the year, or is perennial with occasional floods, the nature of the action is the same, and there is no difference in the roundness of the stones.

I had the good fortune once to witness an example of this. It was in the island of Tenerife, in October 1883. It was said that there had been no rain for six months, and certainly all the beds of the streams were perfectly dry. During the forenoon clouds collected, and about mid-day there was a thunderstorm with a violent and prolonged deluge of rain. After the rain had been pouring for about half an hour, a strange rattling sound was heard, which every moment grew louder, without it being possible to say what caused it. At length it explained itself by the bed of the stream in front of the anchorage of the ship in which I was, suddenly filling with brown muddy water, descending with great velocity to the sea and charged with vast quantities of stones, sand and mud. The rattling noise was produced by these stones and rocks being driven violently over the rocky bed and against each other by the current, against which nothing could stand. Naturally all the other river-beds in the island fared alike. Their accumulations were washed away into the sea likewise. The rocks of Tenerife are volcanic and produce in weathering, besides kaolin, hydrated oxide of iron and other ochreous substances, which have a reddish colour. The consequence was that for nearly a week the island was surrounded by a fringe of red water. The quantity of rain which fell was so great that when it reached the sea it floated on the top and retained in suspension the fine ochreous mud, which would have been precipitated by mixture with the sea-water had the supply of fresh-water been less abundant.

A still more remarkable instance of this kind is related by the Prince of Monaco. At the end of August, 1901, he

Plate V

Fig. 6 a. San Antonio (C.V.). View of Stream-Bed after Three
Rainless Years.
(From photograph by H.S.H. Prince Albert I of Monaco.)

Plate VI

Fig. 6 b.   The Same View on the Same Day after a Violent
Rain Storm.
(From photograph by H.S.H. Prince Albert I of Monaco.)

visited the island of San Antonio in the Cape Verde Group. Here rain is very scarce. It was said that it had not rained for three years. He landed opposite the entrance of the valley of Tarrafal and took a photograph, in which the dry river-bed, encumbered with rocks and stones, occupied the foreground. About noon a cloud-burst occurred from which it was necessary to take shelter for a considerable time. On returning to the landing-place to rejoin his ship, he found that the stream had cut out a broad bed in the stones more than a metre in depth and precipitated the whole of the débris into the sea. He photographed the river-bed after this catastrophe, and the pair of photographs constitutes one of the most important documents in the natural history of denudation. By the kindness of His Highness, I am permitted to reproduce these photographs. Fig. 6*a* represents the view in the morning. Dr Portier, the well-known bacteriologist, is standing on the gravel plain in front of the larger boulder, and the whole of his body is visible over the smaller boulder in the immediate foreground, which is nearly covered by the gravel. Fig. 6*b* represents the same view in the evening, after the stream had done its work of distributing the accumulation of débris due to three years' *chemical and gravitational denudation*. All that is now seen of Dr Portier, over the top of the nearer boulder which has been almost completely undermined by the torrent, is the top of his helmet. The stones in the river-bed in the morning were as round as those occurring in any perennial river, yet it was impossible in this case for the river to have had anything to do with the rounding of them. The sole function of the river in this island is, once every year or so, to clear the bed of rounded stones, not to make them.

It must be observed that, in the island of San Antonio, though rain is rare moisture is abundant, and the valleys are fertile and well cultivated. The moisture condensed on the ground, and retained beneath the surface, aided by the high temperature due to the low latitude, produces ideal conditions for the chemical decomposition of the volcanic rocks of which the island consists, and the consequent production of the best class of soil. This, owing to the rare occurrence of rain, is allowed to remain *in situ*, and contributes to the maintenance of a considerable population. The decomposition of the mineral constituents, which yield to weathering and furnish the soil, undermines the portions of the rock which resist it and removes their support. These then surely gravitate to the lowest level, where their weathering is continued, causing always more and more perfect rounding of the stones, with the production of the equivalent amount of soil in which the stones remain embedded, provided that the process is not interfered with by *running* water.

The "Crumble" Formation.—I never accepted with enthusiasm the teachings of the ultraglacial school of geology, but it was only during the course of the cruise of the "Challenger" that I became convinced that ice is not required to produce denudation. I found in all countries within the tropics that the rocks were decomposed to a depth of 20 or 30 metres, the resulting material often remaining *in situ* with such a fresh appearance that it was difficult to imagine that it could be anything but the unaltered rock.

It was only necessary, however, to touch it with a stick, or even with the fingers, for it to crumble into fragments of all sizes down to sand and clay. In my journal almost every new island or place visited is logged as consisting of "the usual *crumble* formation." There could be no doubt about the cause of it. It furnished convincing evidence of the powerful decomposing action of the heat and high vapour tension which characterise the tropical atmosphere. Fig. 7 represents the Chilian town Autofagasta. The configuration of the mountain slopes illustrates the effects of chemical and gravitational degradations in a rainless district. The landscape so produced consists of a succession of *taluses*. The débris thus accumulated on the mountain side would be cleared out if it was the side of a valley and a glacier passed along it. Applying this knowledge to the consideration of the conditions in my own country, which were referred to ice action, such as the enormous quantities of gravel in the Spey valley, I came then to the conclusion that the glaciers could

Plate VII

Fig. 7. Chemical and Gravitational Degradation of Rock on Chilian Coast.

only be responsible for clearing the débris out of the valleys and distributing it over the plain, that the production of the débris itself might be attributed to the occurrence of the last of the usually postulated warm interglacial periods, and that the existence of the débris furnished the best evidence of the reality of a previous warm period. It is not necessary for the climate during this period to have been anything like as warm as that of the equator; all that is required is moderate warmth and moisture.

I have attributed the presence of fragments of rock, whether great or small, and to a considerable extent their rounded form, to the chemical action of the moisture derived from the atmosphere. Water penetrates, by its own weight, into any cracks or joints which may occur in the rock, and into minute discontinuities of substance by capillarity. It spreads and extends its decomposing influence along the surfaces which lend themselves most readily to the process of soaking. The warmer the rock is, the more energetic is the decomposing action. On the tops of high mountains rock surfaces, exposed to the direct rays of the sun, acquire often, even in winter, a high temperature, and when they are covered by snow, they are protected from excessive cooling by radiation. All rocks have surfaces of relatively imperfect continuity. These are found by water, which enters easily provided the temperature of the surface of the rock is not such as to convert it into ice. Far-spreading decomposition is then only a question of time, and, after it has spread the falling asunder of the parts of the rock under the influence of gravity is a certainty. There is no necessity for assistance by any other agency. **The chemical action of atmospheric moisture and the tendency of every part of a mountain or rock to yield to gravity when not adequately supported, suffice to account for all the degradation of rock which we observe.**

I have not been able to discover the author of it, but it is a very old and generally accepted doctrine, that the rock fragments which are found so frequently covering the tops of mountains are split off from the parent rock by the energy liberated by water freezing in its interstices. I know of no detailed description of the process, nor have I met or read of any one who has actually witnessed a rock being split in this way.

Discontinuities in the rock are postulated in order to admit the water which is to be frozen, but no detailed specification is furnished of how the opening is to be closed and the freezing is to be effected after the water has entered. But it has been shown above that if the water gains admission, it will in time disintegrate the rock by chemical action, and gravity without assistance will complete the degradation. Therefore freezing is unnecessary in order to account for the facts. Moreover, the covering of a mountain top by fragments of its own rock is a common occurrence in latitudes where frost is rare or absent.

In discussing the natural history of ice, it has been necessary to some extent to include that of ice and steam. In its physiographic relations we have found that ice is as efficient a preserver of mineral matter as it is of vegetable or animal matter in the everyday relations of life. We have also found that the substance which chemists indicate by the symbol H_2O has the most destructive action on mineral matter when it has passed from the gaseous state in the atmosphere to the liquid state on the surface of the earth, and when this takes place in regions where the atmospheric temperature is high. Chemical decomposition accompanied by disintegration is the result, and it is followed by degradation under the all-pervading influence of gravity. This is the primary process in the wasting of land substance, and we have called it *chemical and gravitational degradation.* The conditions which favour the exhibition of chemical and gravitational degradation are high temperature, abundance of moisture, and scarcity of water. These conditions are met with together in the tropical regions of both continents. The word "tropical" is here used in its restricted sense and excludes "equatorial." The equatorial climate is not only moist but very wet. It also is productive of energetic chemical and gravitational degradation, but its primary effect is profoundly modified by the secondary effect of running water. This modification is equivalent to intensification, in so far as the running water, by removing the products of decomposition, and thus *denuding* the rock, exposes fresh surfaces to the decomposing agency.

In its simplest and most perfect form the effect of chemical and gravitational degradation, which has been active through all the ages that a distinctly tropical climate has existed, is to be witnessed on the prairie and the pampa.

Here the wayfarer finds himself on a surface as feature-less as the ocean, but equally spherical; and he may use his horizon for the purpose of finding his position astronomically with almost as much confidence as he would at sea.

Water finds its level quickly; sea-ice, slowly; land-ice, more slowly; and land, in a time which, though very great, is still far from infinite.

No. 7. [*Extract of paper read before the physical section of the Schweizerische Naturforschende Gesellschaft at its meeting at Basel on 6 September* 1910, *and printed in its Verhandlungen,* I. *p.* 330.]

BEOBACHTUNGEN ÜBER DIE EINWIR-KUNG DER STRAHLUNG AUF DAS GLETSCHEREIS

AUSZUG.

Das Hauptresultat der Einwirkung der Strahlen auf den Gletscher ist die weisse Oberflächenschicht, welche eine Dicke von 1—2 Meter hat. Wenn man diese Schicht entfernt, so kommt man auf blaues Eis. Wenn die so erhaltene blaue Fläche den Sonnenstrahlen ausgesetzt wird, so erhält man in sehr kurzer Zeit ein Aetzbild der Kornstruktur. Die Linien in diesem Bilde bedeuten die Räume zwischen den Körnern, wo das Eis, wegen vorhandener Verunreinigungen bei einer etwas niedrigeren Temperatur schmilzt als die Masse des Kornes. Setzt man die Bestrahlung fort, so dringen die Strahlen in das Eis hinein und schmelzen das die Zwischenräume begrenzende Eis, selbst einige Dezimeter unter der Oberfläche. Auf diese Weise werden die Körner durch Rinnen isoliert, welche dem Schmelzwasser Abfluss verschaffen und die weisse körnige Oberfläche hervorbringen, auf welcher es sich so leicht gehen lässt. Könnte man nach dem Entfernen der Äusseren weissen Schicht und Bloslegen des blauen Eises, dieses der Einwirkung der Konvektionswärme allein aussetzen, so würde das Eis mit einer glatten Oberfläche weiter schmelzen wie man es an den Wänden der dunkleren Räumen der Grotte beobachten kann. Ein solcher Gletscher würde kaum gangbar sein.

Nach dieser natürlichen Erklärung des weissen Eises der Oberfläche entstand die Frage, giebt es weisses Eis im Innern der Gletscher? Um diese Frage zu beantworten muss man in's Innere der Gletscher kommen können. Zu diesem Zwecke bediente sich der Verfasser der künstlichen Grotten, welche in den meisten grossen Gletschern der Schweiz zu treffen sind und speziell derjenigen des Morteratsch Gletschers.

Als Resultat seiner Beobachtungen in dieser Grotte ist er zu der Ansicht gekommen, dass weisses Eis im Innern der Gletscher vielleicht nicht ganz abwesend ist, weil das Eis durchscheinend ist, dass es aber auf keinen Fall in solchem Masse vorhanden sein kann um eine durchgehende Bänderstruktur mit dem blauen Eis zu bilden.

Aber es lässt sich auf anderer Weise Auskunft über den Zustand des vor Strahlung geschützten Eises verschaffen. Es handelt sich darum, die Eismasse möglichst vollständig in einem Medium, welches für die das Eis auflockernde Strahlen undurchdringlich ist, einzuschliessen. Ein solches Medium ist das Wasser. Die Polarmeere, namentlich die Antarktischen, sind dicht gedrängt mit Eisbergen, welche bei rezenter Herstellung tafelförmig sind und eine Fläche oft von vielen Quadrat-Kilometer bedecken. In diesem Zustand sind sie sehr stabil. Wenn sie aber mit der Zeit dilapidiert und zerstückelt sind, so können die kleineren Berge recht unstabil werden. Irgend ein Stoffverlust kann dann eine solche Störung des Gleichgewichtes hervorrufen, dass der Berg teilwiese oder ganz umkippt. Der Zuschauer hat dann vor sich einen ultramarin blauen Berg in Mitten von unzähligen weissen Bergen. Ein solcher Eisberg ist im Challenger beobachtet worden. Die Farbe des so vom Wasser entblössten Eises war intensiv blau, und nichts weisses, als Schichten oder Flecken, liess sich darin sehen. Etwas Aehnliches, aber in sehr reduzierten Grössen- und Farbenverhältnissen, kann man an einem heissen Tage an der Mergelinsee beobachten. Bei Stoffverlust wälzen sich die kleinen Eisberge um, wobei ein Teil der rauhen weissen Oberfläche unter Wasser geht und ihr Platz von der glatten Oberfläche des durchsichtigen Eises, welches vorher unter Wasser war, genommen wird. Es folgt also, dass das süsse Wasser der Alpen einen ebenso kräftigen Schutz gegen die eisauflockernde Strahlen liefert als das Salzwasser des Oceans. Natürlich kann das Wasser diesen Dienst nicht leisten ohne die Strahlen selbst zu absorbieren und dadurch entsprechend erwärmt zu werden.

Der Kapitalversuch, zuerst von Hugi im Jahre 1822 gemacht, findet so eine einfache und natürliche Erklärung. Wenn ein Stück frisches Eis aus der Höhle geholt und in eine starke Sonne gesetzt wird so fällt es, nach nicht langer Zeit, in einen Haufen Körner zusammen. In dem Eisstücke war schon etwas Wasser (die Mutterlange des Kornes), in den Räumen zwischen den Körnern vorhanden. Dieses Wasser muss jene Schutzstrahlen absorbieren. Im Masse der Erwärmung des Wassers giebt es die Wärme an das anliegende Eis ab, welches dementsprechend schmilzt. Während einem Aufenthalt von 20—30 Minuten in einer starken Sonne kommt auf diese Weise genügend Wärme in das Innere des Eisstückes um durch intergranulare Schmelzung das ganze Stück zu disartikulieren und in einzelne Körner aufzulösen.

Der Schmelzpunkt des Eises ist sowohl vom Drucke, als von der Natur des Mediums in welchem es schmilzt, abhängig. Ist die Natur des Mediums konstant, so ändert sich der Schmelzpunkt des Eises nach den Veränderungen der Natur des Mediums in welchem es schmilzt. Zur genügenden Deutung der Naturerscheinungen des Eises muss der Wirkung eines jeden dieser beiden Naturgesetze gehörige Rechnung getragen werden.

No. 8. [*From The Scottish Geographical Magazine,*
1912, *Vol. XXVIII, p.* 169.]

IN AND AROUND THE MORTERATSCH GLACIER[1]: A STUDY[2] IN THE NATURAL HISTORY OF ICE

MY first acquaintance with the Swiss glaciers was made in the summer of the year 1867, on a short walking tour, which brought me, as it does most tourists, to Grindelwald. As a tourist I visited the lower glacier and penetrated into the artificial grotto, which was not illuminated ; indeed it was so dark that though one heard the music of a zither, it was impossible to distinguish the person who played it.   Possessing at the time no acquaintance with ice except such as is to be obtained from books, I learned very little from this visit. It is true that I saw a great deal, but I may say that I observed nothing the recollection of which I could take away with me. The ice-world was like a new planet to me, and my attitude towards it resembled that described by Forbes when, looking

---

[1] An address read before the Society in Edinburgh, by Dr W. S. Bruce, on behalf of the author on January 24, 1912.

I wish to take this opportunity to thank Dr Bruce for the readiness with which he undertook, at very short notice, to deliver my lecture when I was disabled by illness, and to congratulate him on the success with which he accomplished the task.—J. Y. B.

[2] Previous studies are to be found in the following :—'Some Observations on Sea-Water Ice' (*Proceedings of the Royal Society*, 1874, xxii. p. 431); 'On Ice and Brines' (*Proceedings of the Royal Society of Edinburgh*, 1887, xiv. pp. 129–149, also in *Nature*, xxxv. p. 608, and xxxvi. p. 9); 'Chemical and Physical Notes' (*Antarctic Manual*, 1901, pp. 73–108); 'On Ice and its Natural History' (*Proceedings of the Royal Institution of Great Britain*, 1909, vol. xix. pp. 243–276); 'Beobachtungen über die Einwirkung der Strahlung auf das Gletschereis' (*Verhandlungen der Schweizerischen Naturforschenden Gesellschaft*, Basel, 1910, i. p. 330). (See above, pp. 25, 130, 233, 280.)

back after the completion of his famous work on the *Mer de Glace* in 1842 and the following years, he says :

> "I cannot now recall without some degree of shame the almost blindfold way in which until lately I was in the habit of visiting the glaciers. During three different previous summers I had visited the Mer de Glace, and during two of them, 1832 and 1839, I had traversed many miles of its surface; yet I failed to mark a thousand peculiarities of the most obvious kind or to speculate on their causes....We are not aware in our ordinary researches in Physical Geography, or the natural sciences in general, how much we fall back on our general knowledge and habitual observations in pursuing any special line of inquiry, *or what would be our difficulty in entering as men upon the study of a world which we had not familiarly known as children*[1]."

However, notwithstanding the unfamiliarity of the scene, I observed a good deal round about the glaciers, and I perfectly remember that the tongue of the glacier in which the grotto was driven was in the main valley immediately below the village of Grindelwald, and the same was the case with the upper glacier, the lower end of which protruded far into the Grindelwald valley. Both of these glaciers were in the average state which they had preserved during the previous hundred years, as could be ascertained from their representations in the albums of prints prepared for the use of tourists in the pre-photographic days.

These albums were got up with great care and often at great expense. The prints are generally beautifully executed, by well-known artists of the time, and their design is usually very true to nature. They seldom bear the date of their publication, because at that time views in Switzerland changed very little and a good album was expected to last through several generations of tourists.

In September 1909 I had the curiosity to re-visit Grindelwald, taking with me photographs of some of these prints, and I made it my business to find the positions from which the views had been taken, and from these I took photographs of the same views.

[1] *Travels through the Alps of Savoy, etc.*, by James D. Forbes, 1845, p. 59.

Plate VIII

TWO VIEWS OF THE LOWER GRINDELWALD GLACIER TAKEN FROM THE
SAME POSITION.

Fig. 1. Engraved in 1777.
By recollection, the glacier was in this state in September 1867.

Fig. 2. Photographed in September 1909.
The degree of coincidence of the views may be judged by the church
which appears in both and necessarily occupies the same site.

Change in the Lower Grindelwald Glacier between 1777 and 1909.

Fig. 1 is a photograph of an illustration of the Lower Grindelwald glacier which was published in 1777[1]. Fig. 2 is a photograph taken by myself in September 1909 from the same spot from which the drawing for the engraving of 1777 was taken. The position of the church in both pictures gives a means of judging of the degree of coincidence of the two views. Although the present church is a different building from that which existed in 1777, both of them occupy the same site. The picture of 1777 represents the glacier as no more extensive or more voluminous than I remember it to have been in 1867.

The picture made in 1909 shows no ice in the main valley, nor is there any visible in the lateral valley until a considerable elevation is reached, and the source of the glacier in the *Eismeer* is not far distant.

Change in the Rhone Glacier since before 1856.

Fig. 3 is a photographic reproduction of the print of this glacier in an album entitled *Voyage Pittoresque de la Suisse*, published by H. F. Leuthold in Zürich. It bears no date, but the copy which I possess was brought from Switzerland in the autumn of 1856. The print of the Rhone glacier is signed by the artist *Bäntli*, which may afford a clue to the date of the original sketch. In the work published in 1777, above mentioned, there is a picture of the same glacier, and both the extension and the volume of the glacier appear to be identical in the two pictures. Fig. 4 is from a photograph taken by myself in September 1909 on the spot close to the Grimsel road, whence the sketch for the print must have been taken.

When I passed by the Furka road for the first time, in 1883, the lower glacier, though already diminishing from year to year, still covered about one-half of the area of the low

[1] *Tableaux Topographiques, etc., de la Suisse et de l'Italie.* Raoult: Paris, 1777, vol. i. plate no. 172. Drawn by Besson.

ground which is shown to have been covered by it in the album. Now the lower glacier has disappeared entirely, and the ice-fall is diminishing also. The word *wastage* hardly expresses the meaning; we are here face to face with devastation, and nearly the whole of it has taken place in no more than fifty years.

At the date of my visit to Grindelwald (1867) there was no idea of a period of continuous diminution of the glaciers being before us, nor did one fully realise it until many years later. There does not appear to be any record in historical times of a retreat of the Swiss glaciers such as we can witness at the present date.

It is very important to keep this fact in view when we read the works of the veterans of the first three-fourths of last century and when we compare them with those to be found in modern literature. The old men, like Hugi, Agassiz, Forbes and even Tyndall, knew nothing of decaying glaciers.

Oscillations, no doubt, there were, but all these glaciers, in the main, held their own. Taking one year with another, the wastage of the summer was made good in the course of the year. These were the glaciers which they describe in their works, and if we compare them with equally faithful descriptions of glaciers in recent years, it is difficult to recognise that they refer to the same thing. If, as has been said above, the old men never saw a decaying glacier, it may be said with equal truth that none of the present generation of observers has ever seen a robust one.

M. Vallot, who has charge of the observatories of Mont Blanc, pointed out a few years ago that the sinking of the glaciers in the Chamonix district amounted then to fully one-eighth of the total amount of sinking since the last ice age.

But the recent shrinkage of the glaciers has taken place in fifty years; *therefore, at the same rate the whole shrinkage from the extension of the glaciers in the ice age to their present state may have taken place in four hundred years.*

M. Vallot[1] has also shown that assuming the glaciers of

[1] *Annales de l'Observatoire du Mont Blanc*, iv. 123.

Plate IX

Two views of the Rhone Glacier.

Fig. 3. From a print dating before 1856.

According to an engraving of 1777 it presented the same appearance at that date. According to recollection at least one-half of the lower glacier still existed in 1883.

Fig. 4. From a photograph taken in September 1909.

the ice age to have been at least 500 metres thicker than the present glaciers in the Chamonix district, they must have travelled many times faster than the present ones do. Applying the formula derived from his work on the *Mer de Glace* to the case of the great glacier which in former times filled the valley of the Arve, he shows that, allowing it to have a total thickness of 1000 metres at its exit from the valley at Geneva, it may have had a rate of travel of 4·5 metres per day; that of the *Mer de Glace* at the present day is only 0·35 metre per day.

If the velocity of the ice radiating from the centre of snow and névé in the Alps at that spot was anything like 4 metres per day, and the thickness of it in the valleys was anything like 1000 metres, it is evident that when the decrease began, and the supply of snow and névé ceased to be able to feed a drain of ice from the mountains to the plains on the colossal scale which these figures indicate, the flow of ice would continue for a long time with very little abatement before the length of the glaciers would be much reduced. But as the supply at the fountain-head has been reduced, the persistence of the rate of flow of the glaciers must have the effect of stretching the ice in their upper reaches, and this would cause a diminution of their thickness without the necessity for melting to intervene. The melting, however, would proceed all the same, in proportion to the energy of insolation in the locality and to the superficial area of ice exposed to it.

The total effect of melting and stretching combined would be to accelerate the disappearance of the ice in the earlier part of the period of decrease, while the final rate of disappearance could not fall below that which it is at present.

Consequently, it may with certitude be affirmed that, when the ice of the ice age really began to disappear, the first half would do so at a greater rate than the second half, and that the time required for its reduction to a state such as that which it exhibited in the middle of last century may have been less than four hundred years.

Shrinkage of the Morteratsch Glacier in the last few years.

The studies in and around the Morteratsch glacier which I have carried on continuously since the year 1906 enable me to illustrate the wastage of ice which this glacier has suffered in the years since that date.   For this purpose I have chosen two photographs, Figs. 5 and 6, which represent the entrance to the grotto in September 1907 and 1911 respectively.   The camera used was a 5″ × 4″ Kodak, and the exposures were made by hand, the camera being held in the ordinary way. Both photographs are taken from the same spot at an angle of the foot-path and 3·75 metres nearer the glacier than a rather dilapidated cairn or *Steinmännchen*.   The comparison of these views shows well both the vertical shrinkage of the ice and the apparent horizontal retreat of this part of the western flank of the glacier.   The latter strikes the eye at once by the perspective of the figures, and it has been confirmed by exact measurement.   The vertical shrinkage in the interval between the autumns of 1907 and 1911 is shown in a striking way by the great increase of the amount of the slope of Munt Pers uncovered by the subsidence of the upper surface of the glacier owing to the liquefaction of its mass.

By far the greatest annual subsidence took place in the very hot summer of 1911, but in each of the years the loss was considerable.   The average annual amount was from 2·5 to 3 metres and in the summer of 1911 it was between 4 and 5 metres.

The apparent horizontal retreat of the ice-foot under the centre of the grotto between the dates at which the two views were taken was 35·6 metres, or at the average rate of about 9 metres annually.   From this it is evident that the western flank of the glacier near its lower extremity has lost in each of these summers on an average a shell of ice 9 metres thick, while the maximum loss on the upper surface of the glacier has not been greater than 5 metres and the average not more than 3 metres.   In summer the ice on the surface always receives a very full day's direct insolation ; the ice on the

Plate X

VIEWS OF THE ENTRANCE OF THE MORTERATSCH GROTTO, BOTH TAKEN FROM
THE SAME POSITION AND SHOWING THE SLOPE OF MUNT PERS BEHIND.

Fig. 5. Photographed 17th September 1907.

Fig. 6. Photographed 5th September 1911.
The wastage of ice due to superficial melting is shown by the increased exposure
of the slope of Munt Pers. The lateral wastage is indicated by the perspective of
the figures; and, by measurement, it amounted to 35·6 metres in the four years.

flank receives very little, and that only in the afternoon. On the other hand, it is fully exposed to all the reflection and reverberation from the steep rocky slope which rises directly from the bottom of the valley and close to the ice-foot. It is, however, difficult to admit that this heat of low intensity can really be so much more efficacious in melting ice than the direct rays of the summer sun. A glance at the two photographs shows that the ice-flank is very steep and faces westward. This position not only limits the amount of insolation which it can receive, it also affords it considerable protection against loss of heat by radiation into space. Every one knows that even on a hot summer day when the surface of the glacier is seamed by countless rills of running water and the sun-weathered ice-grains crumble beneath the boot, if the sun goes behind the mountains the fountain of the waters is closed almost immediately and the ice becomes hard and unyielding. Surface melting has ceased for the day. It is not impossible, but I have not checked this by observation, that the protection from cold which the flank enjoys may increase the amount of ice which it loses by melting in the year. But whatever may be the explanation, the distance, 35·6 metres, is exact. It depends on measurements made with the same 15-metre tape and involves no greater possible error than two or three centimetres. In September 1906 I fixed very exactly the position of the northernmost extremity of the glacier. In September 1909 I placed myself again exactly on the same spot, and found the ice-face only 5 metres distant. *At the end of the glacier the rate of apparent annual retreat of the ice was less than 2 metres, while on its western flank it was nearly 9 metres.*

Dependence of the Melting Temperature of Ice on the Nature of the Medium in which it Melts.

From observations which I made on sea-ice in the Antarctic Ocean in the "Challenger," which were followed up by researches in the laboratory after returning home, and from the study of much interesting work on the grain of the glacier then being done by Professor Forel of Lausanne, and others, I believed that my work on the lowering of the freezing-point of water in the presence of even minute quantities of foreign matter dissolved in it, and the consequent depression of the melting-point of pure ice when in contact with such water, would find a fruitful practical application in the study of the glaciers.

The account of these researches was published under the title, 'On Ice and Brines,' in the *Proceedings of the Royal Society of Edinburgh* for 1887.   (See above, p. 130.)

The principle which guided this research was the following: If the crystalline body which is formed when a non-saturated saline solution is partially frozen is pure ice, then pure ice of independent origin, such as snow, must, when mixed with the same saline solution, and heat is supplied, melt at the same temperature, when the concentration is the same.

The experiments undertaken with this end showed conclusively that the crystals formed by freezing a non-saturated saline solution are at least as free from salt as freshly fallen snow is, and that therefore the saline matter, from which they cannot in practice be freed, exists in solution and belongs to the adhering brine.

It was not until this had been established, in 1887, that it became legitimate to say: "The freezing-point of *water* is lowered by the presence of salt dissolved in it," instead of saying: "The freezing-point of a saline solution is so much lower than that of pure water."

The former of these statements expresses the fundamental principle of cryometric chemistry. "In fact *the temperature at which ice melts in nature depends on the medium in which it melts as well as on the pressure to which it is subjected. If the pressure is constant, it varies with the nature of the medium; and, if the nature of the medium is constant, it varies with the pressure.*"

By a curious coincidence, the study of the freezing-point of saline and other solutions in very high dilution (infinite dilution was aimed at), was taken up at this time and prosecuted with great zeal by the rising pioneers of modern physical chemistry, and their labours furnished abundant material confirmatory of the conclusions at which I had arrived from my own experiments. The *Zeitschrift für Physikalische Chemie*, which was founded at this date (1887), is much occupied during the first eight or ten years of its existence by reports of researches and speculations in this field ; one result of which was that the amount by which the freezing-point of water was lowered by the dissolution in it of a minute proportion of a salt was claimed as a trustworthy basis of arriving at the molecular weight of that salt.

The purest water occurring in nature is at the best only a dilute, but not an infinitely dilute, solution of foreign ingredients. It would naturally be expected that physicists and chemists who have made a study of land-ice since the year 1887 would recognise not only the possibility but the certainty that the ice and the water which they meet with on glacier, névé, or snow-field, must, under constant conditions of pressure, melt and freeze at a temperature inferior to the melting and freezing point of pure H_2O.

We search their writings on this subject in vain for any recognition or application of this fundamental principle.

Explanations which ignore an important law of nature must be defective and may be misleading.

The Source of the Impurity in the Intergranular Water.

The principal source of the impurity in the water which moistens the granular surfaces of ice, névé and snow is rock débris, which may take the form of stones, sand, dust or mud; and it is to be found everywhere, even in what is apparently clear ice.

This has often given me a great deal of trouble when photographing the ice of the walls of the grotto. In summer, when all the ice in the grotto is melting, its surface is contaminated with dirt, principally mud with some sand. When the outer superficial layer of the ice melts, the water runs away apparently quite clear, and the mud which has been liberated by the melting of the ice in which it was confined, seems by preference to remain adherent to the ice, partly as a fine semi-transparent layer and partly as small patches, perhaps of the size of a bean. If the surface of the ice be rubbed over with the hand it becomes completely obscured with the mud, and it costs some trouble to get it clean enough to be able to take a photograph through it. The mud is undoubtedly the product of the *kaolinisation* of the felspathic rocks of the neighbourhood, and the sand is some of the quartz which has been liberated by this purely chemical process of disintegration, which is accompanied by the production of carbonates and of other slightly soluble salts[1].

It is never questioned that the water of a stream which in time of flood may have passed from the clouds over the soil into the stream bed in a few hours, has taken into solution some mineral matter, and river water is therefore always accepted as being *prima facie* impure. No one who has been placed in the situation of having to melt snow on the mountains to quench his thirst is likely to forget his dissatisfaction on discovering that the water produced was muddy. The sand and dust blown on to the surfaces of snow and névé get completely mixed up with the mass which, throughout a large part of the year, is in a constant state of alternate melting and freezing, and the mineral matter which thus becomes incorporated with the snow, névé and ice persists in this state at any rate for many years. The certainty that some of it will pass into solution and influence the melting temperature of ice in contact with it cannot be contested.

Usefulness of the Artificial Grottos, which are met with in all frequented Glaciers, for the Study of Glacier-Ice in its Primary State.

The motive which drove me in the first instance to visit the Swiss glaciers was to study the glacier-ice rather than the glaciers themselves. For this purpose it was necessary to have entry to the inner mass of the glacier. At first

---

[1] See Bischof, *Chemische und Physikalische Geologie*, Bd. i. Kap. i., vi.

I confined myself to the natural cave or *Antrum*, from which the glacier stream proceeds, and left the artificial grotto maintained for the delectation of the summer visitors on one side. I had a feeling, which I imagine was not quite peculiar to myself, that the use of this pleasure-grotto for scientific purposes would be a little *infra dignitatem* of the student. However, one or two visits to it convinced me that the artificial grotto was the only place where the study of glacier-ice in its primary state was possible. Having arrived at this conviction, the economic reasons for the construction and the maintenance of the grotto troubled me no more. I reflected that if an astronomer can legitimately accept and use an observatory furnished by one millionaire, the humble student of glacier-ice may use as his laboratory the grotto maintained by the subscriptions of, it may be, a million tourists.

Intergranular Melting of Ice and Freezing of Medium.

If the ice in the interior of the grotto in a glacier be studied in summer it will be perceived that melting does not take place exclusively at the surface; the interior of the ice appears to be wet also. The internal ice receives heat from without only by conduction, which suffices for the maintenance of intergranular moisture. The only other source from which it can obtain heat is the transformation of work. Helmholtz mentions this as a possible source of heat available for the fusion of ice in the interior of a glacier. He says[1]: "I should like here to direct attention to the fact that a not inconsiderable amount of heat due to friction must be produced in the larger glaciers. In fact calculation shows that when a mass of névé descends from the *Col du Géant* to the source of the *Arveyron*, it is possible for one-fourteenth part of it to be melted by the heat produced by the mechanical work."

The heat which can be obtained from both of the above sources is insufficient to produce any considerable liquefaction of the ice as a whole. Nevertheless the intergranular moisture

[1] *Populäre Vorträge*. Braunschweig: Vieweg, 1865, p. 133.

is a very important item in the economy of the glacier. It is in fact the mother-liquor of the grain and provides the medium in which the activity of the crystallisation and dissolution of its ice develops itself.

Every act of crystallisation in this medium furnishes crystals of ice and a mother-liquor. The impurity which was uniformly distributed in the original liquid is concentrated in the mother-liquor and eliminated from the crystal. The melting temperature of the pure ice when in contact with the mother-liquor is lower than it would be if in contact with absolutely pure water. It is this great natural law which secures the preferential melting at the intergranular surfaces. If the mass of the ice of two contiguous grains has exactly the temperature $0°$ C., and the liquid between the adjacent surfaces contains so much dissolved impurity as to reduce its freezing temperature to, say $-0.001°$, then there is established a temperature gradient from the ice to the liquid which permits the ice actually in contact with the liquid to melt at the lower temperature, and at the expense of the heat in the mass of the grains. This example, if considered attentively, will show not only that, with the limited amount of heat available in the interior of a glacier, some melting must take place between the grains, but also that the total amount of ice so melted can never, at any one time, amount to very much. It is however enough to facilitate the flow of the ice in the dark; and, when the ice comes to the light, by its power of arresting certain of the solar rays, this liquid, even when very sparingly present, plays an important part, inasmuch as it serves to initiate or induce the differential action of the sun's rays, and exposes to their action in the interior of a mass of ice a working surface far greater than that furnished by its outer surface.

Effect of Solar Radiation on Glacier-Ice.

The principal effect of the action of the sun's rays on the glacier is the production of the white surface layer. This has a thickness of from one to two metres, and, at a certain

distance, it is only with difficulty distinguished from a surface
of snow. Its whiteness is due to the same cause as is that of
a field of snow, namely, discontinuity in the discrete masses
of ice which compose the masses in both cases. The snow is
produced and deposited in very fine crystals which press but
lightly on each other and, when freshly fallen, it includes so
much air that a measured volume of it when melted does not
produce more than one-eighth or perhaps only one-tenth of
the volume of water. The snow-field with its dazzling white
surface is produced synthetically by deposition from the
atmosphere. The nearly equally dazzling white surface of
the glacier is produced analytically by the action of the sun's
rays on blue glacier-ice. The blue glacier-ice has been produced out of the snow of the snow-field by coalescence,
involving fusion and solidification under the control and
direction of the laws of crystalline conformity. It is easily
exposed by chipping away the outside layer at the surface of
the glacier. If the blue surface so obtained is exposed to the
sun's rays, it in a very short time shows a delicate etching,
the lines of which represent the outcrop of the contiguous
surfaces of the individual grains. These surfaces were already
bathed with the intergranular liquid above referred to, which,
by absorbing certain heat rays which are transmitted by the
ice of the solid grains, procures the heat which is then spent
in the further melting of the contiguous ice-surfaces. As the
liquid which bathes these surfaces contains the greater part
of the soluble impurities, pure ice, when in contact with it,
melts at a lower temperature than that at which it melts
when moistened with pure water. Therefore the internal
melting of the ice of the glacier by radiant heat proceeds by
the melting of the outside surface of each individual grain.
In this way the grains in the immediate surface layer of the
glacier rapidly get separated by gutters, by which the water
of the melted ice is drained away, leaving the individual
grains separated by air-spaces. These have the same optical
effect as those which separate the individual crystals of a
mass of snow ;—they produce whiteness. This rough surface
of easily crushed grains is what enables the mountaineer to

travel the glacier so easily. Were it possible, after the removal of the white outer layer and exposure of the blue ice, to submit this to the action of heat of convection alone, the ice would melt with a smooth surface, as can be observed on the inner walls of the grotto in summer. A glacier having such a surface would be barely passable. After the individual grains have been thus laid bare they themselves experience a further disintegrating effect under the continued action of the sun's rays, which is the same in kind as that produced on massive glacier-ice. As the sun's rays analyse a piece of glacier-ice, separating its constituent grains, so they continue the analytical process on the isolated grain or crystal, and separate it into its constituent lamellæ, which are situated normally to the principal axis of the crystal. Both of these analytical separations are due to the power of the sun's rays to detect the impurities which occur even in the purest crystal, and to eliminate them by discriminating fusion.

Having found that the whiteness of the ice on the surface of the glacier is a secondary feature and is due to its exposure to solar radiation, the question naturally arises, does white ice occur in the interior of the glacier ? In order to answer this question it is necessary to be able to penetrate into the interior of the glacier.

The Morteratsch Grotto.

As has been already indicated, the most convenient place to study such a subject is the interior of the artificial grotto, which is to be met with in all frequented glaciers.

The Morteratsch glacier is one of these ; and at the date of my first visit, in the summer of 1893, there was a very fine grotto, driven in the western flank of the tongue of the glacier, which at that date extended more than a hundred metres further down the valley than it does now. I visited it almost daily, and at each visit I had the opportunity of observing phenomena which I did not expect. In the following February (1894) I returned to the Engadine in order for the first time to see a glacier and an ice-grotto in winter ; the phenomena were still stranger to me than those that I had witnessed in summer: perhaps because the descriptions of glacier-ice in winter were rare and not very informing. The glacier restaurant was but little frequented and no one visited the grotto, the entrance of which was snowed up. I had, however, very little difficulty in clearing a way into it, and I was astonished at the difference in its appearance as compared with that which it presented in the previous summer. The roof had sunk considerably and the pillars and longitudinal buttresses which supported it were much sheared in the direction of the entrance. There were also fewer air-bells in the ice than in summer, and the outer surface of the ice as well as the inner surface of the air-bells was dry; the temperature of the ice at a depth of three or four centimetres was $-4°$ C., which was also the temperature of the air in the grotto at the time. It must be remarked that, not being visited, *the grotto had been practically lying in darkness for some months, and the features of the ice in these circumstances would certainly be different from those encountered in summer, or in winter when it is continuously open to the light of day.*

In the summer of 1906 I began to use photography in my work, and it has naturally proved a great assistance. Since the above date I have visited the Morteratsch grotto at least twice, and in the last years four times in the year. *The critical dates for the glaciers, as for the seas and lakes, are the equinoxes and the solstices.*

Granular Features of the Ice in Summer.

At the entrance of the grotto the ice is exposed partly to the direct rays of the sun and partly to its indirect radiation from the sky. Under these combined influences the ice at the entrance contracts the texture and general appearance of that on the upper surface of the glacier.

The occurrence of continuous white ice depends principally on the receipt of a considerable proportion of the whole radiation of the sky, so that the surface and the ice immediately below the surface may be disarticulated, and the water produced by melting may find, in the interstices so formed between the grains, gutters for efficient drainage. The proportion of radiation so received from the sky by the walls of the grotto diminishes rapidly as the distance from the entrance increases and the occurrence of continuous white ice diminishes at the same rate. At the entrance of the grotto it is only on the roof itself or in its immediate neighbourhood that continuous clear ice is to be seen, and this depends on the fact that these surfaces are by their position protected from direct radiation from the sky, and are exposed only to the much less powerful radiation of the ground in front of the grotto.

The delineation of the grain on the surface of the walls of the grotto ceases at a distance of four or five metres from the entrance. In certain places, however, where the ice is directly or normally exposed to the light from the entrance, the grain may be beautifully delineated on it, even though it may be half-way up the grotto. Fig. 9, from a photograph made 19th September 1907, illustrates the revelation of the grain of the ice on the outer end of one of the ice-walls which receives direct daylight; this, however, is much subdued by the distance, about fifteen metres, of the ice-face from the entrance. This attenuated daylight acts like a very dilute acid in etching. The effect corresponding to the strong acid is the profound disintegration of the ice on the surface of the glacier produced by full exposure to the direct rays of the sun and the hemisphere of the sky.

Granular Features of the Ice in Winter.

The winter visits were particularly interesting. The condition of the Morteratsch grotto turned out to be different in each of the last four winters in which I visited it. Fig. 7 represents it as photographed in January 1907. At that date the most striking as well as the most picturesque feature was the abundant deposit of rime or hoar-frost which penetrated into the innermost parts of the grotto. It covered the ice completely on the roof and on the walls down to a height of

Plate XI

THE MORTERATSCH GROTTO IN WINTER.

Fig. 7. Taken on 16th January 1907.
Shows the great abundance of hoar-frost on the roof
and its absence on the walls.

Plate XII

THE DELINEATION OF THE GRAIN OF THE GLACIER.

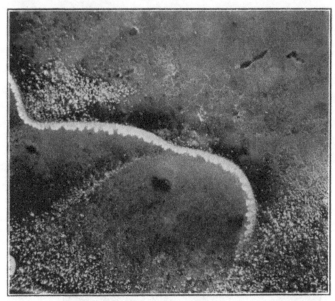

Fig. 8. Shows the delineation of the grain in Winter by the condensation of moisture in the form of hoar-frost on the roof of the grotto.

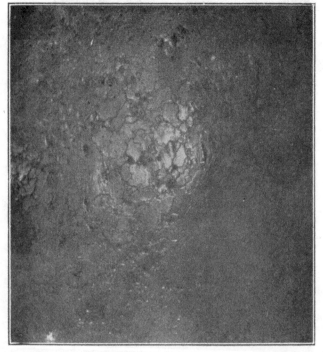

Fig. 9. Shows the revelation of the grain in Summer on a pillar in the interior of the grotto which received subdued daylight from the entrance.

about 1·4 metres above the floor of the grotto where it ceased. Below this line it was completely absent. The deposit of hoar-frost was especially fine in the middle chamber of the grotto, shown in Fig. 7, where it furnished a wonderful picture of ice-flowers. In the galleries it was less abundant. It settled, as may be seen in Fig. 8, in the shell-like cavities of the roof as a comparatively thin coating and in great abundance on the sharp ice-edges which separated these cavities. In these cavities the delineation of the grain of the ice produced by the condensation of the hoar-frost upon it was particularly beautiful. The phenomenon was not repeated in the three following winters, and I am not able to give a sufficient reason why it was so abundant in the one winter and was practically absent in the others.

It is to be remarked that in winter the ice of the walls of the grotto has a temperature below its melting-point, it is hard and polished like glass, and its surface bears no delineation of the grain.

During this season such delineation is met with naturally in the interior of the grotto only as the result of the condensation of moisture as hoar-frost, and artificially in the immediate neighbourhood of the lamps used for illumination. Each of these lamps is a small local centre of disarticulation because it supplies not only the heat of convection which raises the temperature of the surface of the ice in its immediate neighbourhood to 0° C., but also the radiant energy which penetrates the surface and produces intergranular melting. The texture of the ice in immediate proximity to such a lamp simulates on a small scale that of the surface of the glacier which is due to the strong light of the sun. When I first visited the Morteratsch grotto only petroleum lamps were used, each of which melted a considerable dome out of the roof. Later these were replaced by acetylene lamps which produced greater wastage on the walls. Now the grotto is illuminated by electric incandescent lights which produce disarticulation but little melting.

Disintegration behind the Surface of the Walls.

For a distance of two or three metres from the entrance the superficial disintegration of the ice can easily be followed into its mass producing disarticulation of the grains and intergranular fusion along with contraction. In the study of this feature the stereoscopic camera is particularly useful. It renders services which are not to be obtained from the simple camera. A stereoscopic positive represents the ice exactly as *the two eyes* see it, and being permanent can be studied later. Moreover, there is often too little light in parts of the grotto for the eye to be able to see anything in the short exposure which suits it. The camera, whether stereoscopic or not, can be set up and exposed for any length of time which may be judged necessary in order to obtain a picture, and this, if stereoscopic, is all that is required for study.

The visible capillaries in the ice are tubes filled with vapour of water or air which lay themselves along the upper edges of the grains in the course of the intergranular melting of the ice. When these tubes become smaller and contract they often suffer segmentation and form a series of spherules.

If the external surface of the wall remains water-tight we can have the case of a grain in the middle of the massive ice, which *floats* in the water of fusion which surrounds it. As a rule, however, the grains are so closely packed that the grain which would float is prevented, and, in place of doing so, is able only to exercise the corresponding pressure on the grain above it.

In the interior of the grotto the ice has on the whole a smooth surface and the discontinuities in the ice behind it, which proceed from the disintegration by radiation, appear to be arranged more or less in layers. When a grain loses a superficial sheet of ice by melting, the water which appears occupies a smaller volume than did the ice which has disappeared, and the space vacated during the operation is occupied in the first instance by aqueous vapour of low pressure. If the ice-grain under consideration is situated near the surface of the wall of the grotto where disintegration is more advanced, the wall may not be water-tight and air penetrates easily into the space left vacant in the act of fusion, and we have really *air-bells*.

Let us imagine that the mass of ice consists of grains which are all of equal size and have the regular form of cubes. During intergranular fusion, a trough forms itself round each grain which is illuminated, and it is filled with water to the extent of nine-tenths of its volume, while the remaining tenth is filled with air or vapour.

If the tubes of an upper layer do not lie conformably on those of the lower layer, an apparent stratification is at once produced, and it would have a very regular character. But in the real glacier-ice, as we meet with it in nature, the grains are quite irregular in form ; and, in order that they may be capable of completely filling space, they must necessarily be of very different sizes. When radiation produces intergranular fusion in such a mass, the apparent stratification produced must be irregular. This kind of stratification, crude though it may be, is always produced in the white surface layer of the glacier, and may be recognised in lines on the rough exterior of the glacier.

When the winter returns, the intergranular water which has accumulated during the summer begins to freeze so soon as the cold has begun to affect the inner walls of the grotto ; but this takes some time, during which there is neither melting nor freezing in the grotto. When, however, it does begin to take effect, the outer surface of the wall of the grotto is cooled down below 0° C., and this temperature propagates itself into the mass of the ice as an isothermal surface, which may be considered a vertical plane. Isothermal planes corresponding to lower temperatures follow in its wake, and first the intergranular water freezes at a temperature which depends on its purity, leaving, however, always a relatively impure liquid nucleus often of microscopic dimensions. While the intergranular water is being frozen, it attaches itself to the grain in conformity with its crystalline orientation, so that the new ice added is not to be distinguished from the pre-existing ice of the grain. If there has been no leakage, then, when all the water has been frozen, there will remain only a cicatrix or scar which will contain the microscopic liquid nucleus consisting of brine of low freezing-point. In the glass-blower's art we meet with scars of this class frequently. When two pieces of glass tube are joined together before the blowpipe, the joint is almost always characterised by a linear scar which, when examined under the microscope, is found to consist of a string of minute spheroidal discontinuities or air-bells. In the course of my work the most perfectly *healed* ice, or that in which the scar was least visible, was met with in the Morteratsch grotto in February 1894 ; but there, as we have already seen, the freezing and cicatrisation probably took place in the dark.

The features in winter are well shown in Figs. 10 and 11, which were photographed on 5th March 1910, while the winter still lasted. The ice-wall shown in Fig. 11 lies inside of that shown in Fig. 10, which it slightly overlaps. White ice in its interior is therefore less abundant than in the ice represented in Fig. 10, which is close to the entrance of the grotto. The forms assumed by the white ice in Fig. 10 are remarkable and unusual. As it was near the end of winter, the ice-walls shown in these illustrations were frozen to a depth of several decimetres.

It is the custom for the proprietors of glacier grottos to have them repaired and enlarged every spring, so soon as the frost has come out of the ice. In the innermost part of the grotto, where the addition is always made, this happens very quickly, partly because the winter cold of the outside is much attenuated before it reaches the inner terminal chamber of the grotto, and partly because lights of one kind or another are maintained there for the benefit of sightseers, and this interferes with the cooling of the ice.

At the beginning of June 1910 I visited the new gallery which had been driven in the spring in ice which is far removed from external radiation, whether arriving by way of the galleries of the grotto or by transmission through the thickness of the ice above. I found the ice in general very free from air-bells. Only where the miner had struck the ice

Plate XIII

DISCONTINUITIES WHICH CAUSE THE APPEARANCE OF WHITE ICE
IN THE WALLS OF THE GROTTO IN WINTER.

Fig. 10. Shows a very remarkable picture of these discontinuities
in the north wall about two metres from the entrance.

Fig. 11. Overlaps Fig. 10 a little and shows the comparative absence
of White Ice one or two metres further in.

with his pick could one perceive bunches of white ice where the blow of the pick had produced dislocation of the grains. It was only at places on the wall from which the entrance of the grotto was directly visible, that the beginning of the secondary production of white ice could be perceived.

As the result of my observations in the artificial grotto, I have arrived at the view that *the discontinuities which are the characteristic of white ice may perhaps not be entirely absent in the interior of the glacier, because the ice is translucent, but that the white ice which is due to them is certainly not present in such proportion as to permit of its forming with the primary blue ice a ribbon-structure extending throughout the mass of the glacier.*

The Case of a Glacier which has never been Exposed to Light.

Let us imagine that we have a glacier which has always existed in complete darkness. Were it possible to illuminate it for a moment it would appear as an intensely blue mountain. If the illumination were continued, the deep blue colour would quickly fade and before very long would be replaced by dazzling whiteness. Viewed in section the glacier would then appear as a blue internal mass contained in a white external wrapper. Indeed this is exactly the appearance which any of the existing glaciers would present if cut in two. An upper and lateral superficial layer, one to two metres thick, of white ice covering a mass of solid blue ice ; it is the condition of equilibrium of a glacier on the surface of the earth. We cannot detect this when we merely look at a glacier from the outside. *It is only by penetrating into the inside of the glacier that we become aware that white ice is a secondary feature characteristic of the surface.*

While quarrying the ice of a glacier fresh surfaces are continually being exposed, which, in proportion as they are exposed to the daylight, begin to change their colour from blue to white. It might be thought that we are thereby placed on the horns of the dilemma,—if we wish to see if there is white ice in the inside of the glacier we must illuminate the ice; but if blue ice is sufficiently illuminated it becomes white ice; therefore the experiment is not decisive. But the sun-weathering of fresh ice demands time, whereas the human eye, if prepared, can decide whether ice is blue or white in a moment of time. When a fresh ice-surface is about to be exposed it is of the greatest importance that the observer be on the spot: but this usually depends on luck.

Water a Protective Medium, Opaque to the Rays which Disintegrate Glacier-Ice.

There is, however, another way of arriving at the condition of ice which is protected from the disintegrating effect of radiation, namely, to enclose the mass of ice to be examined as completely as possible in a medium which is opaque to, or transmits only feebly the rays which produce the granular disarticulation of ice. *Such a medium is water.*

The polar seas, especially those of the southern hemisphere, are crowded with icebergs. When they are of recent origin they are of tabular form and cover a surface of many square kilometres.

In this condition they resemble flat boards which, when they float on their sides, have great stability. But in the course of time they begin to get dilapidated and to break up into smaller units which may arrive at the state in which their horizontal and vertical dimensions differ little from each other. Such a berg is about as stable in one position as in another; in other words it is unstable. Any loss of substance can bring about a disturbance of equilibrium which may cause the berg to turn partially or completely over before it finds a new position of equilibrium. The spectator who has the luck to assist at an act of this kind, sees before him an intensely blue berg floating in the sea surrounded by countless white bergs. On the "Challenger" we had this good fortune. On one of those wonderfully beautiful days of the Antarctic summer the ship was engaged in the work of a "station," surrounded by icebergs of the usual form and appearance. By the shifting of the position of the ship or by an independent alteration of

Plate XIV

Fig. 12. The Mergelin See.

The waters are retained by the ice of the Aletsch Glacier
from which small icebergs are frequently detached.

Fig. 13. On the Morteratsch Glacier.

the grouping of the bergs, a deep blue berg suddenly appeared amongst the others. The colour was pure ultramarine, and nothing white, whether as patches or layers, was visible. The weather being very fine, work was continued all day in sight of this beautiful and striking object which excited the admiration and the curiosity of every one on board. When the work was done the ship steamed round the blue berg and found its other side white; so that the original berg had only half turned over. If, therefore, the ship had been working during the whole of the day on the other side of the berg this marvellous object would have been close to her without being seen, and indeed without the possibility of the existence of such an object being suspected. It was a piece of extraordinary luck. None who saw the berg ever forgot it.

To any one who is familiar with the ice conditions of the Antarctic it is evident that an event of this kind must be rare. As it is also very seldom that a ship finds itself in these latitudes, it must be a still greater rarity for a freshly-capsized iceberg to come under human eyes.

The Mergelin See and its Icebergs.

It would hardly be expected that anything similar could occur in European, and still less in Swiss, waters. Yet, something perfectly similar, differing only with regard to size and depth of colour, may be seen in any summer in a small but famous lake in Switzerland, namely the *Mergelin See*, which is confined in its little valley by the Aletsch glacier, which closes its outlet. This glacier is continually shedding portions of its substance which then float about in the lake, giving an exhibition in miniature of the colossal icebergs which crowd the waters of the Antarctic Ocean.

The shore of this lake (Fig. 12) is the best place that I know of for studying the natural history of glacier-ice. The bergs and smaller lumps of ice which float in it may go through all the possible metamorphoses of glacier-ice in the course of a single midsummer day with a powerful sun. I have spent many days in this study on its banks and

nothing could be more fascinating. The best season for this
work is early in July, because in this investigation the physical
agency used is the energy of the sun's radiation. This reaches
a maximum at noon on the date of the summer solstice,
21st June, but for quite a month on each side of this date
there is hardly any diminution, either of the sun's meridian
altitude or of the length of the day, and these are the factors
which determine its power of doing work. Consequently the
early part of July is quite as good as any date in June;
moreover, before the beginning of July there is no hotel
accommodation on the mountain, and the observer must make
his visits to the lake from Fiesch or other station in the
Rhone valley.

While studying the bergs and other fragments of the
glacier-ice stranded on the margin of the lake, I noticed that
the bergs which were afloat were seldom at rest, and this was
particularly remarked on days when the sun was most powerful.
On observing more closely the bergs which exhibited this
unrest I saw that it was accompanied by subdivision of their
mass and alteration of their equilibrium. This act is called
"calving" by the whale-fishers and causes one of the principal
risks to their ships if they approach the icebergs too closely.
In the waters frequented by whalers the calving of the bergs
is not an everyday occurrence. On the *Mergelin See* where
the mass of even the largest berg is comparatively small, this
act can be observed frequently on a hot day. If it be attentively observed it will be found, if the "calf," the portion of
ice which has been shed by the berg, comes, as it usually
does, from the part of the berg under water, that it is quite
transparent and has a perfectly smooth surface. If the parent
berg be now observed, it will be found that it has altered its
"trim." This alteration of trim may amount to only a slight
"list" or it may assume the dimensions of a complete capsize.
As a rule the alteration of trim is considerable, and the
alteration which it produces in the appearance of the berg is
due to the submergence of a part of the ice having the rough
dead-white granular appearance acquired by continued exposure in air to the direct rays of the sun, and to the emergence

of a part of the ice possessing a smooth surface and perfect transparency, properties which it was able to preserve in spite of the apparently perfect freedom with which the sun's light penetrates the water and renders visible to the eye the submerged features of the berg. From this it follows necessarily that the particular rays which bring about the disintegration of the glacier-ice are absorbed by water, while, on the other hand, they are transmitted abundantly by ice.

Even the largest berg in the *Mergelin See* is a very small object compared with the blue berg seen in the Antarctic, and with its insignificant mass the proper colour of its ice is much attenuated ; it is represented by a delicate bluish tint.

We see then that equal protection is afforded to the ice from the disintegrating rays of the sun by the fresh water of an Alpine lake and by the salt water of the ocean. Naturally the water can afford this protection to the ice only by absorbing these rays itself, and in doing so it must be warmed correspondingly.

Hugi's Fundamental Experiments.

We now have the information which enables us to follow in detail and to explain the capital experiment first made by Hugi in the year 1822.

When a piece of fresh glacier-ice is fetched from the *Antrum* or the grotto, or from some place inside the glacier where the ice is protected from radiation from the sky, and is then exposed to the rays of a powerful sun, in a very short time it crumbles together and falls into a heap of individual glacier-grains.

We have seen that the ice in the interior of the glacier is always at the melting-point of the ice on the intergranular surfaces, which is influenced by the relative purity or impurity of the water occupying the intergranular spaces. As the purity of the ice-crystal which remains solid is always greater than that of the water in contact with it, the mass of the grain is protected by the lower temperature at which its superficial layers melt. But no matter what the melting temperature may be, no melting takes place without an adequate supply of heat.

The fresh piece of glacier-ice, which we imagine to have been fetched from the *Antrum*, is a piece of the internal ice of the glacier, inasmuch as it has not been exposed to the direct action of sky-light. We have seen that the amount of heat, small though it be, which is transmitted through the ice by conduction, or which may be generated by the transformation of work, is sufficient to keep the internal ice of the glacier always wet. But this moisture is not promiscuously diffused through the mass of the ice: it is strictly localised and confined to the interstitial spaces separating the individual grains. When the piece of ice is exposed to the sun's rays, this intergranular water, which, chemically speaking, is the *mother-liquor of the grains*, interrupts the passage of the rays and acts upon them in the same way as does the water of lakes or seas. It absorbs the disintegrating rays which are allowed to pass by the solid crystalline ice, and these provide the equivalent amount of heat in the water, which, being everywhere in contact with the grain-surfaces, melts the equivalent amount of ice. A lump of ice of such size that a person can carry it without difficulty, after exposure to a powerful sun for from twenty to thirty minutes, appropriates out of the rays passing through it sufficient heat to produce in its interior so much intergranular fusion that the whole lump gets disarticulated and is resolved into its individual grains before the mass itself has suffered sensible diminution of bulk.

If we attempt to make the experiment on a warm summer day when the sky is overcast with clouds, the disarticulating effect produced by the diffused sky-light is insignificant, and the lump of glacier-ice will melt as a whole, preserving its smooth exterior surface, although the form of the individual grains may be delineated, because the intergranular surfaces, being in contact with relatively impure water, will melt at a lower temperature. Consequently the smooth melting surface of the block will always exhibit a channelled delineation of the grains, although it may melt completely without falling into grains. If the sky, without being completely overcast, is cloudy, and the temperature of the air comparatively high, the rate of melting as a whole may be so great that the lump may melt away by the outside before the amount of radiation available has been able to supply sufficient heat to the interior to produce the intergranular fusion which permits complete disintegration.

In very cold winter weather, when the temperature of the air is continuously below the melting temperature of ice, exposure to the sun disarticulates the grains as perfectly as it does in summer. But so long as the temperature of the ice remains below the freezing temperature the grains hold together. The appearance of the lump shows that it has been in fact disarticulated. The moment its temperature reaches 0° C. disintegration takes place even in the dark, and the lump becomes a heap of grains without there having been any melting. Hugi's experiments in this direction are classical. They were made during his *Winterreise ins Eismeer* in January 1832, a date which marks one of the most important epochs in the development of our knowledge of the natural history of ice.

Recapitulation.

A glacier is a granular crystalline mass of ice, which, in summer, is throughout at its melting temperature. In winter it has the same temperature, with the exception of a superficial shell of small thickness, which follows qualitatively the atmospheric changes of temperature, subject to some retardation of epochs. It does not, however, begin to do so until the atmospheric changes have proceeded so far that the surface of the glacier is cooled below its melting temperature, and this epoch is affected by the date when the glacier receives its winter covering of snow.

The granular surfaces are moistened with water derived from the ice of the grain. This liquid contains the greater part of the soluble impurities chiefly derived from the rock-débris which is intimately mixed with the ice. In contact with it the melting temperature of the ice is lowered, independently of any alteration of pressure. But the supply of heat available for melting ice in the interior of the glacier is very small; consequently very little of it melts. What does melt, however, is sufficient to maintain everywhere a liquid nucleus between the grains. This compact but granular mass is *the primary ice of the glacier, its colour is blue, and it persists unaltered in obscurity.*

Among the rays emitted by the sun, there are some which are absorbed by water with production of heat, although they are transmitted abundantly and without alteration by ice. When the primary ice is exposed to the sun, these rays penetrate its mass and are absorbed by the intergranular liquid, the temperature of which would be raised, but, as it is contained in a space bounded by two parallel and almost contiguous surfaces of ice, the rise of temperature is checked and an equivalent amount of ice is melted instead.

When a volume of ice is melted, its place is taken to the extent of ninety per cent. by liquid water, and to the extent of ten per cent. by gaseous water of very low pressure. The latter strikes the eye as a discontinuity in the ice: such discontinuities, when present in adequate proportion in a transparent substance, produce the impression of whiteness.

We have seen that such discontinuities are found in the artificial grotto. When the ice-wall in which they have been developed is at the entrance of the grotto, or not more than about two metres distant from it, the disintegration produced by the radiation from the sky is so considerable that the ice-wall does not remain water-tight, and not only does the ten per cent. gaseous discontinuity fill itself with air, and for the moment become really an air-bell, but the air continues to enter, and permits the ninety per cent. of water to run out, when the hundred per cent. replacement of the volume of ice by air is achieved. In this way the apparent discontinuity caused by the direct effect of radiation in the intergranular spaces has been increased ninefold by its indirect effect in bringing about drainage by destroying the staunchness of the ice-wall and causing it to leak. By our studies in the artificial grotto it has thus been established, not only that the drainage caused by the thoroughness of the disintegration of the ice exposed to the full light of day is a dominant feature of the ice within a distance of a couple of metres from the external surface of the glacier, but also that the leakiness of the ice-wall ceases almost abruptly at this distance from the entrance, so that, except in the immediate vicinity of lamps, or where the ice has been exposed to recent mechanical disturbance, the walls of the grotto are practically water-tight.

When the surface of primary ice exposed to the daylight has been obtained by clearing away the rough white ice from an area of the upper surface of the glacier, the transformations which have just been described take place very quickly.

The arched surface of the glacier facilitates drainage, and the melting of the ice on the granular surfaces enlarges the interstices between the grains, so that through them the water escapes as soon as it is produced. In the course of a day or two, according to the intensity of insolation, the compact layer of blue primary ice becomes converted into a disconnected, easily crumbling layer of white ice, which is not distinguishable from that of the undisturbed surface of the glacier. Consequently—*the white ice which forms the outside layer of every glacier is a secondary formation, and is derived from the primary blue ice of the interior by the process of sun-weathering, that is, by disintegration under the influence of solar radiation.*

As the discontinuities which produce white ice are dependent on solar action, it is unlikely that they can be present in the interior of the glacier in such proportion as to permit of their forming with the primary blue ice *a ribbon-structure* throughout the mass of the glacier, as is frequently postulated. This is borne out by the observations made during a series of years in the interior of the Morteratsch Glacier.

If the water of the intergranular spaces, which is present in the primary ice in so small a quantity compared with the mass of the ice which contains it, is able, by absorbing the disintegrating rays, to melt the ice of the grain, it must, when present in large quantity, *protect the ice immersed in it from this action*. This is proved by the behaviour of icebergs in the sea-water of the Antarctic Ocean, and in the fresh water of the Mergelin See.

No. 9.   [*From the Philosophical Magazine*, 1895, *S.* 5, *Vol. XL*,
pp. 153–172.]

ON THE USE OF THE GLOBE IN THE STUDY OF CRYSTALLOGRAPHY[1]

THE use of the globe in Crystallography is twofold. It enables us to study the character and follow the details of the form of the solid projected radially on its surface. Also all the measurements which can be made on the solid itself can be as conveniently made on its projection on the sphere, and all calculations and developments connected therewith can be made by simple graphical construction and measurements on the sphere. The black globe, with the divided circles belonging to it, is a calculating machine adapted to the solution of all the problems to which the analytical methods of spherical trigonometry are usually applied.

To the student of crystallography, of astronomy, of mathematical geometry, and of geometry of three dimensions generally, the globe and its circles fill the same place as the drawing-board and scale do to the engineer and surveyor. A globe which is to be used for geometrical constructions should be quite free and unencumbered with the fixed axis usually met with in those intended for geographical or astronomical instruction. Also the divided circles with which measurements are made should be capable of being applied directly to the surface of the globe, so as to avoid errors of parallax. The globe which I have found most suitable for the purpose, and I have used it for a number of years, is one published by Mr E. Bertaux, of 25 Rue Serpente, Paris. It has a diameter of 22 centimetres, and has either a black surface for drawing on with slate pencil or chalk, or a white

---

[1] Read before the Chemical Society, December 6, 1894.

surface for drawing on with lead pencil. To this is adapted a system of divided circles of the same radius as the sphère, called the *métrosphère*, which is the invention of Captain Aved de Magnac, of the French navy. The métrosphère consists of one complete circular band of brass, the upper edge of which is a great circle of the sphere. It is divided into degrees throughout one-half of its length, numbered from 0° to 180°. At right angles to this circle a semicircular band of the same radius passes across from one side to the other. The edge of this semicircle, which is turned towards the graduated half of the complete circle, springs from 0° and 180° respectively, and it coincides with a great circle of the globe. The combination of circle or equator and semicircle or meridian bridging its diameter resembles a crown. At the apex of the crown or pole of the equator a movable quadrant is pivoted. It can traverse the whole of the part of the sphere enclosed by the divided part of the equator and by the meridian, and it can be clamped anywhere in the divided part of the equator. The quadrant is divided into degrees, as is also the meridian. When in use the metrosphere rests on the globe, so that there is complete contact, and it can be shifted all over its surface. It is possible by its means to draw and measure any arc or angle on the surface of the globe, and consequently to solve graphically all problems of spherical geometry with an accuracy which depends only on the dimensions and workmanship of the globe and metrosphere. It is convenient to have a scale of chords of arcs of great circles of the sphere, so that arcs may be measured or laid off with a pair of compasses. The real usefulness of the globe is not to be learned by theory or precept, but by actual experience in the solution of problems, whether the study be astronomy, or navigation, or geography, or solid geometry. Within the bounds of a paper like the present it is possible only to show the general direction in which it is of use in the study of polyhedra, of which crystals are a limited class.

The fundamental data for the determination of a crystal are the angles which its faces make with each other. These are measured with the goniometer. Let us follow the process as applied to a *polyhedron of any number of plane faces, arranged in any way so as to completely enclose the space.* The polyhedron or crystal is attached to the goniometer, by means of which we can measure directly the angle included between the normals to any two faces: alongside, we have our globe with its metrosphere, on which we propose to enter the results obtained with the goniometer. We imagine the polyhedron situated inside the globe and concentrically with it ; then the normals to its faces are radii of the sphere, the surface of which they meet in points, which are called the poles of the faces. The angle included between any two such radii is measured by the great circle arc between the two corresponding poles, and it is equal to the angle between the two faces as given by the goniometer. Hence the angles between the faces as measured on the goniometer can be transferred directly to the globe, on which they are laid down as arcs.

We begin with the globe clean. Any point on it is chosen as the pole of the first face. From this point a great circle is drawn in any direction. When the angle between the first and the second faces has been measured, it is laid down on the globe as an arc of the same number of degrees of the great circle just drawn from the starting-point. The poles of the two faces, Nos. 1 and 2, are in the extremities of this arc, which is the *base line* of the survey of the crystal ; and, by triangulation from it, the poles of all the other faces can be laid down so soon as the angles between them have been measured on the goniometer or otherwise determined.

Consider a third face, No. 3, adjacent to Nos. 1 and 2, but not parallel to either of them. Let angles be measured between it and Nos. 1 and 2, and let these angles be transferred to the sphere as arcs. From pole No. 1, as centre, with the arc corresponding to the angles between Nos. 1 and 3 as radius, describe a circle on the sphere : similarly, with No. 2 as centre, and arc equal to the angle between Nos. 2 and 3 as radius, describe another circle. These circles cut each other in two points similarly situated on opposite sides of the arc 1 to 2. The "bearing" of the third face from the first two decides at once which of the two points of intersection is the pole of No. 3 on the sphere. The poles of all the other faces are obtained by an exactly similar construction. When this has been done, we have on the globe a number of points which form a complete catalogue of the faces of the crystal. Similarly, the arcs connecting each pair of poles furnish a catalogue of the inclination of every single face to every other. Every face of the crystal cuts the sphere of projection in a circle having the pole of the face as centre. Let us take the poles of two adjacent faces, Nos. 1 and 2, and with a pair of compasses draw a circle round each of them with a radius which is greater than half the arc between the two poles. These circles cut each other. Let the points of intersection be joined by the arc of a great circle. That arc is the projection on the sphere of the edge produced by the intersection of the two particular faces, the normal radii of which we have assumed to be equal.

When the lengths of the radii are taken in any other ratio than that of equality the position of the edge is shifted, but its direction remains the same. It is always perpendicular to the plane containing the normals to the faces, which form it by their meeting. Therefore the great circle which is the projection of the edge is at right angles to the great circle drawn through the poles of the two faces forming the edge, and the direction of the edge is the direction of the axis of this great circle. If we imagine the edge to be carried parallel to itself until it reaches the centre of the sphere, it will coincide with the diameter which is the axis of the great circle in which the poles of the faces lie. Let the points where this axis pierces the surface be marked on the globe. They fix the direction of the edge of the two faces, and of all parallel edges.

When this construction has been repeated for every pair of adjacent poles on the sphere, we have the projections of all the edges as arcs of great circles. And if they have been all carried parallel to themselves to the centre of the sphere and their extremities then marked, we have another series of points which catalogue the edges just as the poles do the faces of the crystal. One diameter of the sphere represents all the edges of the crystal which are parallel to it, therefore the number of the diameters which have been thus entered on the globe is the number of the different or independent edges of the crystal.

We have here considered the edges formed by the meeting of adjacent faces, or the actual edges occurring on the crystal or polyhedron under measurement. But when every face is marked on the globe by its pole, it is equally easy to determine the direction of the edge which would be formed by the meeting of two faces which are remote from one another; so that, just as we were able, by measuring the arc between every pair of poles, to catalogue the inclination of every face to every other whether adjacent or remote, so we are able to lay down and catalogue the direction of every edge made by the meeting of every face with every other however they may be situated relatively to each other.

Again, the diameters representing the edges parallel to them all cut one another in the centre of the sphere. The arc connecting the similar extremities of any pair of such diameters measures the angle included between them, which is equal to the *plane angle* included by them as edges of the face which they assist to delimit. In this way we obtain a catalogue of all the plane angles occurring on the faces of the crystal; and they are derived by a simple graphical construction from the observed inclinations of the faces. Further, the diameters representing the edges which bound one face all lie in the same plane; therefore the extremities of such diameters lie in one great circle. By drawing great circles through all the groups of diameters lying in the same plane, we get a group of great circles, each of which represents the intersection with the sphere of a plane which passes through the centre of the sphere and is parallel to the face of the polyhedron, which is bounded by edges parallel to the diameters which lie in the plane, and the extremities of which are connected by the great circle. We have thus a catalogue of the faces of the crystal which are inclined to one another, and the inclination of any two of the great circles measures the inclination of the pair of faces. Also the diameters marked as meeting the surface of the sphere in the great circle supply the number and direction of the edges which bound the particular faces (where it must be remembered that parallel edges are represented by the same diameters) and the plane angles of the face are given by the angles between the diameters. The representation of the faces of the crystal by great circle planes and that of the edges by diameters, all of which necessarily meet in the centre, facilitates the choice of a suitable system of crystallographic axes.

If we consider the poles of three adjacent faces, Nos. 1, 2, and 3, and draw small circles round them, the radii of which are equal, and of such a length that each circle cuts the other two, then, as before, we have the projection of each edge represented by the arc of the great circle connecting the intersection of each pair of circles. These three arcs cut one another in a point inside of the triangle formed by the intersection of all three small circles. This point is the projection or pole of the *corner* formed by the meeting of planes 1, 2, and 3. If this corner be carried parallel to itself to the centre, its bounding edges will coincide with their parallel diameters, which thus form representative parallel central corners. These diameters meet the sphere in the extremities, which have been already fixed. If these points be connected by arcs of great circles, they determine a spherical triangle whose area is a measure of the corner.

By a well-known rule the excess of the sum of these angles above two right angles divided by four right angles gives the area of the triangle as a fraction of the surface of a hemisphere. Corners delimited by more than three edges can be specified in the same way by splitting up the polygons which subtend them into triangles. The secondary figures thus described on the surface of the sphere are always different from the primary ones. Thus, the corners of the cube, when collected at, and radiating from the centre of the sphere, delineate the regular octahedron, which in its turn, when similarly treated, delineates the cube. It is a form of inversion.

It has thus been shown how the globe can be profitably used along with the goniometer. We have put no restrictions on the form of the polyhedron under measurement, and we have been able to render a complete account of the faces and their inclinations to each other, the direction of the edges produced by their meeting, and the plane angles produced by the meeting of pairs of edges. We have also been able to draw a complete and accurate representation of the poly-hedron as projected on the surface of the sphere, which can be studied in detail. A collateral advantage which the student gains by using the globe in the study of crystallo-graphy, or of solid geometry generally, is the excellent mental discipline which it affords. A very short training of this kind develops enormously the sense of direction. The greatest advantage is reaped by those who combine the use of the globe with the analytical treatment such as given by Miller in his treatise.

The globe offers great advantages for demonstration. It is very easy to draw correctly the projection of a crystal, especially when it belongs to one of the more symmetrical systems. In the regular system, for instance, which has the greatest number of simple forms, the edges between the faces of any of these forms are found at once when the poles of the faces have been laid down by drawing arcs of great circles at right angles to the great circle containing any pair of poles and midway between them. In this way an accurate representation of the crystal is quickly obtained. The curvature of the faces is not greater than is frequently met with in nature. Combinations of the simple forms can then be studied with advantage and in great detail. Further, hemihedral forms are as easily drawn as holohedral ones ; more especially, twins, according to any law, can be *com-posed* and drawn on the globe as easily as simple forms. In short there is no operation in the geometry of crystals which cannot with the greatest ease be performed by the use of the globe.

In bringing this matter before the Society I have not been concerned to produce anything very new or very original.

It has been my object to draw the attention of other chemists to a method of studying Crystallography which I myself have found profitable. It is true that I had never seen a globe used in the study of Crystallography, nor had I met with any suggestion of its applicability. I looked through all the available crystallographic literature without finding any indication of it. When the projection on the sphere is mentioned, it is always with a view to dealing with it according to Miller's method by spherical trigonometry.

In May 1893 I gave a demonstration to the Philosophical Society of Cambridge (in connection with globes generally) of the suitableness of black globes for studying crystals. When I began to prepare this paper I made a further thorough search through the literature, because I could not believe that the person who first had the idea to project the crystal on the sphere had done so with any other view than to study it when he had got it there. I could not meet with a copy of either Neumann's *Beiträge zur Krystallonomie*, or Grassmann's work *Zur Krystallonomie und geometrischen Combinationslehre*, which are alluded to in the beginning of Miller's *Treatise on Crystallography*, and I suspected that one or both of these authors might have recommended the use of the globe itself. A few days ago, through Messrs Mayer and Müller of Berlin, I procured a copy of Grassmann's book. Its full title is *Zur physischen Krystallonomie und geometrischen Combinationslehre.* Von Justus Günther Grassmann, Professor am Gymnasium zu Stettin : Stettin, 1829. It is the first number of the first volume of a comprehensive work entitled *Zur Mathematik und Naturkunde*, which the author proposed to complete by degrees. Nothing further was, however, published; but the single number is a sufficiently remarkable work. It is worthy of note that he was obliged to publish it at his own expense, as it found no acceptance at the hands of the Scientific Societies or the Journals of the day. This is no doubt the cause why it produced so little effect in its time and is so difficult of access now.

Of Neumann's *Beiträge* also only the first fascicule appeared, and I have not yet been able to see a copy of it. I am informed that it is excessively rare. In Grassmann's work, however, the use of the globe is expressly recommended for presenting clearly to the eye and mind the combination of directions out of which he develops the systems of Crystallography, and he states that one can make use of this instrument (the black globe) for solving by graphical construction all the problems of Crystallography, just as astronomical problems can be solved by the use of the celestial globe.

My expectation was therefore confirmed that the originator (if he was the originator) of the idea of projecting a crystal on a sphere, actually carried it out on a globe on which he made graphical constructions for the solution of problems and for the illustration of his subject. But Grassmann was not a crystallographer, he was a mathematician, and he deals with crystals mainly as affording interesting examples in nature illustrating a branch of pure mathematics, *die Combinationslehre*. The use of the globe in Crystallography proper, that is, starting from the crystal, is not dealt with at much length. His work received very little attention, and would probably have dropped entirely out of sight had it not come under the notice of Miller and suggested to him the treatment of the projection on the sphere by the methods of spherical trigonometry, which is now almost universally employed. It is to be observed that when the analytical work becomes too complicated and difficult on account of the want of symmetry of the crystal, there is no way of dealing with crystallographic problems except the geometrical, and the handiest geometrical method is the one very shortly described in this paper. For this reason also the use of the globe as a help in the study of Crystallography cannot be too strongly recommended.

POSTSCRIPT, 24*th June*, 1895.

Examples.—In giving examples of the use of the globe in dealing graphically with the relations of the faces and edges of a crystal, it is necessary, to avoid unnecessary diffuseness, to make some conventions as to nomenclature. We call the *independent* faces and edges of a crystal, those that are inclined to one another; so that parallel faces and edges are represented by one independent face or edge. When we have placed the poles of all the faces of a crystal on the globe, we can represent *all* the faces by drawing suitable small circles round the poles, and we represent all the *independent* faces by drawing great circles round the poles. If we adopt the latter process one crystal is represented by a group of great circles, the plane of each great circle being parallel to the face or faces which it represents, and the points of intersection of any pair of great circles or their nodes mark the extremities of the diameter which is parallel to and represents the edge made by the pair of faces, and all other edges parallel to it. The diagram produced on the globe by following the latter process will be called the *central representation*; that obtained by the former, the *radial projection* of the crystal or polyhedron. The radial projection of a crystal when constructed with due regard to the length of the normal radii of the faces, and, consequently, to the exact position as well as to the direction of the edges, has the great advantage of affording to the eye a bodily presentation of the crystal, with all its irregularities of development. In the central representation the variability of the normal radii and, consequently, of the size of the faces, which distinguishes the crystal from the polyhedron, is effaced, and only the geometrical properties remain. The one process or the other will be adopted according to the purpose in view.

We shall designate the great circles representing the independent faces by numerals, 0, 1, 2, 3, etc., and the independent edges by the numbers of the two great circles which produce, by the meeting of their planes, the parallel diameter. Thus, the edge made by the meeting of faces Nos. 0 and 1 corresponds to the diameter made by the meeting of the planes of great circles 0 and 1, and it is designated edge (0, 1); similarly we have edges (0, 3), (1, 5), (2, 4), and the like. The position and inclination of a diameter is fixed when the point where it meets the surface of the sphere is known; for it necessarily passes through the centre. It meets the surface in the node of the two great circles to which it is common. Hence the inclination or direction of the edge is given by the position of the corresponding node.

The nodes of intersection of the great circles will be designated by the numbers of the circles which meet in them, the opposite nodes being distinguished by a dash ('). Thus circles 0 and 2 meet in the nodes (0, 2) and (0, 2'), No. (0, 2) being the node which occurs first in azimuth, when travelling from node (0) along the equator in the direction which it has been agreed to call positive. Circles Nos. 1 and 3 meet in the nodes (1, 3) and (1', 3'); the former of these is situated in the hemisphere above circle No. 0, which is taken as the principal hemisphere of construction. Node (0, 1) being distinguished from the others by being made zero of azimuths, is called node (0), the node opposite it is called node (0, 1'). When the faces are represented by great circles, the inclination of any two is equal to the angle made by the plane of the one great circle with that of the other. They will be designated as angle (0, 1), (1, 2), and the like, the greater or smaller of the two supplemental angles being chosen according to the circumstances of the case. In order to fix the position of the great circle, another element is required besides its inclination to the great circle of reference or equator; this is given by the azimuth of its node on great circle No. 0. This also fixes the position of their common diameter and the direction of the corresponding edge. Hence the specification of a great circle representing a face includes the specification of the edge which it makes with the fundamental face of reference No. 0, and also that of its equatorial node. The plane angle made on any one face by the meeting of two other faces in it is equal to the angle between the diameters formed by the meeting of the planes of the representative great circles. Thus, the plane angle formed in face No. 0 by faces No. 1 and 2 meeting in it is equal to the angle between the diameters formed by the intersections of the planes of Nos. 1 and 2 with No. 0, and it is represented by the arc of great circle No. 0 contained between nodes (0, 1) and (0, 2). Such angles will be designated by the numbers of the three faces which are concerned in their production; the face containing the angle coming first. Thus angles (0—1, 2), (2—1, 3) mean the plane angle on face No. 0 made by faces Nos. 1 and 2 and that formed on face No. 2 by faces Nos. 1 and 3. The great circle corresponding to the fundamental face of reference will always be called No. 0, and the fundamental edge (0, 1) is the diameter made by the intersection of the planes of great circles Nos. 0 and 1. One of the nodes of these two circles, which will be called node 0, is chosen as the zero of arcs of azimuth, which are measured along great circle No. 0, from 0° to 360°. Points lying outside of great circle No. 0, which is the equator of our system, are further fixed by their altitude above or below it. The *original position* of the metrosphere means the one which it had when great circle No. 0 was drawn and node 0 was marked: then the equator of the metrosphere coincided with the equator of the diagram on the globe, and the zero of azimuths on the metrosphere corresponded with that on the globe.

In central representation, positions on the sphere are expressed by the azimuth from node (0) in which a great circle passing through the pole of circle No. 0 and the point cuts circle No. 0, and the arc on this circle contained between its intersection with circle No. 0 and the point. This is the altitude of the point and it is + or − according as it is to the right hand or the left of the equator, when moving in the positive direction of the measurement of azimuths. The coordinates of a point will be expressed shortly in the form $(\theta°, \phi°)$, where θ is the azimuth and ϕ the altitude. The azimuth will always come first. In central representation a face is specified by giving the angle which the plane of the great circle which represents it makes with the plane of the equator, or great circle No. 0, and the azimuth of one of the nodes of intersection of the two circles. An edge is specified by azimuth and altitude of the node marking its extremity in the positive hemisphere. As it is a diameter of the sphere its direction is fixed.

As the measurements with the metrosphere used may vary to the extent of half a degree, angles and arcs whether observed or calculated are stated in degrees, and half or quarter degrees. It is impossible to illustrate examples without the aid of the globe itself, and the reader should have one before him in order to make for himself the construction described. The black or white globe with metrosphere mentioned in the paper is, on the whole, the best that is to be had in the market. The sphere Lejeune, to be had of the same publishers, is much more elaborate and admits of arcs being laid down and measured with greater accuracy and precision than the metrosphere of de Magnac, but it is less handy and costs four times the money. The handiness of the metrosphere is its great recommendation. One metrosphere can be used with any number of globes; so that separate details can be worked out on separate globes and combined on others, thus avoiding the risk of mistakes due to overcrowding. As extra globes for the metrosphere cost only ten francs each, the expense is small compared with the convenience afforded.

In the absence of a globe especially constructed for drawing, an ordinary terrestrial or celestial globe may be used with profit. It should be fully mounted, with complete brass meridian and wooden, or metal, equator, and should be furnished with a flexible brass quadrant to be attached to the meridian in the pole of the observer or elsewhere. If now the brass meridian be brought into such a position that the axis of rotation of the globe, which already lies in the plane of the meridian, comes to lie also in that of the equator, and the quadrant be attached to the meridian in the pole of the equator, we have an exact representation of the metrosphere, excepting that the circles in the ordinary mounting stand a little way off the globe, while those of the metrosphere touch and rest on it. The principal great circles such as the equator, ecliptic, and prime meridian are always clearly marked on a globe, and they are valuable as circles of reference. One of them is taken as circle No. 0. Great circles can be easily shown on the ordinary globe by pieces of twine or thin tape joined up at their extremities by an elastic band. If the tapes or threads are of different colours, the constructions made with them are very plain and easily followed. A point may be marked on a varnished globe by attaching a small piece of gummed paper near its position, and marking the exact position on the paper with a pencil. Small circles, if they are wanted for themselves, are difficult to deal with ; but it is usually only intersections of pairs of them that are wanted and they are dealt with as points. If the globe is kept fixed with circle No. o in the plane of the equator, points on its upper hemisphere can be laid down in altitude and azimuth by means of the flexible quadrant. It will be seen, therefore, that an ordinary globe, even though its surface be not prepared for drawing on, can be used for the purposes of crystallography. The terrestrial globe is to be preferred to the celestial, because on the latter the meridians are those of celestial longitude which meet in the pole of the ecliptic, while the axis of the globe passes through the poles of the equatorial.

As a first example let us consider the cube, because its details are so simple and so familiar that they can be easily followed without drawings and models. If we place it on the goniometer and measure the inclination of its contiguous faces, we remark that they are all right angles, and we mark the poles on the globe in the way already described. Let us now draw small circles round each pole, the radius of the small circles being sufficiently great for neighbouring circles to intersect. If we then draw great circle arcs through each pair of intersections, we shall, after obliterating the small circles and other superfluous lines, have not only a correct projection of the cube on the sphere, but also a striking bodily presentation of it. All the geometrical details can be studied on it. No face is represented by a great circle ; therefore, in specifying positions we have to take a great circle in a plane parallel to one of the faces for our equator of reference, and any point on it as the intersection with another face for the zero of reckoning. We find, on examining our projection, that there are only three independent faces, and we can at once construct the central representation of the crystal by drawing great circles round the poles of the faces.

Place the metrosphere on the globe and draw the complete

equatorial circle representing face No. o of the crystal.
Mark on it the nodes of the meridian. The one corresponding to o° on the equator of the metrosphere is node No. o,
from which azimuths are measured. The diameter connecting
nodes No. o and No. (o, 1′) represents edge No. (o, 1) which
face No. 1 makes with No. o. From the specification of
the crystal the angle of inclination of these two faces is a
right angle. Therefore in its *original position* the meridian
of the metrosphere coincides with great circle No. 1. Let it
be drawn. The third independent face, No. 2, which we
find on the crystal is inclined to each of the other two at a
right angle. With the metrosphere in its original position,
clamp the movable quadrant at 90° of azimuth, the quadrant
coincides with circle No. 2. It is generally more convenient
to bring the meridian to coincide with the great circles to be
drawn, because then the complete semicircle is drawn at
once, and both nodes are marked at the same time. For this
purpose the metrosphere is rotated from its original position
round the axis of the equator, which remains coincident with
circle No. o, until division 90° of azimuth coincides with
node (o, 1′). The meridian now corresponds with the position
of circle No. 2. Let it be drawn. Circle No. 2 obviously
cuts No. 1 at right angles, which can be at once verified by
measurement of the arcs between node (1, 2) and nodes (o, 2)
and (o, 2′). If we attempt to place any of the three remaining
faces, we find that its place is already occupied by one of
the circles, o, 1, or 2, to which it is parallel. If we now wish
to specify the crystal in terms of its central representation on
the sphere, we have, for the inclination of the faces o to 1,
−90°; o to 2, −90°; and 1 to 2, −90°; and, for the position
and direction of the edges :—edge No. (o, 1) azimuth o°,
altitude o°; edge (o, 2), −(90°, o°); and edge (1, 2) azimuth
indeterminate, alt. 90°. The plane angle made on any face
by any other two is equal to the arc on the corresponding
circle contained between the nodes made by it on meeting the
two other corresponding circles. In the present case they are
found by measurement, as they are seen by inspection to be
right angles.

In representing the cube by great circles parallel to its faces, we have divided the hemisphere into four equal and similar triangles. If we regard this diagram as the spherical projection of a polyhedron where *all* the faces, whether independent or not, are represented—where in fact the faces have been delineated by drawing small circles round their poles, and arcs of great circles through the intersections of these circles to give the edges, we find that we have here the radial projection of the regular octahedron. The transference of the faces of the cube parallel to themselves until they coincide with planes of great circles, has the effect of transferring the corners of the cube parallel to themselves to the centre. If the extremities of each set of three edges which go to form a corner are connected by arcs of great circles, the diagram produced on the globe is the one which we have been considering. As was pointed out in the paper, the corners of the cube when radiating from the centre of the sphere delineate the projection of the octahedron on it. This fact can be expressed by saying that the central representation of the cube is identical with the radial projection of the octahedron.

Models exhibiting the central grouping of corners, and the forms thereby produced, are quite easily made out of cardboard or stiff paper, and are very instructive.

Let us consider the regular tetrahedron. The normals to its faces, four in number, are found to be inclined to each other at an angle of $109\frac{1}{2}°$ across an edge. The poles of these faces when placed on the globe form a group of four points symmetrically arranged, each being separated from its neighbour by an arc of $109\frac{1}{2}°$. Draw the great circles of which these points are the poles, and call any one of them No. 0. Take its intersection with any other and call it node (0), the opposite node is (0, 1′), and the great circle which intersects No. 0 in these nodes is No. 1, diameter (0, 1) being parallel to and representing the edge made by Nos. 0 and 1. Another pair of nodes are found in azimuths 120° and 300°. Their diameter represents the edge (0, 2), and the great circle represents face No. 2. Similarly the nodes of circle No. 3 are in azimuths 240° and 60°. The inclination of the faces is equal, and we find that the altitude of circles 1, 2, and 3 is 70½°.

The edges (1, 2), (2, 3), and (3, 1) are found by measurement of the positions of nodes (1, 2), (2, 3), and (3, 1); node (1, 2) in (330° and 54¼°); node (2, 3) in (90°, 54¼°); and node (3, 1) in (150°, 54¼°). The theoretical figures are here given because the observed ones have been mislaid. The errors, which did not exceed 1½°, occurred in the quarter sphere of azimuth over 180°. This is always the case, because the metrosphere is never exactly true to the globes, and the errors show in proportion as it is shifted from its original position.

If we consider, as we did in the case of the cube, what form gives in radial projection the same diagram on the globe as the central representation of the tetrahedron, we find that it is the combination of the cube and octahedron "in equilibrium." The radial projection of each of its edges is an arc of 60°, and its principal sections are regular hexagons.

In the central representation of the tetrahedron let one of the great circles be suppressed. There remain three, and they constitute the central representation of the rhombohedron, which consists of six rhombs with plane angles of 60° and 120°.

When this diagram is looked on as a radial projection, it is that of an octahedron in rhombohedral position, the two basal faces being equilateral triangles, and the six prismatic faces being isosceles triangles in which each of the equal sides is double of the base.

Let us now consider the pentagonal dodecahedron as it occurs in nature as pyrites, and is designated by Miller π {012}, and by Kopp $\frac{1}{2}(2a : a : \infty a)$. We shall consider one face, which shall be represented by circle No. 0, and the five contiguous faces, Nos. 1, 2, 3, 4, and 5, which by their intersections give No. 0 its pentagonal shape. From the specification or measurement of the crystal we have the inclination of the normal of 0 to the normal of four of the other faces 66° 25·3′ or 66½°, and to the fifth 53° 7·8′ or 53°.

Let No. 1 be inclined to No. 0 at 53°, then Nos. 2, 3, 4, and 5 are all equally inclined at an angle of 66½°. Place the poles of these faces on the globe. Draw the great circles of which they are the poles, then we have great circles Nos. 0, 1, 2, 3, 4, and 5 respectively parallel to and representing the faces of the same number. We may also proceed directly to draw the great circles. Place the metrosphere on the globe : describe great circle No. 0 coincident with its equator, and mark the nodes (o) and (o, 1'), also the pole of No. 0. Clamp the quadrant at 90° of azimuth on the equator and incline the metrosphere round the common axis of the equator and meridian until the angle between the meridian and circle No. 0 is 53°. The meridian is now in the position of great circle No. 1. Let it be drawn, and let its pole be marked. The two circles cut one another at an angle of 53°, and the diameter (0, 1) represents the singular edge of pentagons Nos. 0 and 1.

By specification the inclination of faces 1 and 2 is 66½°, and equal to the inclination of 2 and 0 ; we have now to place a third great circle on the globe, which shall cut both the others at an angle of 66½°. Clamp the movable quadrant at an azimuth of 66½° on the equator. Bring the quadrant to coincide with circle No. 0 and let it slide along it. Then, if it is carried over the whole semicircle of No. 0, the meridian must somewhere coincide with the position of circle No. 2. While the quadrant is being slid along No. 0, the pole of the meridian is describing a small circle parallel to No. 0, which may be drawn on the globe. Now bring the quadrant to coincide with circle No. 1 and let it slip along it, marking the small circle parallel to No. 1 which the pole of the meridian describes. The two small circles cut each other in one point in the hemisphere of construction. Now, with the quadrant coincident with No. 0, bring the pole of the meridian (90° of azimuth) to coincide with the intersection of the two small circles ; the meridian coincides with circle No. 2. Similarly let the quadrant coincide with circle No. 1, and the pole of the meridian with the intersection of the small circles ; then the meridian will be found to coincide with the circle No. 2 already drawn. In actual construction the coincidence was within half a degree. The method just described for finding the pole of No. 2 is inconvenient in practice. When the quadrant has been clamped at $66\frac{1}{2}°$ it is slipped along No. o to what appears to be a likely position, and the position of the pole of the meridian corresponding to it is marked on the globe. It is now applied to No. 1 at a likely place, and the position of the pole of the meridian corresponding to it is marked. If the first experimental position of the pole of the meridian be called a and the second b; then, from the pole of circle No. o with radius equal to the distance of a draw a small circle, it is necessarily parallel to No. o; and from the pole of No. 1 with radius equal to the distance of b draw another small circle, which is necessarily parallel to circle No. 1. These are the same small circles as those above described, and the great circle described from their intersection as pole is necessarily identical with No. 2 already drawn. In making the construction on the globe the pole of No. 2 was found in this way. Small circles can be drawn with the metrosphere, but they are much more easily and accurately drawn with a pair of compasses. The position of the circle as drawn was proved by measuring the inclination of its plane to those of Nos. o and 1. The angle (2, 1) was found to be exactly $66\frac{1}{2}°$, and the angle (2, 0) was 67°. The arc or circle No. o contained between the nodes made by its meeting with Nos. 1 and 2 is equal to the plane angle on face No. o made by faces Nos. 1 and 2 meeting in it. By measurement on the globe it was found to be 101° instead of $102\frac{1}{2}°$. Similarly the arc between the nodes (1, 2) and (0, 2) is equal to the plane angle on face No. 2 formed by the meeting of faces o and 1 in it. It was found, on measurement, to be 122° instead of $121\frac{1}{2}°$. In the same way the arc between nodes (1, 2) and (0, 1), representing the plane angle in No. 1, was found to be 103° instead of $102\frac{1}{2}°$.

Great circle No. o being our equator of reference, and node (0), one of its intersections with circle No. 1, being our zero of azimuths, we have the positions of the faces and edges entered on the globe as follows :

Face No.	1	2			
Edge No.	....	....	(0, 1)	(0, 2)	(1, 2)
Azimuth...................	0°	102°	0°	77½°	288°
Altitude	53°	66½°	0°	0°	52°

These constructions were made with no greater care than is absolutely essential in carrying out any quantitative graphical work, and the plane angles agree within 1½° of the theoretical values. With the very greatest care, errors of half a degree cannot be insured against with a globe of 22 centimetres and a metrosphere divided into whole degrees. With a larger globe and divided circles, especially adapted to this kind of work, greater accuracy could be obtained; but when the errors do not exceed one degree, the angles obtained are sufficiently exact to be used in the construction of models, which always forms the crucial test of the student's work on the globe or in crystallographic composition.

By an exactly similar construction we lay down circle No. 3 representing face No. 3, and on measurement we find for the positions of circle No. 3—(151½°, 65½°), and edge No. (2, 3), (25°, 61°). When we now come to face No. 4, it is inclined to No. 0 at 66½° and to No. 3 at 53°; we have therefore to find a great circle which cuts No. 0 at 66½° and No. 3 at 53°.

The operation presents no more difficulty than when the angles were equal. The quadrant of the metrosphere is set to 66½° and applied to circle No. 0; then it is set to 53° and applied to No. 3; experimental poles being marked in each case giving the radius of the small circles to be drawn from the poles of Nos. 0 and 3 respectively. The intersection of these circles marks the pole of the great circle sought, No. 4. The quadrant is now set either to 53° or 66½°; if to 53°, then it is applied to circle No. 3 and slid along it until the pole of the meridian is in the intersection of the two small circles. The meridian of the metrosphere then coincides with circle No. 4, which is accordingly drawn. When the quadrant, set to 66½°, is applied to circle No. 0 and slid along it till the pole of the meridian is again in the intersection of the two small circles, the meridian ought to coincide with the circle No. 4, already drawn. This construction was carried out and the resulting great circles fell less than half a degree apart. The coordinates of the plane are (209°, 66°). Circle No. 5 is inclined at an angle of 66½° to both its neighbouring faces and is laid down exactly as No. 2.

After laying down the great circles Nos. 0, 1, 2, 3, 4, and 5, representing the fundamental face No. 0 and the five neighbouring ones, which by their intersections make No. 0 a pentagon, the positions and inclinations of faces and edges were taken off with the metrosphere. The measurements are along great circle No. 0 from one of its intersections with No. 1 as 0° of azimuth, and altitudes above or below it.

No. of face or great circle	1	2	3	4	5
Azimuth of node	0°	102°	151½°	209°	282°
Greatest altitude	53°	66½°	65½°	66°	66°

These are laid down directly from the data supplied for the crystal, and the small differences of the values thus found on the globe from those intended to be placed on it will give an idea of the precision to be obtained with the particular globe and circles. In the following table we have values derived graphically from this construction for the position and direction of edges formed by the meeting of any two of the faces

Node No.	1, 2	2, 3	3, 4	4, 5	5, 1
Azimuth	288°	25°	90°	156°	250°
Altitude	52°	61°	63°	61½°	51°

Node No.	1, 3	1, 4	2, 4	2, 5	3, 5
Azimuth	343°	109°	55°	270°	127°
Altitude	23½°	22°	44°	24½°	43½°

represented on the globe by great circles. As before, the edges are represented by their parallel diameters, and these are designated by the nodes in which they meet the sphere. The azimuth and altitude of each of these nodes is given in the table. As the diameters also pass through the centre of the sphere, the edge is specified both as to direction and situation.

The first five of these edges actually occur; the last five are the remaining possible edges which would be formed by the faces produced. They are shown on the globe exactly the same as the others, because, if a number of great circles be drawn on a sphere, every one of them bisects every other.

In the next table we have the plane angles on face No. 0 made by pairs of its contiguous faces. Here also we have actually occurring and possible angles.

Nos. of contiguous faces	1, 2	1, 3	1, 4	1, 5	2, 3
Plane angle observed ...	103°	30°	27°	100°	107°
,, calculated...	102½°	30°	29°	102½°	106½°
Excess......................	½°	0°	− 2°	2½°	½°

Nos. of contiguous faces	2, 4	2, 5	3, 4	3, 5	4, 5
Plane angle observed ...	50°	22°	123°	50°	108°
,, calculated...	50°	25½°	121½°	50°	106½°
Excess......................	0°	−3½°	1½°	0°	1½°

If we suppose faces Nos. 1, 2, and 4 to be produced, they meet in three edges forming a corner. The three plane angles forming this corner are readily found on the globe. They are the angles contained between the edges (1, 2), (2, 4), and (4, 1) on the faces 4, 1, and 2 respectively. Hence we have only to measure the arcs on great circles 4, 1, and 2 included between their intersections with circles 1 and 2, 2 and 4, and 4 and 1 respectively; and by measurement they were found to be :—plane angle on face 1, 108°; on face 2, 72°; and on face 4, 75½°. Similarly, if we produce faces 1, 3, and 5 to meet in a corner, the plane angles then are:—on face No. 1, 107°; No. 3, 76°; and No. 5, 74°. This is a group of similar faces to the last. If faces 1, 3, and 4, a different group, be prolonged they meet in a corner with plane angles on No. 1, 59°; on 3, 77°; and on 4, 103°.

The interdependence between the plane angles of the faces of a crystal and the inclinations of the faces and the edges can be very conveniently studied on the globe.

Consider a face No. 0, as represented by its parallel great circle on the sphere, which shall be the equator of reference. It is cut by any other face, No. 1, along a diameter one extremity of which is taken as zero of azimuths. Let the plane angle to be formed on No. 0 by faces Nos. 1 and 2 be 112°. Lay off on the great circle No. 0 an arc of 112° azimuth; the face No. 2 cuts No. 0, on the diameter of which this point is an extremity.

Let the plane angles of 1 and 2 be equal and 112°; we have to find their inclinations to No. 0 and to each other, and the direction of the edge which they make with each other.

Set a pair of compasses to a span of 68°, which is the supplement of 112°, and with this radius describe small circles from adjacent extremities of the diameters (0, 1) and (0, 2). These circles cut each other in one point in the hemisphere. The position of this point is found by measurement to be azimuth 238°, and altitude $48\frac{1}{2}°$. Through this point and the zero of azimuths (0°, 0°) draw a great circle which marks the position of the circle representing face No. 1. Also through the points (238°, $48\frac{1}{2}°$) and (112°, 0°) draw a great circle which is parallel to face No. 2. The arcs intercepted on each great circle by the other two are 112°, and they represent the plane angles of the three faces meeting in the corner; with the quadrant clamped at 90°, place the meridian on the points of intersection of No. 1 with Nos. 0 and 2, and the altitude of No. 1 is found to be 54°. By similar measurement the altitude of No. 2 is 54°. Therefore the faces 1 and 2 make angles with face No. 0 of 54° or 126°, according as the inside or the outside of the solid is considered. When the metrosphere is placed so as to measure the inclination of Nos. 1

and 2, it is found to be $53\frac{1}{2}°$. The azimuth and altitude of the point of intersection of Nos. 1 and 2, which have been found by measurement to be 238° and $48\frac{1}{2}°$, give the position and direction of the edge made by Nos. 1 and 2. The altitude $48\frac{1}{2}°$ is also the inclination of the edge made by two of the faces to the third face.

Let the plane angle on No. 0 be 120°, and the plane angles on Nos. 1 and 2 be 108° and 90° respectively. By exactly the same construction we find, on measurement, that the faces which fulfil these conditions are inclined 0 to 1 at $73\frac{1}{2}°$, 0 to 2 at 58°, and 1 to 2 also 58°. The edge formed by Nos. 1 and 2 meets the sphere in 210° azimuth and 58° altitude. The plane angles here assumed are those of the square and the regular hexagon and pentagon.

If the edge made by Nos. 1 and 2 lies in azimuth 220° and altitude 50°, the base angle being 120°, to find the inclination of the faces and the plane angles of 1 and 2. Great circles are drawn through the points (0°, 0°) and (220°, 50°) giving circle No. 1, and through (120°, 0°) giving No. 2. The angles which they make with each other are then measured and found to be:—0 to 1, 61°; 0 to 2, 50°; and 1 to 2, $53\frac{1}{2}°$. The plane angles are:—on No. 1, 118°, and on No. 2, 97°, the angle on No. 0 being given 120°.

The resources of the globe are inexhaustible; but the above examples may suffice for the purpose with which this paper was written; namely, to inform some, and to remind others, of the usefulness of the globe as an instrument of research.

No. 10. [*From the Proceedings of the Cambridge Philosophical Society*, 1901, *Vol. XI, Pt I, pp.* 37–74.]

ON A SOLAR CALORIMETER USED IN EGYPT AT THE TOTAL SOLAR ECLIPSE IN 1882 [1]

WHILE engaged in discussing questions connected with the physics of the ocean, I found the want of definite knowledge of the amount of solar heat which really reaches the surface of the land or sea in a form which can be collected, measured and utilised. There was no lack of actinometrical observations, but I found it impossible from them to obtain the data that I sought. The aim of most observers has been to arrive by more or less direct means at what is known as the solar constant, that is, the quantity of heat which is received in unit time by unit surface when exposed perpendicularly to the sun's rays outside of the limits of the earth's atmosphere. For my purpose the radiation arriving at the outside of the earth's atmosphere was of no importance. What I did want to know and to measure was the amount of solar radiation which strikes the earth at the sea-level and is there revealed as heat. It is the energy of this radiation which maintains the terrestrial economy. I was not satisfied with the values of it which could be deduced from the experiments which had been made with the view of ascertaining the value of the solar constant, and I determined to utilise the opportunity of a visit to Egypt in company with the expedition for observing the total eclipse of the sun in 1882 to make observations for myself on the amount of heat which could actually be collected from the solar radiation in these

---

[1] See *Proceedings of the Royal Society of Edinburgh* (1882), vol. xi. p. 827.

favourable circumstances.   I determined to use a calorimeter which should depend for its indications on change of state and not on change of temperature.   The ice calorimeter naturally suggested itself; but apart from the fact that in 1882 ice was not so universally procurable as it is now, the indications of the ice calorimeter are apt to be seriously modified by the condensation of moisture from the air.   I therefore determined to make a steam calorimeter in which the sun's rays should be collected by a conical reflector of definite area and thrown on an axial tube which should represent the boiler.

Locality.—The astronomers had fixed on a spot on the banks of the Nile close to the town of Sohag and in latitude 26° 37′ N. for the observation of the eclipse, and experience showed that it had been very well chosen.   The eclipse was total at 8.34 a.m. on the 17th May, 1882, civil reckoning. The maximum duration of totality that was expected was 70 seconds, and in fact it lasted longer than 65 seconds.

The expedition arrived on the 8th May, and I was able to begin work with the calorimeter on the 11th.   As the instrument was new in every way the work of the first few days was mainly directed towards learning the manipulations and finding out and rectifying defects.   The instrument worked at once much more satisfactorily than I could have expected, and the only important alteration which had to be made was to replace the original metal dome as steam space by a glass tube.   This performed the functions of a gauge-glass, a steam space and a guarantee against priming.   It is of course essential that nothing but condensed steam should arrive in the receiver, and with the glass steam dome this can be assured.   Improvements of one kind or another were made every day up to the 15th.   On the 16th, 17th and 18th experiments were carried out with the apparatus in best working order and under very favourable circumstances. They are collected in Table III.   The observations made on the morning of the 17th immediately after the total phase of the eclipse are given separately in Table IV.   The instrument was constructed and was taken out to Egypt for use under ordinary conditions.   Its exposure during the later phases of the eclipse was not originally contemplated, yet the results are full of interest.

The observations were made on the 16th, 17th and 18th May. The sun's declination at apparent noon was 19° 8′, 19° 22′ and 19° 35′ on these days respectively. We take the mean declination for the period as 19° 22′ N.

The latitude of the station being 26° 37′ N. the mean meridian altitude of the sun was 82° 45′ = 82°·75. Table I gives the sun's altitude and azimuth at noon and at every half-hour on each side of noon until sunset. These data were obtained graphically by measurements on the globe. It will be seen that when the sun is more than one hour from the meridian its altitude changes at the rate of about 6°·5 in half an hour. When the altitude has fallen to 45° about 3¼ hours from noon, the water of the boiler has begun to invade the glass steam dome owing to the inclination which it is necessary to give the instrument in order to keep it pointed towards the sun. This does not prevent the instrument acting perfectly well, as will be seen, in Table III, from the observations made on the afternoons of the 17th and 18th, but it is necessary to watch the operation very closely. Moreover, the principal object of the observations is to find the maximum distilling effect of the sun, and this is not likely to occur when it is more than three hours either before or after noon. The period during which, if possible, continuous observations should be made is from 9 a.m. to 3 p.m.

TABLE I.

Hour from Noon	Sun's Azimuth	Sun's Altitude	Hour from Noon	Sun's Azimuth	Sun's Altitude
0	0	82·75	4	95	34·5
0·5	46	79·5	4·5	98·5	28
1	65	74	5	102	21·5
1·5	73·5	67·5	5·5	105	15
2	81	61	6	108	9
2·5	85·5	54	6·5	111	2
3	89	48	6·65	112·5	0
3·5	93	41			

During the three days, the 16th, 17th and 18th May, the conditions were very favourable, and particularly so on the forenoon of the 18th when the sun was intensely hot and the atmosphere motionless.   On none of these three days was the sun at any time obscured by cloud.   The only interference with the sun's rays was by dust, with which the desert air is always charged.   It made itself evident by settling upon the mirror surfaces of the reflectors, where it formed a dust of

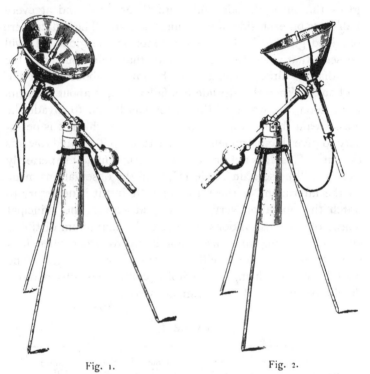

Fig. 1.   Fig. 2.

infinite fineness which could only be seen when regarded edgewise.   The results obtained on the forenoon of the 18th are to be taken as the best.

Figs. 1 and 2 are photographic representations of the instrument showing its general appearance and arrangement. Fig. 3 gives a section of the calorimeter, and Fig. 4 a perspective view.

Construction of the Calorimeter.—Fig. 3 is a principal section of the instrument by the plane which contains its axis OP and that of the earth QS. The dimensions of the parts are most easily specified by their projections on the axis OP and on a line at right angles to it; but as the sections at right angles to OP are all circular it will be sufficient specification of the section of the instrument at any point, say B, to give the distance LB on the axis from one extremity L of the steam tube to B, and the radii of which the projections on the axis are the point B. These radii would be, in order from the axis outwards, $\frac{1}{8}$, $\frac{1}{4}$, 3 and 4 inches.

In the following table the first line contains the points on the axis, the second contains their distance from L, the lower extremity of the steam tube, the third contains the radius of the innermost circle, the section of the steam tube, and the following lines contain the radii of the other circles in ascending order.

Points on the axis ...	L	K	C	B	A	E	F	G	H
Distance of points from L	0	4	16	17	19	19½	20½	22	22½ inches
Radii of the circular		⅛	⅛	⅛	⅛	⅛	⅛	⅛	,,
sections of which each		½	¼	¼	¼	¼	15/32	15/32	,,
point is the centre		1	1	3	5	¼	¼	6¼	,,
and the projection			2¼	4	5½			7	,,

The measurements are given in British units, because these were used in its construction. The construction of the reflector will be described later. The mirrors are carried on arms of sheet brass which spring from a piece of brass tube which fits telescopically over the condenser tube. Their outer extremities are kept in position by being fixed to a flat ring of sheet brass, ¼-inch wide. Each mirror is made of a properly shaped band of "reflector-metal," that is, of sheet copper to which a sheet of silver has been made to adhere by being rolled under great pressure. It is then bent round until its edges abut, when they are soldered together[1].

The main condenser is the tube KC, 12 inches long and 2 inches in diameter. Out of it at the top springs the boiler

---

[1] In the original memoir it was erroneously stated that the silver of the mirrors had been electrolytically deposited on the copper.

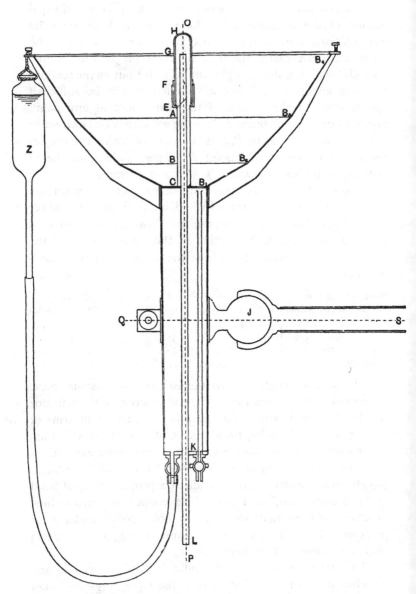

Fig. 3. Calorimeter. Principal Section.

tube $CBAE$, of silver and $\frac{1}{2}$-inch in diameter. At E, and from E to F its diameter is 1 inch, and this carries the steam dome, which is a glass tube closed at one end and inserted into the part EF, where it is fixed with a screw collar and washer. The steam tube passes axially through the whole instrument, terminating just inside of the glass dome. The steam condensed in it runs out at the lower extremity L and is received in a graduated tube in which it is measured or weighed. The glass reservoir Z, which is shown hanging from the outside rim of the reflector, is connected by an india-rubber tube with the bottom of the condenser and the instrument becomes a U-tube, of which the reservoir and india-rubber connection are one limb and the condenser and boiler the other. The instrument is thus easily filled with water and the height at which it stands in the space EF is regulated by means of Z.

When the instrument is going to be set in action it is pointed axially to the sun. When in this position, the tube EF throws a strong circular shadow on the top of the main condenser CB, and concentric with it. With the rotation of the earth the axis moves away from the direction of the sun and the shadow becomes eccentric. The appearance of eccentricity strikes the eye at once, and it is rectified by a slight motion of the instrument round its polar axis. The instrument requires adjustment every two or three minutes.

When pointing truly to the sun all the rays which strike the reflector are reflected on the length AB of the axis. But the boiler tube having a radius of $\frac{1}{4}$-inch intervenes and receives these rays on its blackened surface. The rays reflected from the inner extremity of the inner mirror are reflected on a part of the boiler tube a little below the line BB_2, and those reflected from the outer extremity of the outer mirror are reflected on a part of the boiler a little above the line AB_3. This is due in both cases to parallax.

When the sun's rays strike the surface of the boiler, those that are not thrown back again are absorbed by its blackened surface and passed by conduction through the metal to the water which occupies the space round the steam tube. When everything was at the temperature of the air, and the instrument was pointed to the sun at 2 p.m., the water boiled in 40 seconds, and it continued to boil so long as the instrument truly followed the sun and as the sun was not obscured. This operation had to be stopped when, in order to follow the sun, the instrument had to be inclined at such an angle that the water of the boiler began to trespass too far into the glass dome. The greatest meridian altitude of the sun was 83°, and it was found inconvenient to follow the sun to altitudes less than 45°, so that the instrument was never used in a truly vertical position. This has an advantage which will be appreciated by the chemist, who always inclines a test-tube when he is going to boil a liquid in it. The boiling proceeded with perfect regularity even when the sun was at its hottest, as on the forenoon of the 18th May ; and with the glass dome as steam space everything could be followed minutely. The steam developed in the boiler rises into the dome, from which it finds exit through the inside steam tube GL. In it the steam passes at least as far as B uncondensed, because the temperature of the water boiling outside is slightly higher than its own. But immediately it passes B it is surrounded by water which at first is colder than itself and it is condensed. In this process the steam gives out its latent heat and raises the temperature of the water outside in the condenser correspondingly, and the water produced from the steam runs down the tube and is caught in the receiver. When steam is in presence of water there is no delay in condensation so soon as the temperature of the water is the smallest fraction of a degree below the temperature of saturation. Therefore so soon as the water which moistens the inside of the steam tube has been cooled at all, it instantly condenses steam sufficient in amount to raise its temperature to that of saturation. The result is that the actual condensation of the steam takes place at the upper part of the condenser and immediately below the boiling space. As the instrument is to all intents and purposes motionless and no circulation of water is maintained in it, the hot water remains at the top of the condenser and from it hot feed is supplied to the boiler. While there comes to be a layer of considerable thickness of water at or very near the boiling-point at the top of the main condenser, that part of this water which finds itself forced into the annular space *CB*, if it is not actually at the boiling-point when it enters at *C*, as its inner surface is heated by the full supply of steam as it leaves the boiling water, it cannot fail to attain the boiling temperature before it reaches *B*[1]. Therefore *when the instrument has settled down into steady working, the whole of the heat which reaches the water from the sun is used in transforming water at its boiling-point into steam of the same temperature.* It is essential that the distillation be kept running continuously and the water produced in successive intervals of time weighed or measured. If the meteorological conditions are such that the boiling is interrupted, then it is of no use attempting to make observations, as they would have no value. The reason why I thought it so important to have the apparatus for use with the expedition was that the climate in Egypt in the month of May is very dry and hot and the sky usually cloudless, while the sun also attains a very considerable meridian altitude. Further, of all the results obtained, the one of greatest importance is the maximum. It is necessarily lower than the possible maximum with a perfect instrument under perfect meteorological conditions. But in order to know that we have the maximum we must make many observations, because the conditions

---

[1] It is a *conditio sine quâ non* of the working of the instrument that the water in this annular space, which represents the immediate feed of the boiler, be always, and automatically, maintained at the boiling temperature, and the experimenter who uses it must have this present in his mind. It need not, however, cause him any serious preoccupation. It is amply assured by the very high latent heat of steam. The first gram of steam, in condensing at 100° C., gives out 535 gram-degrees of heat which can raise the temperature of 5·35 grams of water from 0° to 100°. The next gram of steam can, in condensing, raise the temperature of a further quantity of 5·35 grams of water from 0° to 100°. When working under ordinary favourable circumstances, about 1·2 grams steam were made and condensed per minute. Therefore, after the calorimeter had been running for ten minutes, and supposed starting at the temperature of melting ice, there would be an accumulation of sixty grams of boiling water at the top of the condenser, and out of this only twelve grams would be required to make good the water evaporated.

that are apparently the most advantageous are not always so in reality.

It will be seen that besides the tap at the bottom of the condenser which communicates by the india-rubber tube with the reservoir there is one which communicates with the top of the condenser ; it was intended for the removal of the hot water as it was replaced by colder water at the bottom. It was found better to allow the hot water to accumulate at the top, as has been described, and, when the heat threatened to pass too far down, to change the whole of the water and start afresh. As the temperature of the water for an inch or two at the top is at, or nearly at, boiling temperature it loses heat by radiation and convection at a much greater rate than if the water of the condenser were thoroughly mixed and assumed an average temperature. This is an important feature. It is however better with a reflector having the condensing power of the one used to have a larger condenser not only in order to hold more water and so render less attention necessary, but for the mere mechanical purpose of balancing the weight of the reflector. In the instrument used the two were much too nearly of a weight. In designing another, I should make the diameter of the condenser 3 inches, and its length 18 inches. The other dimensions seemed to be in every way suitable. The condensing power of the reflector depends on the ratio of its effective area to the focal surface of the boiler tube. For the same length of focal line the heating surface varies with the circumference or the diameter of the tube, so that if the diameter of the boiler tube were increased from half-an-inch to three-quarters of an inch the condensation of rays would be 32 instead of 48 fold. From experience during the latter part of the eclipse this intensity would be insufficient and it would be necessary to increase the area of the reflector in about the same ratio, we should then be able to collect water at the rate of over 2 grammes per minute, which would be an advantage. The steam tube is large enough for a much greater rate of distillation. But as we can at will alter the diameter of the boiler tube, or its length, or the collecting area of the reflector, the variations that we can make are

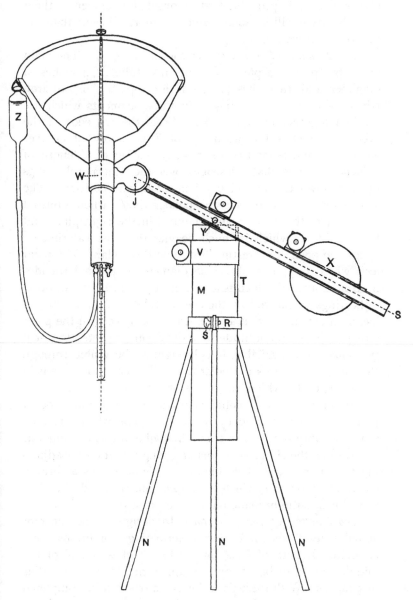

Fig. 4. Calorimeter on Equatorial Mounting. Perspective view.

endless and it is probable that amongst the number of them a combination will be found which is more efficient than the first one that was tried.

The *Equatorial Mounting* is shown in Fig. 4. The main or central part is a piece of stout brass tube M, 4 inches in diameter and 18 inches long. It is supported on an iron tripod NN by an iron ring R in three segments which are pinched together by screws SS. From the top edge of the tube for 6 inches downwards a slot T is cut of rather over an inch clearance, and a collar V, consisting of a length of 2 inches of tube that telescopes over the central tube, slips up and down it and can be clamped in any position. The slot is intended to receive the polar axis JS when adjusted for latitude, the collar being clamped in the right place and the weight X, which is much heavier than the calorimeter, keeps the polar axis resting firmly on the collar. The polar axis which is a tube of 1 inch diameter is pivoted round a horizontal axis Y, which is also a piece of brass tube working in bearings on the top of the central tube. When the tripod has been set up so that the central tube is vertical the polar axis is adjusted for latitude by a quadrant or protractor, or if the pole-star is available, it is brought to be visible through the tube which forms the polar axis. The calorimeter is held by a collar W which surrounds it and is clamped on the condenser tube. This collar is attached to a ball and socket joint J which is carried by a piece of tube which fits telescopically into the polar axis. The ball and socket joint was found to be the simplest means of giving a motion for adjusting the calorimeter for declination. We have thus a form of equatorial mounting which is simple, effective, and cheap, as it is almost entirely made out of brass tubes.

Construction of the Reflector.—In designing the reflector actually used, the following determining conditions were adopted.—Length of focal line to be 2 inches; angle of the middle mirror to be 45° and its upper rim to have a radius of 5 inches. With these data the 45° mirror can be completed at once and the specifications of the outer and inner mirrors follow by a simple graphical construction. The physical

principle involved is that the angles at which rays strike
and are reflected from a mirror surface are equal. The construction is shown in the diagram Fig. 5. On the straight

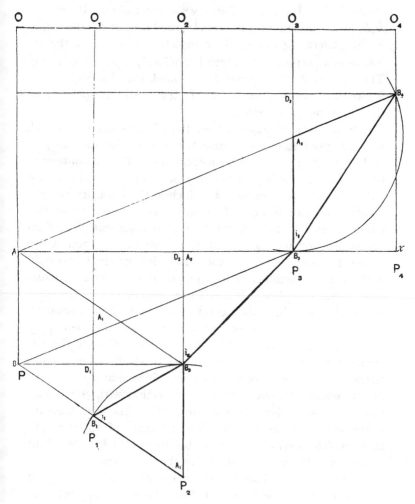

Fig. 5. Geometrical Construction of Reflector used.

line OP, which represents the axis of the instrument, lay off
the length $AB = 2$ inches; this is the focal line. Through
A and at right angles to AB draw Ax, and on it make

$AB_3 = 5$ inches. Through B draw BB_2 parallel to AB_3 and make $BB_2 = 3$ inches. Join B_2B_3. B_2B_3 is obviously the line representing the principal section of the mirror inclined at $45°$ to the axis. Its length $B_2B_3 = \sqrt{8} = 2\cdot83$ inches. If the line B_2B_3 be continued until it cuts AB produced in B_2', then B_3B_2' is the generating line of the complete cone of which the $45°$ mirror is a portion. Its length is obviously $\sqrt{50} = 7\cdot07$ inches. The flat band which, when bent round until its edges abut, forms the reflecting surface of the $45°$ mirror is specified by the following construction.

On a sheet of paper, or on the silvered sheet of metal, describe from the same centre two circles with radii of $7\cdot07$ and $7\cdot07 - 2\cdot83 = 4\cdot24$ inches respectively. The circumference of the greater circle is $44\cdot44$ inches. The upper rim of the $45°$ reflector has a radius of 5 inches, therefore its circumference is $31\cdot42$ inches. The difference between these two circumferences is $13\cdot02$ inches. If $44\cdot44$ represent $360°$ of an arc, then $13\cdot02$ represents $105°\cdot5$. From the common centre of the two circles draw to the outer circumference two radii inclined to one another at an angle of $105°\cdot5$. The construction is then complete. If it has been carried out on the sheet of metal from which the actual mirror is to be constructed, we first cut out the disc of $7\cdot07$ inches radius; we then apply the shears to the point where one of the radii cuts the circumference; we cut along it until we reach the inner circumference, we then cut round this circumference along an arc of $254°\cdot5$, when we arrive at its inner section with the second radius, which is then followed until the outer circumference is reached. The annular disc, less the sector of $105°\cdot5$ amplitude, which remains, is the metal band which, when bent round until its edges abut, forms the $45°$ mirror.

Outer Mirror.—Through B_3 draw O_3P_3 parallel to OP and on it lay off $B_3A_3 = BA = 2$ inches. From A_3 as centre, at the distance A_3B_3 describe a circular arc. Join AA_3 and produce the line AA_3 till it cuts the arc in B_4. Join B_3B_4. B_3B_4 is the line of section of the outer mirror. For, having in view the properties of triangles and of parallel lines, it is clear that the lines O_4B_4 and B_4A make equal angles with the line B_3B_4. But O_4B_4 is the direction of the incident ray at B_4; therefore B_4A is the direction of the reflected ray, and the ray which strikes the outer rim of the mirror is reflected upon the upper extremity of the focal line. In the same way it is evident that the incident ray O_3B_3, which strikes the inner rim of the mirror, is reflected along the line B_3B and falls upon B, the lower extremity of the focal line. Consequently parallel rays which strike the mirror in points between B_4 and B_3 are reflected on AB and strike it in points between A and B which are homologous as regards position with the points in the mirror between B_4 and B_3 which are struck by the primary rays.

In the isosceles triangle $B_3A_3B_4$ the angle A_3 is equal to the angle A_3AB, therefore

$$\tan A_3 = \frac{B_3A_3}{AB_3} = -\frac{AB}{AB_3} = -\tfrac{2}{5} = -0\cdot4 ;$$

therefore $\qquad\qquad A_3 = 111^\circ 48' ;$

and the angle at the base

$$i = \tfrac{1}{2}(180^\circ - A_3) = 34^\circ 6'.$$

Further, the base

$$B_3B_4 = 2 \times AB \cos i = 3\cdot31 \text{ inches.}$$

Therefore the width (m_3) of the outer mirror is $3\cdot31$ inches, and it is inclined to the axis at an angle of $34^\circ 6'$.

The radius of the upper rim of the outer mirror is

$$AB_3 + D_3B_4 = AB_3 + B_3B_4 \sin i = 5 + 3\cdot31 \sin 34^\circ6' = 6\cdot85 \text{ inches.}$$

Further, if the line B_4B_3 be continued till it cuts the axis in B_3', then B_4B_3' is the generating line of the complete cone of which the outer mirror is a portion.

Its length is obviously

$$\frac{6\cdot85}{\sin 34^\circ6'} = 12\cdot23 \text{ inches.}$$

The width of the mirror is $3\cdot31$ inches. Therefore the annular band of silvered metal has an outer radius of $12\cdot23$

inches and an inner radius of 8·92 inches, and the sector to be removed from it has an amplitude of

$$360° \frac{12·23 - 6·85}{12·23} = 158°·5.$$

Inner Mirror.—Through B_2 draw O_2P_2, and on it lay off, below the line BB_2, the length $B_2A_1 = BA = 2$ inches. From A_1 at distance A_1B_2 describe a circular arc. Join A_1B. This line cuts the arc in B_1. Join B_1B_2. B_1B_2 is evidently the line of section of the inner mirror. The demonstration is the same as in the case of the outer mirror.

Also $$\tan A_1 = \frac{BB_2}{AB} = \tfrac{3}{2} = 1·5 \; ;$$

$$\therefore \; A_1 = 56° \; 19',$$

whence $$i_1 = 61° \; 51',$$

and $$m_1 = 2 . AB \cos i_1 = 1·88 \text{ inches.}$$

Therefore the inner mirror has a width of 1·88 inches and is inclined to the axis at an angle of 61° 51′.

Further, if B_2B_1 be continued to cut the axis in B_1', B_2B_1' is the length of the generating line of the complete cone of which the mirror forms a part, and it is the greater radius of the annular disc of silvered metal out of which the band is to be cut. Its length is

$$\frac{BB_2}{\sin 61° \; 51'} = 3·40 \text{ inches.}$$

The inner radius of the annular disc is 1·52 inches, and the amplitude of the sector to be removed is

$$360° \frac{3·4 - 3}{3·4} = 42°·3.$$

The numerical data just worked out and relating to the reflector used are collected in the following table.

The condition that one of the mirrors should be inclined at an angle of 45° to the axis was suggested by the fact that this is the angle of greatest efficiency and by the consideration that it is an angle which is familiar in mechanical workshops,

and is on that account perhaps more likely to be laid off
accurately than another.   There is, however, no particular
advantage in making this restriction, because in designing
a series of mirrors for a reflector one of them is sure to be
inclined at an angle of nearly 45° and to have an efficiency
which is sensibly the same as if the angle were 45°.

No. of Mirror		1	2	3
Description of Mirror..................		Inner	Middle	Outer
Inclination to the Axis	i	61° 51′	45°	34° 6′
Inner Radius ....inches		1·34	3·0	5·0
Outer Radius .... ,,	R	3·0	5·0	6·85
Width of Mirror .... ,,	m	1·88	2·83	3·31
Outer Radius of flat band... ,,	M	3·40	7·07	12·23
Inner do. do. ,,	$M - m$	1·52	4·24	8·92
Amplitude of Sector to be re-} moved ....}	$360° \dfrac{M - R}{M}$	42°·3	105°·5	158°·5

The general problem, to construct the principal section of
a reflector consisting of a series of conical mirrors when the
direction and length of the common focal line are given and
the position relatively to a point on this line, of a point
occupying a definite position in the line of section of one
of the mirrors, is simple.   The diagram, Fig. 6, shows a
construction of this kind.

The line OP is the axis of the reflector, it is also the
direction of the incident rays when the instrument is in use.
Make it the axis y of rectangular coordinates with the upper
extremity of the focal line, A, as origin : ordinates measured
in the direction AB are to be reckoned positive, those measured
in the reverse direction are to be reckoned negative.

Abscissæ are to be measured on a line at right angles to
OP, and they are positive when measured to the right.

Join BB_1 ; and through B_1 draw a line O_1P_1 parallel to OP,
and on it lay off $B_1A_1 = BA$.

Join AA_1 and produce it to a point B_2, so that $A_1B_2 = AB$.

B.   23

Join $B_1 B_2$; then $B_1 B_2$ is the line which represents in section the innermost mirror.

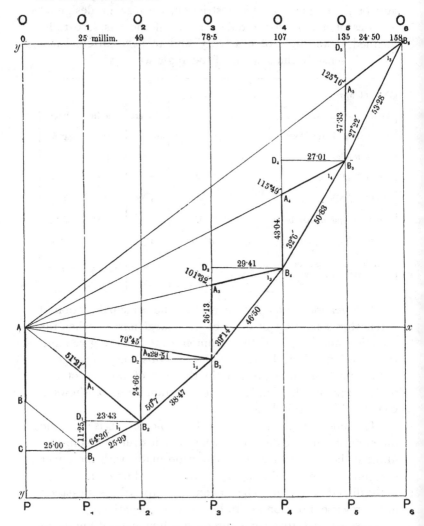

Fig. 6. Geometrical Construction of Reflector, when the position of a point on one of the mirrors, and the position and length of the focal line, are given.

Through B_2 draw a straight line $O_2 P_2$ parallel to OP, and on it lay off the length $B_2 A_2 = BA$.

Join AA_2 and produce it to a point B_3, so that $A_2B_3 = AB$.

Join B_2B_3; then B_2B_3 is the line which represents in section the second mirror of the series.

Through B_3 draw O_3P_3 parallel to OP, and on it lay off $B_3A_3 = BA$.

Join AA_3 and produce it to a point B_4, so that $A_3B_4 = AB$.

Join B_3B_4; then B_3B_4 is the line which represents in section the third mirror in the series; and so on.

It is evident from the properties of parallel lines that the angle which the incident ray makes with the outer extremity of any one of these lines is equal to the angle made with it by the line connecting that point with the upper extremity of the focal line. Therefore all the rays parallel to the axis which strike the outer extremity of a line of section are reflected upon A, the upper extremity of the focal line. In the same way all the rays parallel to the axis which strike the inner extremity of a line of section are reflected upon B, the lower extremity of the focal line. Consequently all the rays parallel to the axis which fall upon intermediate points in the line of section are reflected upon the corresponding points between A and B on the focal line. Therefore all the rays parallel to the axis which strike the reflector are reflected and condensed on the focal line AB.

If the graphic construction is effected on the natural scale, all the measurements, both linear and angular, can be taken from it directly with sufficient exactness to enable the reflector to be constructed. On the other hand, the geometrical construction is so simple that there is no difficulty in arriving at all the values by calculation. It will be apparent from the diagram (Fig. 6) that most of the elements of each section are contained in an isosceles triangle $A_nB_nB_{n+1}$. In it the angle A at the apex of one triangle is derived from the data of the previous triangle. The angle of inclination to the axis of the mirror is

$$i = \tfrac{1}{2}(180° - A),$$

and the length of the base or width of the mirror is

$$m = 2AB \cos i.$$

To find the succeeding values of A let us take A_1 and A_2. A_1 is evidently equal to CBB_1 and to CAA_1; therefore

$$\tan A_1 = \frac{B_1 C}{CB},$$

and these are known from the coordinates of the given point B_1, therefore A_1 is known and the values of i_1 and m_1 follow as above.   Through B_2 draw $B_2 D_1$ at right angles to $A_1 B_1$ and cutting it at D_1.   Then

$$D_1 B_2 = m_1 \sin i_1,$$

and $$D_1 B_1 = m_1 \cos i_1.$$

Then $$\tan A_2 = \frac{CB_1 + D_1 B_2}{AC - (D_1 B_1 + A_2 B_2)}$$

$$= \frac{CB_1 + m_1 \sin i_1}{AC - AB - m_1 \cos i_1},$$

whence A_2 is found.

The values of i_2 and m_2 follow as before, and we have

$$\tan A_3 = \frac{CB_1 + m_1 \sin i_1 + m_2 \sin i_2}{AC - AB - m_1 \cos i_1 - m_2 \cos i_2},$$

whence A_3 is found; and by thus proceeding step by step the elements of all the mirrors in the series are easily obtained.

The diagram (Fig. 6) was actually constructed on the following numerical data:

$$AB = 30 \text{ millimetres},\quad BC = 20 \text{ millimetres},$$

and $$CB_1 = 25 \text{ millimetres};$$

and the values of the different elements as calculated are collected in the following table.   It is to be remembered that $\tan A$ is always the quotient $\dfrac{x}{y}$ of the coordinates of the point A, and that the values of y change sign at the origin.   Thus

for A_1, $y_1 = 50 - 30 = 20$;

for A_2, $y_2 = 50 - 11\cdot25 - 30 = 8\cdot75$;

for A_3, $y_3 = 50 - 11\cdot25 - 24\cdot66 - 30 = -15\cdot91$;

and so on.

In order to make the table complete the specifications of the metal bands for the mirrors are added.

SECTIONS OF MIRRORS.

No. of Mirror		o	1	2	3	4	5
Angle at apex of isosceles triangle	A		51° 21′	79° 45′	101° 32′	115° 49′	125° 16′
Inclination of mirror to axis	i		64° 20′	50° 7′	39° 14′	32° 6′	27° 22′
Width of mirror (millim.)	m		25·99	38·47	46·50	50·83	53·28
Projection of mirror on axis of x (mm.)	a	25·00	23·43	29·51	29·41	27·01	24·50
	$a_0+a_1+\ldots+a_n=\Sigma a=x$	25·00	48·43	77·94	107·35	134·36	158·86
Projection of mirror on axis of y (mm.)	b	0	11·25	24·66	36·13	43·04	47·33
	$\Sigma b=y$	0	11·25	35·91	72·04	115·08	162·41
	$20-\Sigma b=y$	20·00	8·75	-15·91	-52·04	-95·08	-142·41
	$\dfrac{x}{y}=\tan A_{n+1}$	1·250	5·535	-4·902	-2·065	-1·414	-1·115
	A_{n+1}	51° 21′	79° 45′	101° 32′	115° 49′	125° 16′	131° 53′

ANNULAR BANDS FOR MIRRORS.

Outer radius of band	$\dfrac{x}{\sin i}=M$		53·80	101·56	169·72	252·84	345·33
Inner radius of band	$M-m$		25·99	38·47	46·50	50·83	53·28
	$M-x$		5·37	23·62	62·37	118·48	186·48
Amplitude of Sector to be removed	$\dfrac{360°\,(M-x)}{M}=$		35° 56′	83° 39′	125° 55′	168° 43′	194° 22′

TABLE II.—*Meteorological Observations on* 16*th,* 17*th, and* 18*th May,* 1882, *at Sohag in Lat.* 26° 37′ *N.*

Time	Temperature of the air		Evaporation		
	Dry bulb, ° Fahr.	Wet bulb, ° Fahr.	Temp. of water, ° Fahr.	Volume of water, c.c.	Evaporation, millimetres
1882 **May 15** 6 0 P.	82·5		80·0	400	
,, 16 5 30 A.	63·0		51·0	345	2·26
5 45 ,,	63·0		58·0	400	
8 45 ,,	87·0		75·5		
10 0 ,,	90·0		81·0		
11 0 ,,	92·0	63·5	81·0		
12 0 ,,	91·5	63·2	82·0		
2 0 P.	95·0	62·5	82·5		
3 30 ,,	95·0	62·5	79·0		
4 0 ,,	91·5	60·2	76·5		
4 30 ,,	90·0	59·5	73·5		
5 0 ,,	89·0	58·2	67·0		
5 30 ,,	87·0	58·0			
6 0 ,,	85·3	57·3	63·5	145	11·09
6 0 ,,	85·3	57·3	80·0	400	
,, 17 5 40 A.	65·0	50·0	51·5	346	2·25
6 10 ,,	67·0	52·0	60·0	400	
8 0 ,,	74·0	56·5	69·0		
8 10 ,,	73·0	56·0	68·0		
8 16 ,,	73·0	57·1			
8 21 ,,	72·5	56·0			
8 26 ,,	71·2	55·5			
8 30 ,,	71·0	55·0			
9 0 ,,	77·0	58·5	68·5		
10 0 ,,	86·5	64·5	81·0		
11 0 ,,	91·0	66·0	83·5		
2 30 P.	94·0	66·2	81·0		
3 30 ,,	94·0	64·0	78·0		
4 30 ,,	91·0	60·2	73·5		
6 0 ,,	86·0	56·5	62·0	158	10·52
6 20 ,,			80·0	400	
,, 18 6 20 A.	60·5	49·5	55·0	333	2·75
6 30 ,,	60·5	49·5		400	
7 20 ,,	73·5	54·0	69·5		
9 0 ,,	85·0	61·5	82·0		
10 0 ,,	92·0	65·5	87·0		
11 0 ,,	98·5	64·0	89·0		
12 0 ,,	95·0	66·1	88·5		
2 0 P.	105·0	65·0	85·0		
3 0 ,,	103·5	64·5	81·5		
4 0 ,,	97·6	64·0	78·0		
4 35 ,,			77·0		
6 15 ,,	96·5	60·2	68·0	140	11·30

Meteorological Observations and Notes.—The climate at
Sohag is a desert climate tempered by the influence of the
Nile.   This influence extends only a very short distance from
the banks of the river.   As the population is confined to the
banks of the river its benefits are enjoyed by the whole
population.   During the few days in May that the expedition
sojourned at Sohag the sun attained a meridian altitude of
roughly 83°, so that its power differed very little from that
of a vertical sun.   The prevailing wind is from the North
which gives a freshness to the atmosphere while it also
enables the countless sailing craft on the Nile to navigate
its waters against its not insignificant current.

While occupied with the calorimeter I made observations
on the temperature of the air using both the wet and dry
bulb thermometers, and I also measured the evaporation by
night and by day of water exposed freely in a plate raised
about 6 inches above the ground.   A glance at Table II or
Fig. 7 which contain these results will indicate better than
any description the nature of the climate in that part of
Egypt in May.

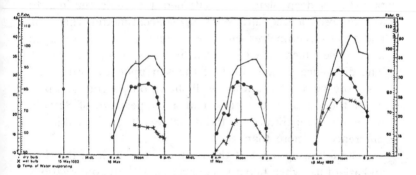

Fig. 7. Meteorological Observations on the 16th, 17th, and 18th May, 1882.

The latitude of the station was 26° 37′ N. so that, on the
ocean, it would be in the heart of the Trade Wind.   In fact
the Trade Wind regions of the ocean are the desert regions
of the sea.   The water at the surface is there drier than anywhere else, that is, a given volume of it contains more salt

and less water than is to be found either on the equatorial or on the polar side of the region. The northerly wind on the Nile is the Trade Wind blowing from colder to hotter latitudes and always increasing its evaporative power. Thanks to this power the temperature of the Nile is lower than it would otherwise be. I took its temperature frequently at all hours of the day, it varied only between 74°·5 and 76° Fahr., while the temperature of the air above it varied from 50° to 105° Fahr. the variations of the wet bulb thermometer are much less, namely from 49°·5 to 66°·2 Fahr. The temperature of the water of the Nile is very nearly the mean of the maximum and minimum temperatures of the air above it. These means were on the

14th,	15th,	16th,	17th,	18th,	mean
77°·5,	74°·5,	79°·0,	79°·5,	82°·75,	78°·65.

The temperature of the Nile is a little below this mean. The large range of temperature in the air is partly due to the cooling effect of the evaporation from the surface of the water.

The evaporation experiments were made on water contained in a deep plate. It contained 400 cubic centimetres and exposed a free surface of 243 sq. centimetres when full, and 230 sq. centimetres when nearly empty. It was set upon a tin cone which raised it about 6 inches above the ground. The difference of the effect of the sun upon the water and upon the sand close to it was well shown on the 16th at 2 p.m. when the water in the plate had a temperature of 82°·5 F. while the temperature of the sand was 134° F., making a difference of more than 50° F.

The following notes of the weather were made at the time and are of use when taken in connection with the observations of the calorimeter.

16th May.—The sun rose in a cloudless sky and there was a very light wind from the west. As the morning wore on, it came round more and more to the north and freshened slightly. At 10 a.m. the wind seemed freshening and came in gusts, retarding distillation. In the afternoon it was calm.

17th May.—The day of the eclipse, the sun rose in a

cloudless sky and the inhabitants of Sohag had already begun
to collect on the banks of the Nile where they remained until
the eclipse was over. The temperature recorded as at 8.30 a.m.
was really observed 50 seconds after totality began, and I had
no difficulty in reading the thermometer although all the
principal stars were shining brightly, along with an unsuspected comet which appeared with totality, between two and
three sun's diameters from the darkened disc and with a
slightly curved tail quite as long as the sun's diameter. This
comet had not been detected before and I understand that it
was never seen afterwards. It was a very striking feature
of the eclipse to those whose occupations enabled them to
look at it. I made a sketch immediately totality was over.
Perhaps the most impressive period of the eclipse is during
the two or three minutes that precede the total phase. Until
a very large proportion of the sun's disc has been obscured
the decrease of light causes no remark, especially from people
who are accustomed to climates where clouds are more common
than sunshine. But when the time comes that every minute
and indeed every second alters by many per cent. the visible
radiating area of the sun and when at last this area is halved
in one second and becomes nothing in the next, the effect
which this sudden extinction of the sun produces is a very
profound one. And this not on man only, but also on the
beasts. There were some turkeys in the camp, and they
went about as usual until the final phase above indicated
began, when they showed every symptom of alarm. When
the sun reappears his light increases as rapidly as it disappeared, and five minutes after totality all interest in the
eclipse has gone. What struck me most besides the comet,
were two so-called protuberances. I say *so-called* because to
the naked eye they look much more like indentations or
notches in the moon's disc and coloured red. This is a subjective effect and due to the same cause as the "black drop"
in the case of the transit of Venus.

Of all the natural phenomena which I have had the
opportunity of witnessing there is none which produces so
powerful an impression as a total eclipse of the sun. In

connection with this it may be recalled that the eclipse of
17th May, 1882, repeats itself after 19 years on the 17th May,
1901, with this important advantage, that in place of seventy
seconds the maximum duration of totality will be six minutes
and a half, and it will occur very nearly at noon at stations in
Sumatra and Borneo.

The 18th May was the hottest day experienced. Perfect
calm reigned until 2 p.m. when a breeze began to blow up
the Nile and continued throughout the afternoon although it
was never very strong. During this forenoon the maximum
results were obtained with the calorimeter and the temperature
of the air reached its maximum 105° F. at 2 p.m. It will be
noticed that the temperature of the wet bulb thermometer
was only 65°, or 40° F. below the dry bulb ; the air was of
extraordinary dryness. One effect of a climate such as this,
where great dryness is associated with very high temperature,
is that, although perspiration is abundant, the skin is never
moist, indeed it is so dry that it has a tendency to crack.
Another remarkable subjective effect of high air temperatures
such as those of the afternoon of the 18th is the notice given
when the temperature of the air passes from below that of
the human body to above it. It is a matter of common
experience that in preparing a warm bath, very slight differences of temperature can be appreciated by the hand when
the water is at or about the temperature of the human body.
With air the conditions are different ; the capacity for heat of
all the air that can at any moment touch exposed portions
of the body is very small and produces no noticeable effect.
But although it cannot do so directly it can do so vicariously
for instance through the metal rim of a pair of spectacles.
The calorimetric work described in this paper necessitated
continuous exposure to the rays of the sun which were being
collected and measured, and in order to protect the eyes from
the intense glare of the sun it was prudent to use neutral
tinted spectacles. The moment the temperature of the air
passed upwards through the temperature of the skin was
signalised by the spectacles feeling hot. Although the temperature rose to 105° F. the capacity for heat of the rim of

a pair of spectacles is too insignificant to cause any inconvenience.

Discussion of Observations.

The observations made with the calorimeter on the 16th, 17th and 18th May are given in detail in Table III and they are represented graphically in Fig. 8.   On the 16th the distillate was received in a cylinder capable of holding over 100 c.c. and graduated into single cubic centimetres.   In the two columns for rates on this day, the one is the rate per minute while 10 c.c. were collected, and the other the rate per minute while 20 c.c. were collected.   On the 17th and 18th a tube graduated into half cubic centimetres and holding 20 c.c. was used.   Time was taken as every 5 c.c. were collected, and the tube was emptied when 20 c.c. had been collected. The readings for every 5 c.c. were made without removing the receiver from the distilling tube.   The portion of 20 c.c. was measured in a truly vertical position and is more exact than the measurement of its constituent portions of 5 c.c., although every care was taken to note the time when exactly 5 c.c. had run without running the risk of losing any of the distillate.

The most important manipulation is attending to the equatorial motion of the instrument.   The observed rate of distillation agrees the more closely with the true rate the more carefully the axis of the instrument is kept pointed towards the sun.   This was controlled by observing the shadow of the steam space on the top of the condenser with which it is concentric.

The position was adjusted every two or three minutes when it was usually put a shade in advance of the true position so as to give it a position correct for the middle of the interval.

The calorimeter ought always to be fed with pure distilled water.   Unfortunately this was not available, and Nile water had to be used.   It contains a considerable amount of earthy carbonates and is apt, after prolonged use, to froth.   With the glass dome, however, this was at once detected, and if it was serious the water was changed.

TABLE III.—*Giving Rates of Distillation observed on the 16th, 17th, and 18th May, 1882, at Sohag, Lat. 26° 37′ N.*

	Apparent Solar Time	Volume of Distillate	Rate for Distillation of		
			10 c.c.	20 c.c.	
1882 May 16 A.M.	h. m. s.	c.c.	c.c. per m.	c.c. per m.	
	9 26 0	0			Cloudless sky, wind very slight
	36 55	10	0·916		from the West, drawing to
	46 55	20	1·000	0·956	North on the Nile
	56 40	30	1·026		
	10 6 50	40	0·984	1·002	
	17 5	50	0·976		Wind seems to be freshening
	29 15	60	0·821	0·893	and comes in gusts, retarding
	39 20	70	0·992		distillation
	49 50	80	0·952	0·972	
	59 20	90	1·053		
	11 8 5	100	1·144	1·096	
	12 0	0			
	21 10	10	1·091		
	29 45	20	1·166	1·127	
	38 25	30	1·155		
	46 50	40	1·189	1·171	
	55 35	50	1·144		
P.M.	12 3 50	60	1·212	1·177	
	12 5	70	1·213		
	19 0	80	1·447	1·330	
	27 10	90	1·225		
	34 10	100	1·429	1·327	
	1 27 0	0			
	35 0	10	1·250		
	42 35	20	1·320	1·285	
	50 0	30	1·349		
	2 15 30	0			
	23 15	10	1·292		
	31 30	20	1·213	1·253	
	40 0	30	1·177		
	48 5	40	1·237	1·207	
	56 35	50	1·177		
	3 4 40	60	1·237	1·207	
	16 55	73			
	22 50	80		1·101	
May 17 A.M.	8 34 0	0			Eclipse total
	51 0				Exposed calorimeter
	58 0				Water begins to "sing"
	9 1 0				Water begins to boil
	3 0				Water boils briskly
	17 0	0			Water begins to distil and to be
	19 30	1	0·400		collected
	21 0	0			
	29 30	5	0·588		Sky quite cloudless
			5 c.c.		
	36 5	10	0·760		
	40 55	15	1·035		

TABLE III.—*continued.*

	Apparent Solar Time	Volume of Distillate	Rate for Distillation of		
			5 c.c.	20 c.c.	
1882 May 17 A.M.	h. m. s.	c.c.	c.c. per m.	c.c. per m.	
	9 45 45	20	1·035	0·806	
	47 0	0			
	51 15	5	1·177		
	56 0	10	1·053		
	59 50	15	1·306		
	10 4 5	20	1·177	1·171	
	5 0	0			
	8 52	5	1·292		
	14 35	11	1·101		
	18 40	16	1·225		
	22 20	20	1·091	1·155	
	23 30	0			
	27 50	5	1·155		
	31 55	10	1·225		
	36 5	15	1·201		
	40 30	20	1·133	1·177	
	41 30	0			
	45 35	5	1·225		
	49 30	10	1·278		
	53 30	15	1·250		
	57 45	20	1·177	1·232	
	58 30	0			
	11 2 45	5	1·177		
	7 0	10	1·177		
	11 15	15	1·177		
	15 35	20	1·155	1·171	
	16 30	0			
	20 25	5	1·278		
	24 30	10	1·225		
	28 25	15	1·278		
	33 0	20	1·091	1·212	
	34 30	0			
	38 50	5	1·155		
	42 55	10	1·225		
	47 0	15	1·225		
	51 50	20	1·035	1·155	
	54 0	0			
	58 25	5	1·133		
P.M.	12 2 30	10	1·225		
	6 30	15	1·250		
	10 40	20	1·201	1·202	The apparatus was emptied and filled with fresh water for the afternoon's work
	2 1 0	0			On exposing to the sun the water
	5 45	5	1·053		began to " sing " in 5 seconds
	10 55	11	1·177		and boiled in 40 seconds
	14 35	15	1·091		
	19 5	20	1·111	1·106	From 2 p.m. to 2.30 p.m. rather
	24 30	0			more wind than before. Sky
	28 50	5	1·155		all day quite cloudless. Wind

TABLE III.—*continued.*

	Apparent Solar Time	Volume of Distillate	Rate for Distillation of		
	h. m. s.	c.c.	5 c.c.	20 c.c.	
			c.c. per m.	c.c. per m.	
1882 May 17 P.M.	2 33 15	10	1·133		North, but not much of it. The weather altogether very much like the 16th only it feels somewhat warmer
	37 55	15	1·071		
	42 50	20	1·071	1·091	
	44 0	0			
	49 5	5	0·984		
	53 40	10	1·091		
	57 55	15	1·177		
	3 2 45	20	1·035	1·066	
	10 0	0			
	14 40	5	1·071		
	19 50	10	0·968		
	24 30	15	1·071		
	29 40	20	0·968	1·015	
	31 30	0			
	37 10	5	0·883		
	42 10	10	1·000		
	47 10	15	1·000		
	52 50	20	0·883	0·943	
	55 30	0			
	4 1 0	5	0·909		
	6 20	10	0·937		
	11 55	15	0·896		
	17 20	20	0·923	0·916	
	19 30	0			
	25 25	5	0·845		
	31 5	10	0·883		
	37 0	15	0·845		
May 18 A.M.	43 55	20	0·720	0·819	Cloudless sky, absolutely no wind
	8 47 30	0			
	51 25	5	1·278		
	55 20	10	1·278		
	59 20	15	1·250		
	9 3 20	20	1·250	1·264	
	5 0	0			
	8 55	5	1·278		
	12 35	10	1·365		
	16 20	15	1·333		
	20 20	20	1·250	1·306	
	22 0	0			
	25 45	5	1·333		
	29 20	10	1·397		
	33 50	16	1·333		
	37 5	20	1·225	1·326	
	39 15	0			
	43 0	5	1·333		
	46 50	10	1·306		
	50 40	15	1·306		
	54 25	20	1·278	1·306	
	57 0	0			
	10 0 50	5	1·306		

TABLE III.—*continued.*

1882 May 18 A.M.	Apparent Solar Time	Volume of Distillate	Rate for Distillation of		
			5 c.c.	20 c.c.	
	h. m. s.	c.c.	c.c. per m.	c.c. per m.	
	10 4 35	10	1·333		
	8 25	15	1·306		
	12 35	20	1·201	1·285	
	15 0	0			
	18 45	5	1·333		
	22 40	10	1·278		
	26 15	15	1·397		
	30 15	20	1·250	1·313	
	32 0	0			
	35 40	5	1·365		
	39 0	10	1·501		
	42 25	15	1·465		
	46 15	20	1·306	1·405	
	47 30	0			
	51 5	5	1·397		
	54 35	10	1·429		
	58 5	15	1·429		
	11 2 0	20	1·278	1·381	The calm was broken by a light northerly breeze lasting about 10 minutes
	4 0	0			
	7 35	5	1·397		
	11 0	10	1·465		
	15 20	16	1·381		
	18 25	20	1·301	1·389	
	21 30	0			
	24 50	5	1·501		
	28 15	10	1·465		
	31 40	15	1·465		
	35 20	20	1·365	1·447	
	37 30	0			
	41 5	5	1·397		
	44 40	10	1·397		
	48 5	15	1·397		
	51 55	20	1·306	1·389	
	53 30	0			
	57 0	5	1·429		
P.M.	12 0 40	10	1·365		
	4 10	15	1·429		
	8 35	20	1·133	1·326	Perfectly calm Light northerly breeze
	1 55 30	0			
	59 20	5	1·306		
	2 3 15	10	1·278		
	7 10	15	1·278		
	11 10	20	1·250	1·274	
	12 30	0			
	17 25	6	1·225		
	20 40	10	1·225		
	25 35	16	1·225		
	29 0	20	1·177	1·214	
	31 0	0			
	35 25	5	1·133		Northerly wind freshening

TABLE III.—*continued.*

1882 May 18 P.M.	Apparent Solar Time	Volume of Distillate	Rate for Distillation of		
			5 c.c.	20 c c.	
	h. m. s.	c.c.	c.c. per m.	c.c. per m.	
	2 40 35	11	1·166		
	44 5	15	1·144		
	48 45	20	1·071	1·128	
	51 0	0			
	55 10	5	1·201		North wind quite fresh
	59 45	10	1·091		
	3 4 25	15	1·071		
	9 0	20	1·091	1·112	
	11 0	0			
	15 25	5	1·133		
	20 10	10	1·091		
	24 40	15	1·111		
	29 10	20	1·111	1·111	
	31 0	0			
	35 15	5	1·333		
	39 45	10	1·111		
	44 25	15	1·071		Wind freshening still
	49 35	20	0·968	1·076	
	52 30	0			
	57 10	5	0·937		
	4 1 30	10	1·155		Water beginning to froth
	6 0	15	1·111		
	11 20	20	0·937	1·062	

By far the most important agent in altering the true rate, as due to the sun alone, is the wind. During the three days we were fortunate in having both calm and wind, so that an idea can be formed of the cooling effect of wind. On the 16th with a calm afternoon the mean rate between 2 and 3 p.m. was 1·221, and on the 17th, when it was breezy, the rate was 1·087 or about 10 per cent. less.

The breezes which occur on the Nile are usually cool and from the north. They did not at any time exceed force 3 of Beaufort's scale. They were never steady, but came in puffs or gusts, so that one 20 c.c. or even 5 c.c. interval would be affected and the subsequent one not.

On the 13th some satisfactory observations were made when the sun though behind a cirrus cloud was still able to keep distillation going. The rate was 0·752 at 10 a.m.

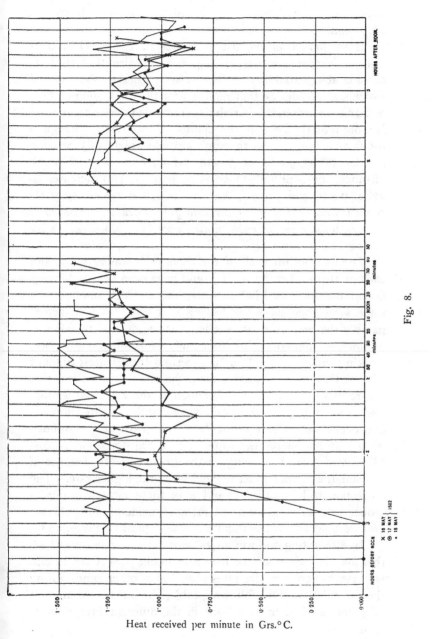

Fig. 8.

Heat received per minute in Grs.°C.

When the sun had cleared the cloud the rate rose to 0·91 at 10.30 a.m.

If we look over the list of figures in Table III or their graphical representation in Fig. 8, we notice that there is considerable variability in the results whether the interval which we consider be that required for the distillation of 5 c.c. or 20 c.c. Further, this variability from one interval to another is more remarkable than the change of rate due to change of the sun's altitude. Yet the sun's altitude which is 83° at noon is only 48° at 9 a.m. or 3 p.m. If we express it in zenith distance, the zenith distance is at noon 7° and increases to 42° at 9 a.m. and 3 p.m. We conclude that *the energy of the radiation received by a surface held perpendicularly to the sun's rays is, within considerable limits, very little dependent on the sun's zenith distance.*

The weather on each of the three days was very fine and each of them taken by itself would have been held to be very favourable for this kind of experiment. Yet amongst the three very good days the forenoon of the 18th was incomparably the best; the sun shone its strongest and the air was motionless; moreover, instrumentally everything was in best working order. Therefore to ascertain the greatest amount of heat that can be obtained from the sun's rays we examine the results obtained in the forenoon of the 18th, and we find that at about half-past ten, 5 c.c. were distilled in three minutes and twenty seconds, being at the rate of 1·501 c.c. per minute. Nearly an hour later the same time is registered for the distillation of 5 c.c., but owing to the greater zenith distance of the sun the former must be held to be the higher rate. The correction to be applied to either of these rates in order to reduce it to its value for a vertical sun is evidently insignificant and we take 1·5 c.c. per minute as the highest rate observed.

In attempting to form an estimate of the extent to which this may fall short of the true rate under perfect conditions, we have to consider the rates observed at other times during the three days and the following table, which gives the amount of water distilled in each hour in the different days, may be used.

If the experiments were to be repeated, I do not know in what particular the conditions of weather, as they were on the forenoon of the 18th, could be made better. Yet if we admit that they might be improved in the proportion that the conditions between 10 and 11 a.m. on the 18th are better than those obtained between the same hours on the 16th, when they were the most unfavourable, the rate would have to be increased in the proportion 57·6 : 80·7 and it would become 1·5 × 1·4 = 2·1 c.c. per minute. This correction is certainly too great when considered as an allowance for faulty weather, and even if held to cover all instrumental deficiencies, such as imperfect equatorial adjustment and others, I believe it will be still much in excess of the truth; moreover, I am

Time	Cubic centims. water distilled		
from to	16th	17th	18th
A.M.			
9 10			78·7
10 11	57·6	71·2	80·7
11 12	70·2	70·5	84·4
P.M.			
2 3	73·0	65·6	71·6
3 4		58·9	65·1

convinced that if the calorimeter furnished steam at this rate it would be in such conditions that it would be impossible to stand by it and attend to it on account of the excessive heat.

Having thus indicated the maximum correction which can be applicable to our observations we return to the consideration of the observations themselves, where we are on the sure ground of experiment.

In the circumstances we may, without sensible error, take the cubic centimetre of water to weigh one gramme. In specifying quantities of heat we do so in gramme-degrees (Celsius) (gr.° C.), or kilogramme-degrees (kg.° C.), as the case may be. Similarly, quantities of work are expressed in kilogramme-metres (kgm.). We take the latent heat of one gramme of steam as 535 gr.° C., and the specific heat of

water as unity, and the mechanical equivalent of heat as
0·425 kilogramme-metres per gramme-degree. Therefore the
heat required to transform 1·5 grs. of water at 100° C. into
steam of the same temperature is 803 gr.° C., and this is the
greatest amount of heat which the calorimeter has recorded
in one minute. On careful measurement of the calorimeter,
especially the reflector, I find that its actual collecting
diameter is 34·3 centimetres, less that of the condenser tube,
5·1 centimetres. So that its collecting area is

$$924 - 20·5 = 903·5 \text{ square centimetres (cm.}^2).$$

Therefore the rays of the sun falling perpendicularly on a
surface of 903·5 cm.² supplied it with heat at the rate of
803 gr.° C. per minute. This is equivalent to 8888 gr.° C.
per square metre; and 8888 gr.° C. suffice for the generation
of 16·6 grs. of steam at 100° C. *Therefore by the use of
ordinary mechanical appliances it is possible under favourable
geographical and meteorological conditions to collect on a square
metre of surface exposed perpendicularly to the sun's rays the
energy of generation of 16·6 grs. of steam per minute.* But
8888 gr.° C. of heat are equivalent to 3777 kgm. of work; and
this work is done in one minute, therefore the agent is working
at the rate of at least 0·84 horse-power.

The agent is the energy of the sun's rays which fall upon
a surface of one square metre, exposed perpendicularly to
them at the distance of the earth. If the sun throws so much
radiant energy that it can be collected and utilised at the
earth's surface at the rate of 0·84 horse-power per sq. metre,
then, as the area of a great circle on the earth's surface is
129·9 × 10¹² sq. metres, the useful energy received by the
whole earth is at the rate of

$$109 \times 10^{12} \text{ horse-power.}$$

Taking the radius of the earth's orbit to be 212 times the
radius of the sun, the radiation of one sq. metre of the sun's
surface is spread over 45,000 sq. metres of the earth's surface;
therefore *the sun must radiate energy at the rate of at least
37,000 horse-power per sq. metre of its surface.*

[1]No allowance has been made for loss.   If we increase the working value of the sun's rays from 0·87 to 1·0 horse-power, and take the earth's mean distance from the sun's centre to be 212 times the radius of the sun, the radiation emitted by one square metre of the sun's surface is spread over, in round numbers, 45,000 square metres of the earth's surface.   Therefore the intensity of the radiation of the sun's surface is equivalent to at least 45,000 horse-power per square metre. This number, especially when used in connection with so very small a surface as one square metre, conveys no definite idea to the mind.   The following consideration may assist in giving definition to our conception.   The specific gravity of solid iron at ordinary terrestrial temperatures is about 7·5 ; therefore one cubic metre of it weighs at the earth's surface 7500 kilogrammes.   Taking the force of solar gravity at the sun's surface to be twenty-eight times that of terrestrial gravity at the earth's surface, one cubic metre of cold solid iron on the sun's surface would exercise a pressure of 210,000 kilogrammes. To lift this mass through one kilometre against solar gravity would involve the expenditure of 210×10^6 kgm. of work : and if this amount of work were done in one minute, the engine employed would have to develop 46,667 horse-power.

Further, the heat which is equivalent to 210×10^6 kgm. of work is 494,100 kilogramme-degrees (kg.° C.).   When iron is burned in oxygen so as to form the magnetic oxide, the heat evolved is given by the thermochemical equation

$$Fe_3 + O_4 = Fe_3O_4 + 264\cdot7 \text{ kg.}° \text{ C.}$$

Using this constant, we find that the mass of iron which by its combustion would furnish the above amount of heat, would weigh on the surface of the earth 313·5 kilogrammes, and would occupy a volume of 0·0418 cubic metre, or 1 square metre × 4·18 centimetres.   Therefore the heat required could be produced by burning 4·18 centimetres of liquid iron on a hearth of 1 square metre per minute.   With a supply of oxygen at high pressure this would not seem to be an

[1] Extract from an account of this paper given in *Nature* (1901), vol. lxiii. p. 548.

insurmountable task. This is put forward only as an illustration, and in no way as an explanation of the source of the heat of the sun.

With this caution, however, I should like to call attention to a coincidence.

The specific heat of Fe_3O_4 is 0·1678, and its molecular weight 232, whence the water value of the gr. molecule is 38·93 grs. The molecular heat of combination is 264,700 gr.° C. Dividing this number by 38·93 we get 6800° C. as the temperature of the Fe_3O_4 produced. Adding 273, we have 7073° C. as the absolute temperature which may be produced. In a recent work[1] Scheiner gives 7010° C. as the most probable effective absolute temperature of the sun.

Whilst the maximum value recorded by the calorimeter is the most important for the determination of the sun's heating power, the other values obtained are of use for testing the working of the instrument. The principal disturbing element is wind. During the forenoon of the 18th there was an almost complete absence of wind. We take the observations of that forenoon, neglecting those that show a diminution of intensity as noon is approached, because the sun's heating power cannot diminish as noon is approached. They are collected in the following table. In the first column a is the mean time corresponding to the mean rate of distillation under d. Under b we have the sun's zenith distance at this time, and under c the secant of this angle, so that $c = \sec b$. Under d are the mean rates of distillation, in c.c. per minute, for quantities of 20 c.c. collected. Under e are these rates reduced to their value per square metre per minute, $e = 11·06\,d$. Under h we have the values of e corrected for the obliquity of the sun's rays. For this purpose the formula given by Herschel in his *Meteorology* is used[2]. It is, using the letters in our table,

$$e = f(\tfrac{2}{3})^c \text{ whence } f = \frac{e}{(\tfrac{2}{3})^c}.$$

[1] *Strahlung und Temperatur der Sonne*, von Dr J. Scheiner: Leipzig, 1899, p. 39.

[2] *Meteorology*, by Sir John Herschel, Bart.: Edinburgh, 1861, p. 10.

In this equation $\frac{2}{3}$ is the transmission coefficient of the atmosphere ; therefore f is really the solar constant expressed in cubic centimetres water evaporated per square metre per minute.   Under g it is given in gr.° C. per square centimetre per minute, for $g = 0.0535\,f$.   Calculating back, we obtain the observed values of the intensity of the sun's rays at the sea-level corrected only for the sun's zenith distance,—$h = \frac{2}{3}\,f$. This is expressed in cubic centimetres water evaporated per square metre per minute, and it is unaffected by the value employed as the transmission coefficient of the atmosphere, because this is eliminated in the process of reduction.

a	b	c	d	e	f	g	h
a.m.							
8.55	44°	1·390	1·264	13·97	24·57	1·314	16·38
9.12	39°	1·287	1·306	14·43	24·35	1·303	16·30
9.29	35°	1·221	1·326	14·65	24·06	1·287	16·11
10.39	20°	1·064	1·405	15·53	23·92	1·280	16·02
11.28	10°	1·015	1·447	16·00	24·16	1·293	16·18

The numbers in this table show that the values of the heating effect of the rays of the vertical sun, deduced from observations made when the sun was at zenith distances ranging between 10° and 44°, are practically identical.

If we take one horse-power per square metre as the intensity of the rays of the vertical sun at the sea-level, their intensity outside of the atmosphere is 1·5 horse-power per square metre, using Herschel's value for the transmission coefficient.   This is equivalent to 15,882 gr.° C. per square metre, or 1·588 gr.° C. per square centimetre per minute.   In round numbers we obtain 1·6 for the value of the solar constant.   While it is possible that this value may be a little too low, reasons are given in the paper for believing that the values commonly received, which lie between 3 and 5 gr.° C. per square centimetre per minute, are much exaggerated.

Observations during the Eclipse.

The calorimeter was directed to the sun as soon after totality as possible. At 8 hr. 34 min. the sun was totally eclipsed; at 8.51 the calorimeter was directed to the sun but

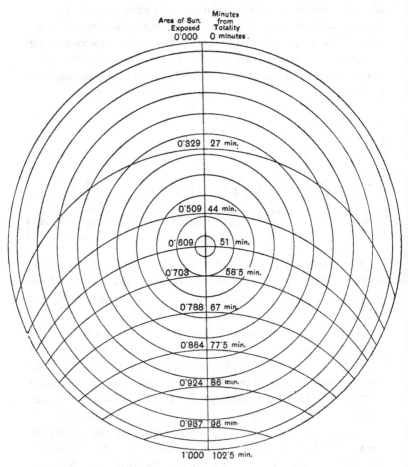

Fig. 9. Diagram of exposed surface of the Sun at successive epochs after Totality.

no boiling took place. At 8.58 the water began to "sing"; at 9.1 it boiled; at 9.3 it was boiling briskly, but it was not till 9.17 that the first drop of distillate fell into the receiver.

By 9.19·5 1 c.c. had passed, and between 9.21 and 9.29·5 5 c.c. passed.

The observations made at this time are collected in Table IV. —In the first column is the apparent solar time of each observation, in the second column is the volume of distillate collected at that time, in the third column is the mean time of collecting each portion, in the fourth column is the date stated in minutes after totality, in the fifth column is the average rate of distillation in c.c.'s per minute during the interval, and in the sixth column is the percentage of the sun's disc exposed.

The diagram Fig. 9 shows the progress of the eclipse and the portions of the sun successively uncovered.

From Table IV we see that when distillation has begun, it increases at a much greater rate than does the exposed sun's surface. This must be so in the early stages, because we see that it is not till 27 minutes after totality and when already 33 per cent. of the sun's surface has been uncovered that the water in the boiler boils, and it takes 16 minutes

TABLE IV.

Apparent Solar Time, A.M.	Cubic centims. collected	Mean date and Interval, A.M.	Minutes from Totality	Rate of Distillation	Amount of Sun's surface exposed
h. m. s.		h. m. s.		c.c. per m.	
8 34 0	0		0		0·000
9 1 0	0		27		0·329
9 17 0	0				
9 19 30	1	9 18 15	44	0·400	0·509
9 21 0	0				
9 29 30	5	9 25 15	51	0·589	0·609
9 36 5	10	9 32 47	58·5	0·759	0·703
9 40 55	15				
9 45 45	20	9 40 55	67	1·034	0·788
9 47 0	0				
9 51 15	5				
9 56 0	10	9 51 30	77·5	1·111	0·864
9 59 50	15				
10 4 5	20	10 0 0	86	1·237	0·924
10 5 0	0				
10 8 52	5				
10 14 35	11	10 9 45	96	1·146	0·987
10 18 40	16				
10 22 20	20	10 18 30	102·5	1·161	1·000

more before any distillate is collected.   Even when 50 per
cent. of the sun is exposed the rate of distillation is only
0·4 c.c. per minute.   After this more weight may be attached
to the observations, but their numerical significance is not
great.   The experiment was not originally contemplated.
The instrument was constructed for use with the strongest
uneclipsed sun that could be found.   Still it shows that useful
information could be obtained by arranging for making trustworthy observations during the progress of an eclipse.   The
provisions which it would be necessary to make are instructive,
because they indicate some of the capabilities and defects of
the instrument.

First of all it must be remembered that the calorimeter is
efficient only when it is running continuously and at or nearly
at its full load.   In the case of a total eclipse there must be
an interval during which the sun cannot keep steam however
large the reflector may be and however great its condensing
power may be.   We have seen that when exposed cold as
soon as possible after the total phase of the eclipse, it was
27 minutes after totality before the water boiled.   One-third
of the sun was then uncovered.   It is therefore reasonable to
suppose that, if the eclipse had happened at noon so that the
first half of it could have been utilised as well as the second
half, the sun would have kept steam in the calorimeter and it
would have continued to distil until two-thirds of the sun's
surface had been obscured.   Then distillation, if it did not
cease, would become so slow that its rate would have no value,
and fifty-four minutes would elapse before one-third of the
sun would again be uncovered during which the calorimeter
would get cold.   *During this interval steam must be kept
artificially.*   This is very easy.   The glass tube which forms
the steam dome is attached to a metal collar which screws
down on a washer.   It can therefore be easily detached.   If
then the steam tube of the calorimeter be connected by means
of an india-rubber tube with a flask in which water is kept
boiling, steam can be passed through the calorimeter at the
normal rate until it is judged suitable to expose it again to
the sun.   There is no difficulty about this.   It might however

be well for use during an eclipse to provide increased reflector power. But it would be necessary to shade it with a diaphragm when used with the uneclipsed sun, and the comparison of the heat of the eclipsed sun with that of the uneclipsed sun would be defective. Fortunately the power of varying the constants of the instrument is so great that one or two trials would suffice to fit it for use during an eclipse.

Although quite insignificant as a natural phenomenon an annular eclipse is better for calorimetric experiments than a total one. On 11th November, 1901, there will be an annular eclipse visible in Ceylon. The annular phase will last over ten minutes and at its greatest 0·875 of the sun's disc will be covered. It is pretty certain that the calorimeter used in 1882 would not keep steam through this phase, but a larger reflector might be used. It would be worth while to have a reflector of such a size that steam would certainly be kept through the whole eclipse, especially during the annular phase when all the radiation is from the peripheral region.

Conclusion.

It is usual for writers on this subject to express the heating effect of the sun's rays in gramme-degrees received by one square centimetre exposed perpendicularly to them for one minute outside the limits of the earth's atmosphere. This is termed the *solar constant*. Expressed thus our maximum rate is 0·89 gr.° C. per sq. centimetre per minute at the base of the earth's atmosphere. If we add 11 per cent. for deficiencies from all sources we have 1 gr.° C. heat received at the sea-level on a surface of 1 sq. centimetre exposed perpendicularly to the sun's rays per minute; and from the conditions under which the maximum rate was observed on the 18th May, I believe that this figure is as likely to be above the truth as below it. If however it is thought that the allowance should be more liberal, we have seen that our maximum rate corresponds to 0·84 horse-power per sq. metre; if we make this one horse-power per sq. metre we have certainly got as much radiant energy as it is possible to collect

at the level of the sea. Further, in speculations connected
with physical meteorology we are not entitled to postulate
a more abundant supply. From this supply falls to be
deducted the energy of evaporation which however is returned
on precipitation, also the energy of storms which to a large
extent are secondary features attending the changing hygro-
metric state of the atmosphere. Notwithstanding the apparently perfect transparency of the atmosphere on the
morning of the 18th we must admit that some of the energy
was lost by absorption in the passage through the earth's
atmosphere ; but the small effect produced by great variations
of the zenith distance of the sun on the rates observed shows
that this effect cannot be great, in fact it is entirely masked
by very slight motion of the air.   Most recent writers put the
value of the solar constant at from 3 to 5 gr.° C. per cm.² per
min., the greater part of which is added to the observed value
in order to compensate for the supposed absorption by the
air.   Thus Scheiner[1] in a recent work writes :—" From what
precedes it is apparent that the values which have been found
for the solar constant do not differ so very much from each
other. The older determinations have without doubt given
too small values, the later ones point with great certainty in
the direction that the solar constant is included between the
amounts of 3·5 and 4·0 gr. cal."   Now even the lower of these
values can be true only if we admit that the atmosphere
absorbs at least as much heat as it transmits.   But we know
by every-day experience the far-reaching effects produced by
what it transmits ; what does it do with the heat that it absorbs ?
Do we see any evidence of work being done at the rate of two
or even of one horse-power per sq. metre, remembering that
the energy of storms is already accounted for ?   If it is
absorbing heat at the rate of 2 gr.° C. per cm.² per minute,
how does it come that the atmosphere is so cool ?   Again,
looking to the length of time that the present state of things
has existed, how has the atmosphere not long ago arrived at

---

[1] *Strahlung und Temperatur der Sonne*, von Dr J. Scheiner: Leipzig, En-
gelmann, 1899; see p. 38.

the state in which it emits as much energy as it absorbs, so that its effective power of absorption would be *nil*?

I do not ask these questions lightly. The subject has occupied my attention off and on for the last eighteen years, and I believe that the only answer to them is that *the value of the solar constant which is now accepted is very much exaggerated.* This view is, I think, supported by the following consideration.

Taking the length of the sun's radius as unity we have in the accompanying table the distance (d) from its centre to the sun's surface, and to the three inner planets, Mercury, Venus, and the Earth, and the squares of these distances (d^2). The squares of the distances represent the area on each planet

Name of Body......................	Sun's surface	Mercury	Venus	Earth
Distance from Sun's centre, d	1·0000	82·0646	153·3466	212·0000
Square of Distance, d^2	1	6,735·0	23,515·0	44,944·0

over which the radiation per unit area of the sun's surface is spread. It will be seen that the area on the earth's surface covered by the radiation from a given area of the sun's surface is almost double that covered by the same radiation on the surface of Venus, and therefore the intensity of radiation on Venus is almost exactly double that on Earth. In other words, the true value of the solar constant at a point on the orbit of Venus is almost exactly double its true value at a point on the Earth's orbit. It is impossible to believe that the cloudless atmosphere of the Earth, the whole mass of which is only 1 kg. per sq. centimetre, can produce an absorbing effect equal or superior to the dissipating effect of such a distance as that separating the orbits of Venus and the Earth.

To conclude, it has been shown that under favourable meteorological and geographical conditions, by the use of ordinary and necessarily imperfect mechanical appliances it is possible to collect from a square metre of surface exposed perpendicularly to the sun's rays 8888 gr.° C. of heat which are equivalent to 3777 kgm. of work per minute or 0·84 horse-power. If we allow 16 per cent. for losses from all causes the result is one horse-power received from the Sun by every square metre of the area included in a great circle of the Earth which is roughly 130 × 10¹² sq. metres and *this figure in horse-power represents the working value of the Sun in its relations to the Earth.* Accepting the value of one horse-power per sq. metre at the distance of the Earth we find by simple arithmetic that *the working value of 1 sq. metre of the Sun's surface must be* 45,000 *horse-power.* It follows that the area of the Sun's surface which we may regard as hypothecated to the Earth's heat service is no more than 2900 square kilometres which would be contained in a circle of 60 kilometres diameter and would subtend an angle of at most one-tenth of a second. As over five hundred millions of such areas are included in a great circle of the Sun, it is clear that the maintenance of the Earth's heat is well assured.

No. 11. [*From Nature, September* 5, 1901, *Vol. LXIV,*
p. 456; *with postscript in* 1911.]

SOLAR RADIATION

SOLAR radiation is a subject which has more than scientific
interest. It is the source of all the energy which maintains
the economy of our globe. It lights and heats the other
members of the planetary system. But, after accomplishing
this, only an infinitesimal proportion of the total radiation
has been used. The remainder, in so far as we know, is
wasted by uninterrupted dissipation into space.

The subject can be regarded and studied from either the
solar or the terrestrial point of view. In terrestrial physics
everything may be said to depend on the energy which, in
one form or another, is supplied by the sun's rays. It is the
revenue of the world, and it is of fundamental importance for
us to know at what rate it falls to be received.

Roughly speaking, the surface of the earth is occupied to
the extent of one-fourth by land and three-fourths by sea.
Therefore at least three-fourths of the surface which the
earth presents to the sun is at the sea-level. Consequently
the rate at which the sun's radiant heat arrives at the
sea-level is the fact which it is of the greatest economical
importance to ascertain.

In considering this problem we have to answer two
questions: What is the best experimental method of determining the heating power of the sun's rays at any place?
and What is the best locality for making the experiment?
Let us take the last first. The energy which a radiation
communicates to a surface is greatest when it strikes it
perpendicularly. At every moment the sun is vertical over

one spot or another of the earth's surface. Therefore our first step should be to choose a locality where the sun passes through the zenith at mid-day.

Before reaching the sea-level the sun's rays have to pass through the whole thickness of the atmosphere. It is a matter of every-day observation that the atmosphere varies in transparency. The second condition is therefore to put ourselves in the position of greatest advantage as regards atmospheric conditions. Clouds and similar visible obstructions are of course excluded. The air should be motionless, the sky should be clear and of a deep blue colour in the regions remote from the sun and should contain nothing that can be called haze, or that interferes with the definition of the sun or other heavenly bodies.

From inspection alone we can only approximately ascertain what are the most favourable meteorological conditions. For this reason it is necessary to multiply observations and never to miss fine weather. In the end we cannot fail to approach nearer and nearer to the exact determination of the maximum heating power of the sun on the earth's surface at or near the sea-level, in so far as the degree of perfection of our instrumental resources permits. This limitation imposes on us the duty to continue observations, not only until the best natural conditions have been found, but also so long as the instruments or experimental methods appear to be capable of improvement. If we suppose for one moment that we have arrived at the point where no further improvement is possible, then the result of our work is the determination of *the rate at which unit area of the earth's surface at or near the sea-level receives heat from the vertical sun in unit time.*

There is no question here of how much is lost on the way from the sun. All that is sought, and the most that is ascertained, is how much arrives. *If we multiply this by the area included in the great circle of the earth we have the amount of radiant heat which we can count on as being supplied to the whole earth in unit of time. This is the constant which is of greatest importance in physical geography.*

When we have ascertained the supply of radiant heat which reaches the earth's surface, we have to inquire what becomes of it. If the heat were to accumulate the world would become uninhabitable. It cannot be doubted that long ago the earth, in this respect, arrived at a condition of equilibrium which is maintained with very slight oscillations. The fundamental principle of this state of equilibrium is that *the heat which the whole earth receives from the sun in the course of a year also leaves it in the course of a year, so that, taking one year with another, the sum of the heat remains the same.*

When we study the details of the annual dissipation of heat we find that the atmosphere, and especially the aqueous vapour in it, performs a very important part. Although practically transparent to the heat-rays passing from the sun to the earth, it is very opaque to those leaving the earth to pass outwards. They are powerfully absorbed and the temperature of the atmosphere is thus raised considerably above that which it would have if it were as transparent to the leaving rays as it is to the entering ones. This has no effect in permanently detaining any of the year's supply, it still disappears in the year, but not before it has produced important climatic effects.

We see in this differential behaviour of the atmosphere towards the incoming and the outgoing rays an example of Kirchoff's law, in virtue of which a body absorbs by preference the rays which it itself emits. It is exceedingly unlikely that any portion of the rays coming directly from the sun proceed from highly heated water or water vapour; we should therefore not expect the water vapour in the atmosphere to absorb them to any appreciable extent. When, however, they strike the surface of the earth, whether it be land or sea, they are abundantly absorbed. The blue water of the ocean transmits the sun's visible rays to a considerable depth. In experiments made by the writer on board the "Challenger," a white surface, about four inches square, was clearly visible at a depth of 25 fathoms. The total length of the path of the incident and reflected ray was 50 fathoms; therefore the

sun's rays which strike the sea have a thickness of at leas
100 metres to work on. When they strike the land, the
direct effect is superficial, but the absorptive power of a
surface of soil is very much greater than that of a surface
of water, and it frequently attains a very high temperature
Even in the driest countries the soil is moist, *and it may be
that, ultimately, the surface of every particle of the soil is
a water surface.* Whether this be so or not, when a land
surface cools, the heat of low refrangibility which it radiates
proceeds to a very large extent from water, and it is accordingly abundantly absorbed by the water vapour in the
lower layers in the atmosphere. In the absence of mechanical
mixture by wind, these layers can lose it only by passing
it on by radiation to higher layers which contain moisture,
whence it ultimately escapes into space. This accumulating
function of the atmosphere provides that *while every portion
of the earth's surface receives heat intermittently it loses it
continuously.*

As the heat of the atmosphere is due to contact with,
or radiation from, the surface, it must be taken from the
supply that reaches the surface of the earth. Further, wind
and all mechanical atmospheric effects are due to differences
of density, and these are produced not only by the thermal
expansion and accompanying rise of temperature of the air,
but also, and *without change of temperature, by the mixture
with it of a lighter gas.* Such a gas is the vapour of water,
and the water which supplies it is at the level of the sea.
*Therefore the sun's heat which arrives at the surface of the
earth at or near the sea-level has to maintain not only the
temperature of the surface of the globe, it has also to maintain all the mechanical manifestations of the air and the
ocean.* This is the ground for asserting, as above, that the
only constant which is of interest in terrestrial physics is
the rate at which the vertical sun heats unit area of the
earth's surface at the sea-level.

The instruments used for measuring the thermal effect of
the sun's rays must fulfil certain conditions. The area of the
sheaf or bundle of rays collected must be accurately known ;

and provision must be made for the exact measurement of the thermal effect produced by them in a given time. The thermal effect produced is measured by a mass of some substance and either by the change of temperature produced in it or by the change of its state of aggregation. Actinometers, such as those of Herschel, Pouillet, Violle, Crova, are instruments of the first kind. The ice calorimeter used by Exner and Röntgen and the steam calorimeter of the writer are instruments of the second kind. The thermal mass of the substance affected is conveniently expressed in terms of the thermally equivalent weight of water, which is called its water value. In the actinometer the change of temperature is either measured by a separate thermometer or the actinometer is itself a thermometer the calorimetric constants of which have been ascertained. In instruments of the second class no thermometer is required; the thermal effect is measured by the mass of water-substance which changes its state in a given time either from ice to water or from water to steam, both being at the same temperature. In the ice calorimeter the quantity of liquefaction is measured by the change of volume, as in Bunsen's calorimeter; in the steam calorimeter the generation of steam is measured by the weight or volume of the distilled water produced. The steam calorimeter was described recently in *Nature* (see above, p. 337), and it is unnecessary to repeat it here. It acted quite satisfactorily in the writer's hands in Egypt in May 1882, and it has since been giving good results in the hands of Mr Michie Smith at the observatory of Kodaikanal in South India, at an elevation of about 7000 feet above the sea. Theoretically, the ice calorimeter is as good as the steam calorimeter, but in applying it to the measurement of the sun's radiant heat it has a practical defect. At the moment before exposure, the ice in the calorimeter is frozen to the inner surface of the metal plate, the outer surface of which receives the sun's rays. The first effect of exposure to the sun is that the ice is detached from the plate. The intervening water introduces perturbations which are not easily allowed for.

The fundamental principle of the actinometer is analogous to Newton's second law of motion; when a body is engaged in the exchange of heat between itself and any number of other bodies, each exchange takes place independently of the others. The rate of exchange in each case depends on the difference of temperature between the two bodies and takes place on the principle that equal fractions of heat are lost or gained in equal times. A body cooling in the air is always subject to at least two quite independent sources of loss of heat, namely, radiation between itself and the surrounding objects and conduction between itself and the contiguous air. Under ordinary circumstances the rate of loss of heat by radiation is subject to but little variation, but that due to conduction is subject to continual variation owing to the varying rate at which the air actually in contact with the thermometer is renewed. It is not to be expected that a body subject to at least two independent sources of loss of heat will cool in the same way as it would if exposed to only one, any more than it is to be expected that a body acted on by two forces will move in the same way as if it were impelled by only one of them. The composition of rates of cooling is like that of velocities in the same straight line; the resultant rate is the nett, or algebraic, sum of all the rates. When the actinometer is exposed to the sun, its temperature rises at first rapidly, and then more slowly until, if the experiment is sufficiently prolonged, it becomes stationary. The temperature is noted at equal intervals of time. The sun is screened off, either after the temperature has become stationary or beforehand, and the temperature is observed at equal intervals during cooling. Whenever the thermometer is at a higher temperature than its enclosure, it is cooling. Therefore when it is exposed to the sun's rays, and its temperature rises ever so little above that of the enclosure, cooling begins; and what is observed in the first operation is, not the rate of heating by the sun's rays, but that rate diminished by the rate at which the thermometer is cooling. Hence, when the two series of observations have been made and tabulated, the rate of rise of temperature when that of the thermometer is, say, 2°, 4° or 6° above that of the enclosure is found. Similarly, the rate of fall of temperature when the temperature of the thermometer is 2°, 4° or 6° above that of the enclosure during cooling is found. Three pairs of rates are thus obtained. The sums of all three pairs of rates should be alike, and each gives a value of the rate at which the temperature of the actinometer would rise when exposed to the sun if there were no cooling. The rule is the same whether the temperature is allowed to rise to the stationary point or not. A distinction is often made between the *static* method, when the experiment is continued until the stationary temperature is arrived at, and the *kinetic* method, when it is interrupted before that temperature is reached. This distinction rests on no substantial difference; at the same time it is convenient to retain the designations to distinguish the manipulative processes.

Were the protecting enclosures, such as the double spherical shell packed with melting ice, used by Violle, or the thick metal shell used by Crova, perfectly efficient, then it would not be necessary to make a separate cooling experiment in connection with every heating one. The necessity for it is due to the fact that, when the sun's rays are introduced, the temperature of the air in the enclosure no longer is, and it cannot be, at the temperature of the enclosing shell; nor can it remain motionless, as it is when at a constant temperature in the shade. These perturbations, which cannot be avoided, so long as there is air in the enclosure, make it impossible to apply a rate of cooling determined beforehand. It is necessary on each occasion to determine the actual integral rate of cooling during the particular experiment.

If the actinometer could be so arranged that the rate of cooling should not be affected by the introduction or exclusion of the sun's rays, the static method could be adopted without hesitation, and the instrument would become a valuable one for continuous self-recording observations. Their value would be mainly relative. The absolute value of the sun's heat radiation, as it reaches the surface of the earth, has to be determined by other means. When it has been ascertained under the most favourable circumstances it does not vary, excepting in the annual cycle of the earth's revolution. The diurnal variation, as shown by registering actinometers, would have a great local importance. Crova, in the long series of valuable observations which he has made since 1875 at Montpelier, has, in fact, put this principle in practice.

Very important observations have been made in the neighbourhood of Chamonix by Violle and afterwards by Vallot. The *Annales de l'Observatoire météorologique du Mont Blanc* contain, ih vol. ii., several interesting reports on the results of these observations. They were made simultaneously at Chamonix and at certain stations on Mont Blanc. The first series of observations was made in 1887 on July 28, 29 and 30, and the instruments used were two "absolute actinometers" of Violle (*Ann. Chim. Phys.* 1879 [5], t. xvii.).

The great advantage of such experiments is that they are made simultaneously at two stations situated at very different altitudes. At the higher of the two the average barometric pressure is 430 millimetres, so that $\frac{33}{76}$ of the whole atmosphere are below the observer, and this portion contains nearly all the aqueous vapour. Above him there is a little more than one-half, and that much the simpler and purer half of the atmosphere. In it aqueous vapour is almost absent. The summit of Mont Blanc is 4807 metres and the station at Chamonix is 1087 metres above the sea. The layer of the atmosphere separating them has, therefore, a thickness of 3720 metres, and it can be visited at any point in its thickness. M. Vallot has acquired a personal acquaintance with this layer of air which can only be obtained by devoting a number of years to living in it and observing it. It is this intimate and continuous acquaintance with so large a proportion of the earth's atmosphere that entitles the observations and conclusions of M. Vallot to especially great weight.

The main results of Vallot's observations are as follows. The ratio between the heat received in the same time by the same area exposed perpendicularly to the sun's rays on Mont Blanc and at Chamonix was found to be 0·82 to 0·85, which agreed well with the proportion found by Violle in 1875. The value of the solar radiation found was, however, much lower than that found by Violle. The maximum values observed by Vallot were 1·56 gr.° C. on Mont Blanc and 1·33 gr.° C. at Chamonix, whilst Violle found 2·39 gr.° C. on Mont Blanc and 2·01 gr.° C. at the Glacier des Bossons in the valley. Violle's observed values are therefore half as great again as Vallot's. No explanation of the cause of this discrepancy is offered, but it is pointed out that the values observed by Crova at Montpelier are more in accordance with Vallot's than with Violle's. They are interesting in themselves and are worth quoting. They relate to the year 1895, the summer of which was very hot.

Intensity of solar radiation observed by M. Crova at Montpelier in 1895, in gramme-degrees per square centimetre per minute.

Season	Means				Absolute Maxima	
	Monthly			Seasonal		
Winter ...	1·02	1·12	1·15	1·09	1·32	January 28
Spring ...	1·20	1·13	1·13	1·15	1·38	May 12
Summer...	1·22	1·14	1·19	1·18	1·42	July 24
Autumn...	1·30	1·20	1·02	1·17	1·41	September 8

The subject was taken up again by Vallot in 1891, and this time he used the mercury actinometer of Crova (*Ann. Chim. Phys.* 1877 [5], xi. 461).

The result of the experiments in 1891 was in the main confirmatory of those obtained in 1887. In the following table the intensities of solar radiation on September 19, 1891, are given as observed on Mont Blanc and at Chamonix :

	Hour	9 a.m.	10	11	Noon	1 p.m.	2	3
Observed intensity of Radiation	On Mont Blanc	1·34	1·30	1·36	1·38	1·34	1·33	1·31
	At Chamonix	1·11	1·16	1·19	1·15	1·16	1·09	1·01
Ratio of Intensities ...		0·83	0·89	0·87	0·83	0·87	0·82	0·77

The mean value of the ratio of the intensities is 0·84, as before. The values of the intensity of radiation are rather lower than those found in 1887.

In the year 1896 Prof. Ångström, of Upsala, made observations on the peak of Tenerife with a special form of actinometer depending on the heating of metal plates. He made observations at three different elevations, namely, at Guimar, 360 metres, Cañada, 2125 metres, and at the summit, 3683 metres. Reduced to a uniform thickness of one atmosphere corresponding to a pressure of 760 mm., the intensity of radiation by the vertical sun was found to be at Guimar 1·39, at Cañada 1·51, and at the summit 1·54 gramme-degrees per square centimetre per minute. These values agree more closely with the values found in 1887 by Vallot than with those of 1891. But the values found by Crova, Vallot and Ångström are all of the same order.

The writer's observations with the steam calorimeter in Egypt in May 1882 were undertaken with the object of ascertaining the maximum rate of distillation near the sea-level under the most favourable circumstances. This occurred during the forenoon of May 18, when the meteorological conditions were as favourable as they could be. The sun shone steadily in a cloudless sky and the air was motionless. The shade temperature reached 40°·5 C. in the course of the day. Time was taken as portions of 5 cubic centimetres were distilled. The shortest time in which this quantity passed was 3 m. 20 s. This is at the rate of 1·5 c.c. per minute, and it occurred twice in the forenoon, namely at 10 h. 37 m.

and at 11 h. 23 m. As the collecting area of the reflector
was 904 square centimetres, this corresponds to 16·6 c.c.
distilled per minute per square metre. If we apply a correction for 20° zenith distance it becomes 17·04 c.c. The
evaporation of 17·04 grammes of water at 100° C. requires
9116 gr.° C. of heat, so that the heat actually collected
and used in making steam was at the rate of 9116 gr.° C.
per square metre or 0·9116 gr.° C. per square centimetre
per minute. Converting 9116 gr.° C. into work at the rate
of 0·425 kilogramme-metres per gramme-degree, we obtained
as the realised working value 3875 kilogramme-metres per
minute or 0·87 horse-power per square metre. The reflector
consists of one mirror inclined at an angle of 45° to the
axis of the instrument. This mirror throws all the reflected
rays normally on the surface of the axial border. The
larger mirror outside and the smaller mirror inside of this
one throw their reflected rays inclined at small angles to the
normal. Taking all the reflected rays together their mean
normal component is 94 per cent. of the total reflected rays.
It is therefore legitimate to increase the above numbers in the
proportion of 94 : 100, giving 0·93 horse-power or 9700 gr.° C.
per square metre per minute. Allowing 7 per cent. for loss,
we have in round numbers for the work-value of the sun's
vertical rays on the surface of the earth at or near the sea-
level 1 horse-power per square metre; the equivalent of this
in heat is 10,590 gr.° C. per square metre per minute, or
1·059 gr.° C. taking the square centimetre as unit of area.

Mr Michie Smith informs the writer that the highest rate
which he has observed at Kodaikanal is 1·754 c.c. distilled
per minute at a height of 7000 feet above the sea. This
result taken together with 1·501 c.c., the maximum amount
distilled per minute on the banks of the Nile, gives a measure
of the absorptive power of the corresponding layer of the
atmosphere in tropical regions.

Considering Crova's summer values as determined at
Montpelier and Vallot's (1891) values for Mont Blanc and
Chamonix, we find that the rate at which the surface of the
earth at the level of the sea receives heat under the most

favourable circumstances from the vertical sun is 1·2 gr.°C. per square centimetre per minute, and if we ascribe to the atmosphere a coefficient of transmission no greater than two-thirds, the value of the *solar constant*, or the heating power which the sun's rays would exert on a surface of one square centimetre exposed to them for one minute at a point on the earth's orbit, is 1·8 gr.°C. As the transmission coefficient is probably greater than two-thirds, the value of the solar constant is probably less than 1·8. Vallot, by giving effect to the rate of absorption actually observed in the air separating his two stations, arrives at 1·7 gr.°C. as the most probable value. These values are in substantial agreement with the older ones, such as those of Herschel and Pouillet; but there is a feeling at present (1901) that not much weight is to be attached to these results, and much higher figures seem to be more readily accepted. In a recent work, *Strahlung und Temperatur der Sonne*, p. 38, J. Scheiner sums up the discussion of this subject by giving 4 as the most probable value of the solar constant.

As we have seen, the heat which arrives at the sea-level has to support the temperature of the land and that of the sea; it has also to supply the energy for all the movements of the ocean; it has to warm and expand the air, and to furnish the latent heat represented by the aqueous vapour in the atmosphere, and it is mainly accountable for winds and storms. All this is maintained on certainly less than 1·5 gr.°C. per square centimetre per minute. But when the above catalogue of functions has been repeated, there is nothing left to be accounted for. If the sun's rays enter at the top of the atmosphere with an intensity of 4 and come out at the bottom of it with an intensity of only 1·5, how is the loss to be accounted for? It represents nearly double the energy which reaches the sea-level and produces such far-reaching effects. If it really entered the atmosphere it must be still there, either as heat or as its equivalent. But we know that the air is not made appreciably warmer by it, and we see no mechanical manifestations which can in any way be put forward as an equivalent. We conclude therefore that there is no excess of heat of this order to be accounted for, consequently values of the solar constant of the order of 4 are exaggerated.

POSTSCRIPT. *February*, 1911.

Since the above was written the generally accepted value of the solar constant has fallen from year to year.

In 1899 Scheiner (*Strahlung und Temperatur der Sonne*, p. 38) wrote: "Aus der im vorigen gegebenen Zusammenstellung lässt sich ersehen, dass die gefundenen Werthe für die Solarconstante nicht allzuweit auseinander liegen. Die älteren Bestimmungen haben zweifellos zu kleine Werthe ergeben, die neueren deuten mit grosser Sicherheit darauf hin dass die Solarconstante zwischen die Beträge von 3·5 und 4·0 Gr.-Cal. eingeschlossen ist. Nach Seite 17 ist anzunehmen, dass infolge der unvollständigen Absorption des Russes oder Platinschwarzes die wahre Solarconstante noch um 5 bis 10 procent zu erhöhen ist; man wird wohl der Wahrheit am nächsten kommen wenn man als wahrscheinlichsten Werth der Solarconstanten die Zahl 4·0 annimmt."

Crova, in his interesting memoir (*Rapports présentés au Congrès Internationale de Physique réuni à Paris en* 1900, iii. 469) arrives at the following conclusions: " Deux notions importantes peuvent être considérées comme acquises dans l'état actuel de nos connaissances :

" 1° Plus l'altitude est grande et les circonstances favorables, plus les courbes horaires sont symmétriques, plus aussi la valeur de la constante solaire est élevée et approchée de sa véritable valeur.

" 2° Il peut être considéré comme démontré que la véritable valeur de la constante solaire est au moins de 3 calories par minute et par centimètre carré, bien que l'on puisse obtenir des valeurs encores supérieures."

Quite recently (1909) Féry quotes 2·40 as the value of the solar constant generally accepted at that date. In his memoir ('Les Lois du Rayonnement et leur Application correcte': par M. Ch. Féry, *Ann. Chim. Phys.* 1909 [8], xvii.

267) he writes : " Depuis une vingtaine d'années la Physique s'est enrichie de données précises en ce qui concerne le rayonnement calorifique. Grâce à l'emploi du four électrique à résistance, qui réalise très approximativement les conditions de l'enceinte isotherme de Kirchhoff, les grandes anomalies signalées par les expérimentateurs qui employaient des *surfaces* rayonnantes noircies pour vérifier la loi de Stefan, ont disparu."

After an interesting discussion of this subject he writes (p. 287):

" D'après la loi du déplacement, la température apparente moyenne du soleil serait

$$\theta = \frac{2940}{0.54} = 5440° \text{ absolus.}$$

En employant cette valeur de T_m et la constante de Kurlbaum rectifiée $= 6.3$, on arrive à la valeur suivante pour la constante solaire,

$$A_w = 6.3 \times (5440)^4 \times \tan^2 \frac{\alpha}{2} = 0.118 \text{ watt,}$$

et en petites calories-minute,

$$A = \frac{0.118 \times 60}{4.15} = 1.70.$$

Ce désaccord de la valeur probable 1.70 de la constante solaire avec la valeur 2.40 habituellement admise, ne peut être dû entièrement aux propriétés sélectives des actinomètres ; je pense qu'il provient en partie d'une correction exagérée de l'absorption atmosphérique."

Here he refers to Crova's statement that the loss from this source amounts to more than one-half. Against this he puts the more recent observations of Wilson (*Proc. R. S.* 1902, xix. 312), who found 0.71 as the coefficient of transmission which gives only 29 per cent. for the absorption. He then proceeds : " Dans ses ascensions de 1906 et 1907, M. Millochau a effectué plus de 750 pointés sur la surface solaire, de Chamonix au sommet, et à cette dernière station à des distances zénithales diverses. Il a obtenu 5550° (abs.)

comme température apparente au centre du soleil, en employant le coefficient de transmission zénithale 0·91 auquel
l'avaient conduit ses nombreuses mesures. Le télescope
de mon système dont il se servait fut réétalonné après ces
mesures au National Physical Laboratory de Teddington....
Pointé à Teddington, voisin du niveau de la mer, par un
jour beau et très sec, sur le soleil, cet appareil a fourni
comme moyenne de huit mesures très concordantes : 5153°
comme température de l'astre.

"Le coefficient zénithal serait pour cette station

$$\left(\frac{5153}{5556}\right)^4 = 0\cdot74.$$

L'absorption ne serait donc que de 26 pour 100.

"Ceci ne doit pas trop étonner puisque le maximum de
l'énergie solaire est dans la partie lumineuse du spectre, où
la vapeur d'eau produit son minimum d'effet."

We see then that in the space of ten years the value of
the solar constant as determined by indirect methods has
fallen from 4·0 to 1·7 owing to the gradual revelation of
their defects.

The solar calorimeter which I constructed and used first
in 1882 furnishes a direct measure of the sun's heating power
by the determination of the mass of water which it can convert into steam. *It furnishes one simple equation for the
determination of one unknown quantity.*

Since I used it on the banks of the Nile, very little above
the level of the sea, it has been used by Mr Michie Smith
at the high-level observatory of Kodaikanal attached to that
of Madras.

The first object, in whatever locality it may be exposed,
is to obtain the maximum heating effect which the sun
can produce *there* under the most favourable circumstances.
During the few days in May 1882, when I used the instrument at Sohag, the sun attained a meridian altitude of
roughly 83°, so that its power differed very little from that
of a vertical sun. The latitude of Kodaikanal is 10° 14′ N.,
and there the meridian altitude of the sun is greater than 80°

for twice 60 consecutive days in the year. The elevation of Kodaikanal is 7000 feet above the sea, therefore the average height of the barometer may be taken to be 585 millimetres and that at Sohag as 760, so that the mass of the atmosphere above the two places is in the proportion 77 : 100. The maximum rate of distillation at Sohag was 1·5 grams per minute. This is equivalent to 16·6, and after correction to the zenith, to 17·04 grams per square metre. At Kodaikanal the maximum observed rate was 1·754 grams distilled per minute. This has no doubt received correction to the zenith, if any was necessary. It is equivalent to 19·39 grams distilled per square metre per minute.

Multiplying these quantities by the numbers representing the latent heat of steam at 100° C. and 93° C., namely, 535 and 542 respectively, we have for the quantities of the sun's radiant heat which have been converted into the latent heat of steam 9116 gr.° C. and 10515 gr.° C. respectively.

We have still to apply the correction to both observations in order to compensate for the fact that the rays reflected from the outer and inner mirrors do not strike the boiler normally. The effect has been computed to be diminished by this imperfection to the extent of 6 per cent.; it is therefore legitimate to increase both values in the proportion of 94 : 100. We thus obtain the quantities 9698 gr.° C. and 11186 gr.ᶜ C.

Conclusion.—Considering only my own observations, the final result of the experiments made with my steam calorimeter is that the calorific value of a sheaf of the sun's rays, having a section of one square metre, is at least 9698 gram-degrees Centigrade per minute at the terrestrial sea-level; and it is certain that heat can be obtained from them there at this rate and in a useful form by mechanical appliances of simple construction.

No. 12. [*From Nature, December* 21, 1905, *Vol. LXXIII,* p. 173.]

THE TOTAL SOLAR ECLIPSE OF AUGUST 30, 1905

IN visiting Spain at the end of August of this year I was actuated by the desire once again, after an interval of twenty-three years, to witness the marvellous and unique phenomenon of a total solar eclipse. It is a sight which cannot be imagined—it must be seen. Happening at a time of maximum sun-spot frequency, it was reasonable to expect a considerable display of protuberances, and I wished to form my own idea of their size by checking their persistence or non-persistence through the phase of mid-totality on a day which otherwise may be taken to have been chosen at random. For this purpose a station on, or very close to, the line of central eclipse was essential. Torreblanca was chosen because it was the station of the Barcelona and Valencia railway which was nearest to the line of central eclipse, lying, in fact, about a mile to the south-west of it.

I observed the eclipse from the railway station, the position of which is lat. 40° 12′ N. and long. 0° 12′ E. (Greenwich). The railway and official time of Spain is that of Greenwich. By the clock at the railway station, mid-totality occurred between 1 h. 18 m. and 1 h. 19 m. p.m. Before the beginning of the eclipse I entered in my note-book half the expected duration of totality, 1 m. 50 s.; when I had observed the second contact, I wrote the time underneath, and, by addition, ascertained the time of mid-totality by my watch. The display of protuberances which appeared just before the moment of second contact, and on the part of the sun's limb which was about to be eclipsed, was, according to all witnesses, exceptionally brilliant. When the time of mid-totality came round I looked for these protuberances. They were absent. Not a trace of them or of any others was visible to the naked eye, and I searched the whole edge of the moon's disc with the greatest attention. Their absence was confirmed by Stephan (*Comptes rendus*, October 9), observing with the best instrumental aid at Guelma.

The sun's true altitude at Torreblanca on August 30, at 1 h. 18·5 m., may be taken as 54°·5. For this altitude the augmentation of the moon's semi-diameter is 14″·3. Adding to this the geocentric semi-diameter, 16′ 21‴·4, as taken from the *Nautical Almanac*, 16′ 35″·7 is obtained for the apparent semi-diameter of the moon as seen from Torreblanca at mid-totality. Deducting from this the semi-diameter of the sun, namely 15′ 50″·7, we obtain 45″ as a sufficient approximation to the width of the annular band by which the disc of the moon overlapped that of the sun. Therefore, to an observer stationed on the central line in this neighbourhood, no protuberances could be visible at mid-totality which had a height less than 45″, and, neglecting the small displacement of Torreblanca from the central line, the protuberances of the second contact, magnificent though they were, could not have exceeded this height.

Eight seconds before second contact I detected the streamers of the outer corona on the western limb of the moon. At this moment there was no trace of the inner corona, which presents to the spectator during the whole of totality the appearance of a bright, luminous ring surrounding the moon.

If we assume that the argument from parallax is applicable to the inner corona, as it is to the protuberances, we have to conclude that, eight seconds before second contact, the light-giving portion of it did not extend further than between 93″ and 94″ from the western limb of the sun. To an observer on the central line, at mid-totality, it is eclipsed to a distance of 45″ from the sun's limb, and this would leave only between 48″ and 49″ as the width of the outer portion, which furnished the unexpected amount of light which persisted through totality. It is clear that if the inner portion, having a width of 45″, had been uncovered, the daylight during totality would have been still more remarkable.

In this respect there was a great contrast between the eclipse of this year and that of May 17, 1882, which I witnessed at Sohag, on the Nile, where a large camp of astronomers of many nations was established. In it, one of the most striking features was the rapid darkening during the last moments before second contact. I have always compared it to what is witnessed when a lecture-room is darkened during the day by quickly closing the shutters of the windows in succession. In 1882 the darkening took place rapidly and completely; and immediately quite a number of stars came out, besides the great comet which revealed itself, all unsuspected, close to the sun's limb, and formed the feature of that eclipse which was most noticed and is best remembered by the spectators. In 1905 the darkening effect was much less striking; but the illustration of the lecture-room holds if we imagine that the shutter of the last window is out of order and has to remain open during the demonstration.

The contrast between the two eclipses is accentuated when we remember that the apparent semi-diameter of the moon, as seen from Torreblanca, was 45″ greater than that of the sun, while on the Nile this excess was only 15″·4. Therefore a width of 45″ of the brightest part of the corona was eclipsed in 1905, as against only 15″·4 in 1882. If, therefore, the uneclipsed coronas had possessed equal efficiency as furnishers of daylight, the darkness during totality ought to have been much greater in 1905 than it was in 1882; but the opposite was the case. Therefore, whatever may be the process by which the inner corona or luminous ring is produced, it was much more active on August 30, 1905, than it was on May 17, 1882.

No. 13. [*From Nature, October* 19, 1905, *Vol. LXXII, p.* 603.]

ECLIPSE PREDICTIONS

IT is always interesting to compare the results of observation with those predicted by calculation. In the case of the recent total eclipse of the sun this is rendered difficult by the want of agreement in the predictions of the two most used authorities, the *Nautical Almanac* and the *Connaissance des Temps*. The discrepancies in the predicted duration of totality and of the breadth of the band traced on the earth's surface by the total phase are made apparent in the following table. It is compiled from the table in the *Nautical Almanac* headed "Limits of total phase of the Solar Eclipse," and the corresponding table in the *Connaissance des Temps* entitled "Limites de l'Eclipse totale et Durée de la Phase totale sur la Ligne centrale." Entries for as nearly as possible the same time in each table have been taken and are placed together:

A B	C		D		E
1905 Aug. 30 G.M.T.	Distance	Bearing	N. A.	C. T.	
h. m.	′	°	secs.	secs.	secs.
C. T. 0 22	113·5 N.	1 W.	(198·4)	206	7·4
N. A. 0 24	101·5 ,,	2 ,,	200·6	(208)	7·6
C. T. 0 35·2	109·5 ,,	2 E.	(211)	219	8·0
N. A. 0 36	102 ,,	11 ,,	211·8	(219·5)	7·7
C. T. 0 50·3	114 ,,	6 ,,	(220·2)	228	7·8
N. A. 0 48	104 ,,	19 ,,	219·1	(227·4)	8·3
C. T. 1 7	116·5 ,,	10 ,,	(223·8)	231	7·2
N. A. 1 8	104 ,,	31 ,,	223·8	(231·2)	7·4
N. A. 1 24	105·5 ,,	37 ,,	220·7	(226·6)	5·9
C. T. 1 24·9	116·5 ,,	12 ,,	(220·2)	227	6·8
C. T. 1 43·1	115 ,,	14 ,,	(209·2)	215	5·8
N. A. 1 44	106 ,,	44 ,,	208·4	(214)	5·6

Column A contains the authority, *Nautical Almanac* (N.A.) or *Connaissance des Temps* (C.T.).

Column B contains the time (G.M.T.) for which each prediction is made.

Column C contains the calculated distance (in nautical miles) and the bearing of the northern limit of totality from the corresponding southern limit.

Column D contains the duration of totality on the central line as predicted by the one authority and (in brackets) as interpolated from the prediction of the other.

Column E contains the differences of these pairs of values.

It will be seen that, for stations in Spain and the adjacent Mediterranean, the duration of totality on the central line was predicted by the French authority to be from seven to eight seconds longer than by the British authority. In the same region, the width of the band of totality is from ten to eleven nautical miles greater by the French than by the British prediction. The orientation of the line connecting the two limits of totality also differs considerably in the two tables.

It is reported that at Sousse and Gabes, two towns in Tunisia, the eclipse was partial, while a total eclipse had been predicted for them. The prediction for these places would surely rest on French authority; we are therefore entitled to conclude that the mistake has been made by the French calculators. An excessive estimate of the width of the band of totality would almost certainly be accompanied by an excessive estimate of the duration of totality, and the table shows that both estimates are considerably greater in the *Connaissance des Temps* than in the *Nautical Almanac*[1].

---

[1] See Contents, p. xxix.

No. 14. [*From Nature*, May 9, 1912, *Vol. LXXXIX*, p. 241.]

THE SOLAR ECLIPSE OF APRIL 17, 1912

THE study of the article by G. Fayet in the *Revue Scientifique* of March 30, an account of which was given in *Nature*, convinced me that, with favourable weather, the solar eclipse of April 17 would prove to be interesting, although its totality was extremely doubtful.

I went to Paris on April 16 and put up at the Gare du Nord. At 10 a.m. on April 17 I took a suburban train from that station to Eaubonne, which was on the central line, as shown in the map given in Fayet's article. I arrived there about 10.35, and after looking round I took up a station at a seat by the roadside in front of the school. When I arrived the boys and girls were being dismissed, by order of the Minister of Education, so that they might see the eclipse. Some of them came round me while I was looking out for the first contact. They were much interested, and were very well behaved. They all had the red glasses supplied by an advertising firm, but they had a curiosity to see the sun through my glass, which showed it in its natural colour, and they were delighted with the effect.

During the first hour they played about, because watching the gradual encroachment of the moon on the sun was tedious. When the diminution of daylight made itself felt they began to gather round me; so I told them they must not look at me but look at the sun and moon, and notice all that they saw, and that, while the eclipse was going on, they must not speak to me or to each other. "*Ne pas parler?*" "Yes," I said, "*ne pas parler*—you must look and do nothing else." They retired into the back part of the playground and stood in a group, and they no doubt looked, for they were quite silent.

When the central phase was over and I had taken my glass away from my eyes, they rushed up in a body and surrounded me, and I asked them if the sun had become quite dark or if there had been some light all the time. Their opinion was divided. From this I concluded that the sun had been at no time completely obscured, for this would certainly have impressed them.

As I was working alone it was useless to try to take the times of contact. Moreover, the second and third contacts, which are the most important, would happen so close to each other that, if I attempted to time them, it would interfere with my seeing what happened. I therefore devoted myself entirely to following the eclipse and observing as well as I could everything that took place. It was certain that "Baily's beads" would be a feature of the eclipse, and I had great curiosity to study them.

The sky was cloudless and the sun very powerful. I had with me an ordinary binocular of low power, which I hoped to find useful when the short-lived central phase, whether total or annular, arrived. In order to be able comfortably to follow the eclipse from the beginning to the end, I had the hand glass which pleased the school children. It was a combination of three coloured glasses measuring 110 × 35 millimetres, so that the sun could be observed through it, using both eyes, whether it was used alone or in conjunction with the binocular. The effect produced by this combination of colours was that the sun, viewed through it, appeared of its natural hue. The density of the coloured medium was such that nothing but the sun's direct rays penetrated it, and the sky, in which the sun appeared to be set, was quite black.

I bought this glass of a hawker in the streets of Barcelona on the eve of the total eclipse of August 30, 1905, and I found it very useful, although the interest of that eclipse centred almost entirely in the total phase, which lasted nearly four minutes, and during it reducing glasses were not required. Very fortunately I was able to lay my hands on it before starting for Paris, and it was indispensable during the eclipse.

The three glasses have an aggregate thickness of 7 milli- metres, and they consist of one green glass 3 millimetres thick and two pink glasses, the external one being 2½ and the internal or middle glass being 1½ millimetres thick. These seem to be pieces of the same glass, and differ from each other only in thickness. They have the same colour, which is very nearly that of a dilute solution of permanganate of potash. The colour of the green glass is a chromium- green, and the colour-intensity of its thickness of 3 millimetres has been successfully compensated by a 4-millimetre thickness of the pink glass. The result of the combination was most satisfactory.

Having noticed that first contact had taken place, I settled down to follow the progress of the eclipse, and I continued making observations every five minutes. The rate of en- croachment on the periphery of the sun became less and less, though the augmentation of the area eclipsed proceeded rapidly.

At 11 h. 35 m. the eclipse appeared to be affecting the general illumination, and I looked round at the school-house to the north of me. The ultramarine blue of the sky was getting darker and more of an indigo. Looking towards the horizon, which was masked by trees, the illumination was becoming decidedly fainter. The indigo colour of the sky spread round and became deeper as the eclipse went on, and, by 11 h. 50 m., there was an impression of approaching nightfall.

At 11 h. 40 m. the invasion of the sun's disc had reduced it to a crescent with very sharp cusps. The limb of the moon still looked quite circular, but at 11 h. 45 m. it became somewhat ogival, and by 11 h. 50 m. this effect was very marked. By this time the schoolboys were observing the sun without coloured glasses.

I now noticed the peculiar appearance of the shadows of the foliage of a tree cast on the ground close to my station. I also noticed that the illuminated surface of the dusty playground became more and more sombre, while the shadows under the trees preserved the same tone, so that, while the illumination of the exposed ground diminished rapidly, that of the ground in the shade of the trees had already nearly reached its minimum of illumination before the eclipse was complete.

After this my attention was confined entirely to what was taking place in the sky. I was now using the binocular with the reducing glass in front of the eye-pieces. With it the rapid diminution of the luminous crescent could be easily followed and the view furnished was very sharp.

As the area of the luminous crescent diminished rapidly before the advance of the dark lunar disc *the colour of its light suddenly changed to red.* This suggested to me that a brilliant display of protuberances might be expected. The tone of the red reminded me, at the moment, of that of nitrous fumes escaping into the air; it was therefore a very pure red. It became visible only after the overwhelming intensity of the relatively white light of the middle of the sun's disc had been screened off by the interposing moon; but it would have been impossible to perceive the red colour, intense though it was, had it not been for the perfection of my reducing glass, which, while it reduced the intensity, pre-served the natural colour of the sun's light.

Before the most striking phenomena of the central phase began to crowd across my view, I noticed the beginning of the phenomenon which most impressed me when witnessing the eclipse of May, 1882, in Egypt. In the last moments before totality the rate of extinction of light was very great, and I compared it with that which would take place in a well-illuminated room when a shutter is rapidly drawn down over the window. In the case of the 1882 eclipse the shutter was drawn quite down, and nocturnal darkness was produced with the appearance, not only of all the principal stars, but also of an unsuspected comet in the immediate vicinity of the sun. In the case of the present eclipse the shutter was at first being drawn down quite as rapidly, but it stopped short, and was almost immediately pulled up again. I have no doubt whatever that if the eclipse had been total, it would have been a very dark one.

The central phase was now close at hand, and the appearance of the luminaries changed so rapidly that it was impossible to time the changes. After the light of the whole solar crescent had become quite red, my attention was attracted to the lower (S.E.) luminous cusp, which seemed to become indented by black bands or teeth. Then the upper (N.W.) cusp showed a similar phenomenon; and, almost in a moment, the black teeth spread over the whole crescent, which then offered a magnificent spectacle. The bands or teeth did not span the crescent always by the shortest path, but they crossed and intersected each other like a crystallisation. There was, however, but little time to study them. Very quickly the dark disc of the moon advanced and pushed the beautiful network over the eastern edge of the sun, which it totally obscured, and, apparently at the same moment, the network reappeared, coming over the western edge of the sun, attached to the black limb of the moon, and at the same time held by the limb of the sun. In a few moments the uncovered crescent of the sun had increased so much that the delicate lacework could no longer bear the tension; it parted and disappeared instantly, while at the same moment the dark limb of the moon recovered its perfect smoothness of outline.

The central phase of the eclipse was over, and I could not say that I had seen either a total eclipse or an annular one, but I had witnessed a very remarkable natural phenomenon.

All the phenomena were so astonishing and followed each other so closely that it was impossible to pay attention to every detail. The two crescents, the disappearing and the reappearing one, seemed to be situated diametrically opposite to each other. I perceived nothing on the upper (N.) or the lower (S.) edge of the common disc, but there might have been a thread of light or a string of minute "beads" on one or both of them; and, consequently, I cannot say if the light of the disappearing crescent passed round the northern or the southern limb of the common disc and so preserved continuity between the departing and the arriving crescents, or if it passed round at all.   All that I *saw* was the extinction of the departing crescent, and, *post saltum*, the illumination of the arriving crescent.

When the moon is in conjunction and the sun is behind it, the mountains cut by a tangential surface cannot be very evident, because *they can only be the summits of the very loftiest peaks.   The valleys are wholly masked.*   My binocular, the magnifying power of which is only twofold, shows the mountains and valleys beautifully when the moon is in quadrature, but during the eclipse it made the edge of the lunar disc appear as a smooth and continuous line.   The mountains were perfectly invisible on it; yet what we take to be their images were enormous.   The phenomenon is not a subjective or an instrumental *spectre*, because it is seen by everybody, with every kind of instrument and without any instrument at all.   It is a reality; it must therefore be due to a substantial cause, and to one which can be shown to be capable of producing the effect.   May not this substance be the often-suggested lunar atmosphere; and, if so, what is its exact specification?[1]

[1] See Contents, p. xxx.

No. 15. [*From Nature, August* 10, 1893, *Vol. XLVIII,*
p. 340.]

THE PUBLICATION OF SCIENTIFIC PAPERS

THE discussion of this important subject has been started *à propos* of physical papers, but the publication of papers in all branches of science is in an equally unsatisfactory state.

Prof. Lodge, in his letter in your issue of July 27, after paying attention to the preparation of useful abstracts of all papers on physical subjects, appearing both at home and abroad, calls attention to what has always appeared to me to be the most important matter for reform, namely, the means and methods of publication of English scientific papers.

There is no complaint more frequently heard abroad than that important papers of English scientific men are almost inaccessible to the foreigner, because it has been the fashion to communicate them to local societies and to rest content with such publication as is secured by their being printed in the Society's *Proceedings* or *Transactions*. If these societies distributed their publications liberally where there are students who ought to have the opportunity of reading them, and without taking account of whether they receive in exchange a publication of an equal number of pages, the evil would be much less. But this is not so. It is notorious—to take, for instance, the Royal Society of Edinburgh, with which I am best acquainted, and which is not by any means the least liberal in the matter of distribution—that unless the author distributes lavishly separate copies of his paper in every quarter where he considers it important that it should be read, it will pass unnoticed, and a worker in the same branch of science will not consider that he is open to any blame for not being acquainted with a paper published in an organ so difficult to procure. I believe that this applies in at least an equal degree to the other two societies mentioned by Prof. Lodge, namely, those of Dublin and Cambridge, and of course it is all the more applicable to societies of less importance. But even the Royal Society itself is open to exception in this respect, for although no fault can be found with the *Proceedings* or *Transactions* as a recognised organ of publication, they are, as a matter of fact, not more readily accessible abroad than the corresponding publications of the Edinburgh Society, and the majority of foreign students never see anything but abstracts of important English papers. The only independent scientific journal of importance is the *Philosophical Magazine*, and though widely known it is not extensively used, and has not grown with the times. The want of means of scientific publication which has been produced by the development of scientific activity in the last twenty or thirty years has been met by an increase in the number of societies, and by a greater development of society publication. The former is probably an advantage, the latter is certainly a disadvantage. The publication of scientific papers cannot be too much centralised in the interests of both authors and readers, and for this purpose a central organ such as indicated by Prof. Lodge is required.

What is at present inefficiently and extravagantly done by a multitude of amateur publishers scattered over the country could at much less cost be efficiently done by a central publishing officer issuing a central organ, in several series, each series appearing in monthly numbers, and the whole run on strictly business lines. Each series should be devoted to a particular science or branch of a science. Thus, there might be several series in chemistry as organic chemistry, inorganic chemistry, physical chemistry and technical chemistry. Physics also would fall into several series, as would other sciences. Each series of original papers would have a parallel one of abstracts of foreign papers on the same subject, and it would be useful to have a separate series, which might be issued weekly or fortnightly, devoted to printing a minute of the proceedings and papers read at the meetings of the various societies throughout the country, to be furnished by their secretaries.

The effect of the realisation of some such plan as this would be the immediate setting free of the large sum of money annually spent by the societies in printing, and the collection of all that is published in *one* organ, which would be an enormous assistance to the student.

Each series would have to be intelligently and liberally indexed, and a separate volume of the indices of all the series published each year. It would then be sufficient for the worker to take in the series devoted to his own branch of science and the yearly index volume, which would prevent his overlooking papers of importance appearing in other series.

This scheme of central publication has occupied my thoughts for some years, and I have from time to time discussed it with my friends, and it has even been brought before one publisher, but without any practical effect.

It is therefore with very great pleasure that I find Prof. Lodge advocating a similar scheme, and I hope that it may be the means of fixing public attention on the present unsatisfactory state of things and of forcing a remedy[1].

[1] See Contents, p. xxxi.

No. 16. [*From Nature, January* 28, 1904, *Vol. LXIX,*
p. 293.]

THE ROYAL SOCIETY

AT the special meeting of the Royal Society held on
January 21, when the constitution and functions of the sectional committees were under consideration, the opinion was
expressed by more than one speaker that the usefulness of
the society in encouraging and advancing scientific work is
not what it might be ; but no very definite suggestions were
made with a view to its improvement.

It seemed to me that the functions of these sectional
committees had a good deal to do with the lack of scientific
enterprise which we observe in the Royal Society, and that
they might with advantage be done away with.

As many of the fellows had left the meeting before I
spoke, and as everything that affects the efficiency of the
Royal Society concerns the public, I crave the hospitality of
the columns of *Nature* to develop as shortly as possible my
views on this matter.

The main function of the sectional committees is to *refer*
papers received by the society from fellows, to some other
fellow or fellows of the society to be certified that they are or
are not fit to be accepted and published by the society.

It is well known that the fellows of the society are *de facto*
chosen by the council after rigid scrutiny and the most careful
inquiry, and the only object of this scrutiny and inquiry is to
satisfy the council that the candidate whom it recommends
is a man of eminence in his own science, and that the work
which he is likely to do will be a credit to the society. So
convinced is the society of the thoroughness and impartiality
with which the council discharges this duty that the confir-
mation of its selection by election has come to be a pure

formality. This being so, it cannot fail to surprise the newly elected fellow, when he proceeds to justify his election by doing work and communicating the results of it to the society, to find that he is now in no better position than he was before he was elected. His work is *referred* in the same way as that of any outsider. His recent selection by the council is ignored by that body or is regarded as having no weight, and it treats him, scientifically, as a perfect stranger.

Furthermore, this *reference*, which amounts to neither more nor less than a secret revision of the title of the fellow to the privileges of the society, is repeated on every occasion when he comes under the notice of the society by offering it work. So long as he is content to be a passive fellow, or at least an inactive one, he is spared this injustice and indignity. It is no wonder then that the fellowship of the Royal Society has come to be looked on as an invitation to repose rather than as an incentive to work.

How different is the state of things which we observe in the parallel society in France, the Academy of Sciences. Its constitution is thoroughly democratic, and all its proceedings are inspired by enlightened self-respect. But we need only contemplate the work which it puts through in the year and compare it with what is turned out by the Royal Society to see that there is something for us to learn by its study.

First and foremost the academy meets fifty-two times in the year, namely, on every Monday, with the exception of Easter Monday and Whit Monday, and then it meets on the following Tuesdays. By the time-table of the current year the Royal Society is to meet twenty times.

Papers by members, or communicated by members of the academy, are not obliged to be sent in before the meeting. The *agenda* of the meeting is compiled at the meeting, each member who has a paper to communicate giving notice of it to the secretary on his arrival in the room, and the papers are taken strictly in the order of their intimation. If the paper communicated by the member is to be published in the *Comptes rendus* of the sitting, it has to be handed in to the secretary at the sitting; the corrected proof has to be returned to the printer on the Wednesday evening, and it is then published without fail on the Sunday. The communication, reading, and publication of a paper presented to the academy is therefore an affair of the inside of a week, and it is a certainty. This promptitude in the putting through of work is due to the fundamental fact that when a man is elected a member of the academy he enters at once into the full enjoyment of all its privileges, and one of the chief of these is the complete confidence of all his fellow-members. When he communicates a paper, whether it be by himself or by someone not a member of the academy, it is accepted without question. The only limitation in the privileges of members is with regard to the space that they are entitled to claim in the *Comptes rendus*. A paper by a member or foreign associate of the academy may fill six pages per number, and his communications in the year may fill fifty pages in all, and this as a matter of right.

It is unnecessary to occupy more space in order to show what a powerful engine the Academy of Sciences is in the production and encouragement of work, or to indicate how easily the Royal Society may successfully rival it. Let every fellow of the society, whether he be on the council or not, have complete confidence in his fellow-fellows and give practical effect to it, and the thing is done. The rest will follow of itself[1].

[1] See Contents, p. xxxi.

No. 17. [*From Nature, May* 12, 1898, *Vol. LVIII, p.* 30.]

NOMENCLATURE AND NOTATION
IN CALORIMETRY

ALL who are engaged in thermal investigations themselves as well as those who have occasion to study the
published work in this department of science, must have
been frequently annoyed by the use of the word *calorie* with
its varying signification. It has been sought to remove the
inconvenience by qualifying the calorie as small or great, and
in other ways ; but on opening a book at any place where
the results of thermal determinations are given, it is in most
cases difficult to discover at once what unit of heat the author
is using.

As different classes of investigation are carried on on
different scales, it is obvious that it is a convenience, if not
a necessity, to have different heat units at disposal. The
unit which is suitable to express the thermal changes in a
beaker in the laboratory would manifestly be inconvenient
when dealing with the daily or seasonal changes in a lake or
an ocean. It is therefore natural and necessary to have heat
units of different magnitudes, but it is neither natural nor
necessary to call them all by the same name, and it is extremely inconvenient not to have a short form of notation
which will show on its face the actual heat unit used.

In the early literature of the equivalence of heat and work
in this country, one unit of heat is universally used ; it is
the pound-degree-Fahrenheit, and in the writings of Joule,
Thomson, Rankine and others of that time it is simply called
"heat unit," as there was no other competing with it. With
the rise and development of thermal chemistry, it was necessary to fashion the compound unit out of the simple units in

common use in chemical laboratories; these are the gramme and the Celsius degree.

The heat given out by one gramme of water cooling by 1° C. at ordinary temperatures is the unit most used in such researches; and it received the name of calorie, sometimes now called small calorie.

For many purposes this unit proved itself inconveniently small, and several larger units have been used, such as the heat given out by one kilogramme of water cooling 1° C. at ordinary temperatures, or the heat given out by one gramme of water cooling from 100° C. to 0° C.; but the name of calorie was retained in connection with them all, and in the specification of a quantity of heat by a number, the nature of the unit was indicated by the syllable cal. or the letter K, neither of which of itself gives any information.

In my own work, and in the study of the writings of others, I have adopted a form of notation which I have found so useful that I propose to lay it before the readers of *Nature*. I do not doubt that others who interest themselves in calorimetric work have been driven to adopt some similar, perhaps the same, perhaps a better form of notation; and I think they will agree with me that some system of self-interpreting notation should be universally adopted without loss of time.

Just as, when dealing with work, we use currently the expressions foot-pound and kilogramme-metre, so in calorimetry it is quite common to talk of a gramme-degree, or a kilogramme-degree; and what I propose is to use no other expression than these compound and self-explaining ones, and, in writing, to express them shortly by g° and k° respectively, to which for clearness the symbol of the thermometric scale must be added, so that they become g° C. and k° C. when Celsius' scale is used, or g° F. and k° F. when Fahrenheit's scale is used.

On this system the expression g° C. would replace the ordinary "cal." and Ostwald's K would be represented by 100 g° C. or 0·1 k° C., or by h° C., to mean hectogramme-degree C. With perfect exactness K would be expressed by g 100° C., but the difference between 100 g° C. and g 100° C. is much less than the probable experimental error in any calorimetric operation. In a table containing a column of quantities of heat expressed in numbers of gramme-degrees-Celsius, the nature of the unit would be indicated at the top of the column by g° C.; exactly as, in a column of temperatures, the unit is indicated by the symbol ° C. or ° F. The original British heat unit is then clearly expressed by lb° F.

A heat unit made up of any unit of weight and any unit of temperature can be perfectly expressed in this system. Thus, if there were any advantage in doing so, we might have g° F., lb° C., k° R. and many others, and their meaning would be at once apparent on inspection.

In oceanographical work, where the heat exchanges between one layer of water and another, or between the water and the air are under discussion, I have found the most convenient heat unit to be the fathom-degree-Fahrenheit, or the metre-degree-Celsius, which are abbreviated for the purposes of notation into f° F. and m° C., respectively. The nature of this unit will be most easily understood by considering an example.

In a paper, 'On the Distribution of Temperature in Loch Lomond in the Autumn of 1885,' read before the Royal Society of Edinburgh, and published in its *Proceedings* for the session 1885–86, I have given, at page 420, a table of the changes in the distribution of heat in the direction of depth, between several pairs of dates, in the Luss basin of Loch Lomond. At a certain depth, indicated by the intersection of the temperature curves, the temperature of the water is the same on both dates. The season being autumn, the layer above this depth has been losing heat, partly to the air above and partly to the water beneath, while the layer below the depth of common temperature has been on the whole a gainer. Thus, taking the dates September 5 and October 15, the intersection of the temperature curves is found at a depth of 16 fathoms; and in the interval of forty days the mean temperature of the water above this depth has fallen by 5°·8 F., from 55°·0 F. to 49°·2 F. The thickness of the layer is 16 fathoms; therefore the loss of heat has been $16 \times 5 \cdot 8 = 92 \cdot 8$ f° F., or 92·8 fathom-degrees-Fahrenheit. The total depth of the lake at the spot was 35 fathoms, therefore the layer of water below the depth of common temperature was 19 fathoms thick. The mean temperature of this layer was 47°·1 F. on September 5, and 48°·9 F. on October 15, showing a rise of 1°·8 F. in the interval. This corresponds to a gain of heat represented by $19 \times 1 \cdot 8 = 34 \cdot 7$ f° F. Assuming that the heat gained by the lower layer has been entirely at the expense of the upper one, we see that the loss of heat of the upper layer, during the interval, has been to the extent of 37·4 per cent. to the deeper water, and 62·6 per cent. to the air. The upper layer of water has thus been passing heat at the average rate of 1·485 f° F. into the air, and into the deeper water at the rate of 0·85 f° F. per day.

It is worthy of remark that the fathom-degree-Fahrenheit and the metre-degree-Celsius are interchangeable in heat calculations, because the fathom is 1·8 metre and the Celsius degree is 1°·8 F.

This is a great convenience, and its usefulness will be apparent by applying it to the above example.

We have seen that, during the interval of forty days, the average transmission of heat from the upper layer of water has been at the daily rate of 1·485 f° F. to the air and of 0·85 f° F. to the deeper water. Writing m° C. for f° F., and considering a horizontal area of one square centimetre, we find at once that the average daily supply of heat from the water to the air has been at the rate of 148·5 g° C., and to the deeper water at the rate of 85 g° C. (gramme-degree-Celsius) per square centimetre of superficial area.

It is unnecessary to provide for special cases where specially suitable units will be chosen as a matter of course; but for ordinary work of constantly recurring type it is important to have a system of nomenclature and of notation, each of which will tell its own story.

No. 18. [*From Nature, August* 17, 1899, *Vol. LX,*
pp. 364–365.]

THERMOMETRIC SCALES FOR METEOROLOGICAL USE

IN the course of some recent work on the meteorology of Ben Nevis, which involved extensive extracting and computing work, I have again had forcibly impressed on me the great advantage which Fahrenheit's thermometer has over that of Celsius for meteorological use, especially in temperate regions.

In chemistry and physics the range of temperature covered is so great that Celsius' scale, which is now universally used, adequately meets every case. The size of the degree and the change of sign at the melting-point of ice do not cause any inconvenience in the laboratory. It is otherwise in the meteorological observatory. There the range of temperature dealt with is very restricted, and the Celsius degree is too large, while the change of sign in the middle of the working part of the scale is simply intolerable. The latter peculiarity is the fruitful introducer of error into both the observations and the reductions, and besides it greatly increases the fatigue of both classes of work.

In view of the agitation to abolish the use of Fahrenheit's scale, and to replace it universally by that of Celsius, it may not be inopportune to direct attention to some of the advantages in securing accuracy and in relieving labour which Fahrenheit's scale offers over that of Celsius when used for meteorological purposes.

In tropical countries it matters little whether one scale or the other is used, except that the size of Fahrenheit's degree is much the more convenient, as the first decimal place is always sufficient. But in Europe and in North America, where the greater number of meteorological observatories is situated, the temperature falls every year below the freezing-point of water. In some localities it passes quickly through this point and remains constantly below, often far below it, returning again in the spring and passing as quickly through it again in the beginning of summer, to remain constantly above it until it drops away again in the fall of the year. In such places, where, however, the population affected is limited, the use of Celsius' scale is not open to very much objection. With the exception of a few days in the fall, and again in the spring of the year, the temperatures are either continuously positive or continuously negative; and during one-half of the year the observer reads his thermometer upwards, while during the other half of the year he reads it downwards. When he has got well into the one or the other half of the year, he will make no more errors than those that he is personally liable to under circumstances of no difficulty. But at and near the two dates when the temperature is falling or rising through that of melting ice the case is very different. If the rise or fall is rapid, his task is comparatively easy, and, after a few unavoidable mistakes, he has succeeded in inverting his habit of reading. But, in those parts of Europe and North America which carry nearly the whole of the population, the temperature in winter is frequently oscillating from one side to the other of the melting-point of ice. If the observer is compelled to use a thermometer which he must read upwards when the temperature is on one side of that point, and downwards when it is on the other side, and if he may be called on to perform this fatiguing functional inversion several times in one day, it is certain that he will suffer from exhaustion, and that the observations will be affected with error.

Were there no other thermometric scale available but that of Celsius, we should simply have to put up with it, and endure the inconvenience of it; but, when we have another scale, one devised primarily for meteorological observations in the North of Europe, by a philosopher who constructed it with a single eye to its fitness for what it was to be called upon to measure, and when, in addition, this scale is still exclusively used in a large proportion of the meteorological observatories of the world, it seems almost incredible that amongst reasonable people, be they scientific or non-scientific, there should be a powerful agitation to abolish the scale which was devised for its work, which excludes error in so far as it can be excluded, and to replace it by one which, besides other defects, introduces, in the nature of things and of men, *avoidable* errors, the elimination of which is the first preliminary of the scientific treatment of all observations in nature.

Every meteorologist in northern countries who makes use of the data which he collects knows that when his temperatures are expressed in Fahrenheit's degrees, he can discuss them at much less expense both of labour and of money for computing than when they are expressed in Celsius' degrees; yet such is the apprehension of even scientific men when brought face to face with the risk of being ruled "out of fashion," that meteorologists who use Fahrenheit's scale, though they fortunately do not give up its use, seem to be disabled from defending it.

What is this stupefying fashion, and can it not be made out friend?

Fahrenheit lived and died before the decimal cult or the worship of the number ten and its multiples came into vogue; but, whether in obedience to the prophetic instinct of great minds or not, it almost seems as if he had foreseen and was concerned to provide for the weaknesses of those that were to come after him. The reformers of weights and measures during the French Revolution rejected every practical consideration, and chose the new fundamental unit, the metre, of the length that it is, because they believed it to be an exact decimal fraction one ten-millionth of the length of the meridian from the pole to the equator. Is it an accident that mercury, which was first used by Fahrenheit for filling thermometers, expands by almost exactly one ten-thousandth of its volume for one Fahrenheit's degree?

Again, how did Fahrenheit devise and develop his thermometric scale? A native of Danzig and living the first half of his life there, he considered that the greatest winter cold which he had experienced in that rigorous climate might, for all the purposes of human life, be accepted as the greatest cold which required to be taken into account. He found that this temperature could be reproduced by a certain mixture of snow and salt. As a higher limit of temperature which on similar grounds he held to be the highest that was humanly important, he took the temperature of the healthy human body, and he subdivided the interval into twenty-four degrees, of which eight, or one-third of the scale, were to be below the melting-point of pure ice, and two-thirds or sixteen were to be above it. Fahrenheit very early adopted the melting temperature of pure ice for fixing a definite point on his thermometer, but he recognised no right in that temperature to be called by one numeral more than by another. The length of his degree was one-sixteenth of the thermometric distance between the temperature of melting ice and that of the human body, and the zero of his scale was eight of these degrees below the temperature of melting ice, and not as is often thought, the temperature of a mixture of ice and common salt or sal-ammoniac. Fahrenheit, as has been said, was the first to use mercury for filling thermometers; and being a very skilful worker, he was able to make thermometers of considerable sensitiveness, on which his degree occupied too great a length to be conveniently or accurately subdivided by the eye. To remedy this he divided the length of his degree by four, and the temperature from the greatest cold to the greatest heat which were of importance to human life came to be subdivided into 96 degrees.

Had he lived in the following century he would have been able to point out that on his scale the range of temperature within which human beings find continued existence possible is represented by the interval 0 to 100 degrees, and there can be little doubt that this would have secured its general adoption. Its preferential title to the name Centigrade is indisputable. Perhaps this may be an assistance to its rehabilitation as the thermometer of meteorology[1].

[1] See Contents, p. xxxiii.

THE METRICAL SYSTEM

IN view of the agitation to induce Parliament to make the use of the metrical system of weights and measures compulsory in this country, it may not be amiss to direct attention to some matters relating to this system which have not recently been brought prominently under the notice of the public.

Laplace, in his *Exposition du Système du Monde*, gives a short and clear account of the reasons which actuated the "Assemblée Constituante" when it referred to the Academy of Sciences the duty of preparing a new system of weights and measures, and of the system which the committee appointed by the Academy produced in response to this remit. As Laplace was a member, and perhaps not the least influential member, of this committee, the account which he gives can be accepted without question. It really forms a digression in the chapter dealing with the figure of the earth and the variation of the force of gravity at its surface. Although the force of gravity is different at different localities on the earth's surface, it is believed to be constant in the same locality. Consequently the length of the pendulum which beats seconds in any locality is a constant length and can at any time be recovered. The idea that this might be made the basis of a metrical system suggested the digression on the subject of weights and measures.

The committee appointed by the Academy to deal with this matter consisted of Borda, Lagrange, Laplace, Monge, and Condorcet, all of them distinguished mathematicians.

Had the system devised by these gentlemen not been final, but subject to revision by another committee of men whose profession brought them into more intimate touch with the every-day wants of the people, the fundamental excellence of the system—namely, the simple relation between the unit of weight and that of length, and all the far-reaching advantages which follow from it, would, we cannot doubt, have been disengaged from the fundamental defect of the system—namely, the inconvenience of the unit of length selected. At the outset, it was laid down as a principle that the unit of length to be chosen should be unlike any existing unit of length; because, if it happened to be the same or nearly the same as that of any country, the jealousy which would thereby be produced in other countries would prevent its adoption by them. It would probably have occurred to a committee of business men that, if the unit selected were identical or nearly identical with that in common use by one people, its early adoption by at least one nation would be assured; and if it had been arranged so as not to be too different from the units of more than one nation the probability of its early adoption by all would be enormously increased. At first, and in the state of international temper which prevailed in Europe at the beginning of the 19th century, no system that was proposed by France would have been generally accepted. Moreover, time is as necessary for moving a people as it is for producing motion in a mass of matter. In the face of practical convenience, jealousies of the kind feared by the committee of the Academy soon disappear; if for no other reason, because the individuals who are jealous die, and their jealousies are not transmitted to their offspring, or at least only in a very attenuated form. If the system had fallen to be revised by a committee which was really representative of the chief occupations and industries of the people, there can be little doubt that it would have directed its attention to this matter; and if it had done so it would quickly have found that there is one calling or occupation where the people of all countries meet on common ground, with

the result that they have adopted in practice a common
system of linear measurements, simply and decimally subdivided, the unit of which is directly derived from nature.
The common ground is the sea. The table of measures
which was and is used at sea runs—100 fathoms are one
cable length, 10 cables are one sea mile; whence one sea
mile contains 1000 fathoms. Every nation used the nautical
mile, and every nation had its fathom. The noon entry in
the log of a ship gave, after the latitude and longitude at
that hour, the distance run since the previous noon, and this
distance was expressed in nautical miles and decimals of the
same, as at the present day. The nautical mile is the length of
one minute of arc (1′) of the meridian, in the latitude where the
ship is. The length of the minute of arc varies within certain
limits according to the latitude. These limits are known, and
they can be appreciated by means of the following table, which
gives the length of one minute of arc on the meridian in
British fathoms of six feet, on different parallels:

Latitude	0°	20°	40°	60°	90°	Mean
Length of 1′, fath.	1007·6	1008·8	1011·75	1015·1	1017·5	1013·3

The French seaman used the *brasse*, which is the equivalent of the British fathom. The statutory length of the
fathom in different countries differed, because it was defined
to be six feet of the particular country, and the length of
the foot differed in nearly every country: but the seamen
of all countries had the same natural standard for it, namely,
the stretch of the arms; consequently, no fathom differed so
much from the one-thousandth part of either the equatorial or
the polar nautical mile that it would have caused any inconvenience if it had been made exactly the one-thousandth part
of any distance within these limits.

Here, then, was a unit of length, the adoption of which
could offer no ground for international jealousy, and it must
be matter of surprise that it was not pressed on the committee by Borda, who had had considerable experience at
sea, and whose name is connected with some of the most
refined instruments used in nautical astronomy. Of course,
he may have done so and been overruled.

It was perfectly recognised by Laplace's committee that for economical purposes the natural unit without an artificial reproduction is useless. The suggestion of the unit is taken from nature, and it is repeated as nearly as possible in a mass of metal. It is the length of this mass of metal at a conventional temperature which is thenceforward the statutory unit. There was, therefore, no difficulty in selecting a convenient length, within the above limits, for the standard nautical mile. If, for instance, the *minimum* or equatorial value for the minute of arc were chosen, and it were divided into one thousand parts, to be called fathoms, each such new fathom would be only three-quarters per cent., or about half an inch, longer than the existing fathom ; if the *maximum* or polar value were chosen, the new fathom would be one and a quarter inches longer than the old one ; while if the length of one minute of arc of the great circle of the sphere, the volume of which is the same as that of the earth, were taken, the difference would be one and one-third per cent., or not quite one inch. Even the greatest of these rectifications is insignificant. When applied to the yard, or half-fathom, it is proportionately less noticeable, and when applied to the foot it would be imperceptible without careful measurement. It must be regretted by everyone that, when so simple a unit was already universally adopted, and decimally subdivided, it was not chosen as the fundamental unit of the system. The folly of the actual selection is sufficiently shown by the fact that not even in France has the metrical system been adopted in navigation, and still less in the subdivision of the day which forms an integral part of the system. Had the nautical mile been selected as point of departure the system would have been identical with that universally used at sea, and with such a field in which to show its usefulness it could not have failed to become universal on land, and that in a few years. Instead of this, after more than a century, it is still (1899) soliciting clients.

While it cannot be questioned that the French ought to have seen that their system was not only academically beautiful but also practically acceptable before they offered it to themselves and to the world, other nations ought to have exercised a little more criticism before they accepted it. Looking to the fact that in France itself it had, after more than half a century, been accepted only in a frag- mentary form, it would have been reasonable to inquire into the cause of this, and the unsuitableness of the fundamental unit would have been at once recognised as the cause of the failure. It could then have been easily rectified, and the only inconvenience would have been the necessity of changing the weights and measures in France, if the metrical system had already been made compulsory in that country. The inconvenience to France at that date would have been insignificant, and the advantage to the world would have been enormous.

When we are asked to say that we think that the use of the metrical system should be made compulsory in this country, it is important that those who ask us to do so should declare clearly whether they wish to introduce the metrical system in its entirety, as formulated by the Academy of Sciences and accepted by the French Assembly, and make it compulsory, or only that fragment of the system which concerns weights and measures as used in commerce. If they really wish to introduce and make compulsory only this fragment, then they should not only clearly say so, but they should formally and specifically renounce the remaining fragment of it which deals with navigation and with the division of the day. It would be intolerable if we adopted the kilogramme and the mètre for our every-day weights and measures and then found that the activity of the agitator was only shifted and took the form of forcing us to alter our division of time and of the circle, which are in perfect harmony with each other and are in universal use throughout the world.

Even if this declaration could be obtained, it is doubtful if it would be worth much. The number of States which have accepted the metrical system and have introduced it in its fragmentary form is large; but the population of these States is less numerous than that of the States which have not accepted it. The principal of these are Japan, China, Russia, Great Britain, with India and the Colonies, and the United States. It is quite an open question whether these great countries, if they made common cause, might not even now make the alteration of the fundamental unit a condition of the acceptance of the system, simply on the ground that, if an alteration must be made, it should be the one which would inconvenience the smallest number of people. When we are asked to forsake our old system of weights and measures of all kinds, we ought to be offered in exchange something which has not already been proved to be a failure. We may freely admit that a universally adopted system of weights and measures would be an advantage without our admitting that that system must be the French metrical one.

A system which is to be universally adopted should be convenient all round, and it should be fitted to be useful for all time. The metrical system as offered to us fails in both these particulars. Now is the time to improve and perfect it, so that we shall hand on to posterity something that we may reasonably expect it to be grateful for. At present, if Great Britain, with the other outstanding nations, accepts the surviving fragment of the French system, we leave to posterity nothing but a heritage of strife over the division of the day into ten hours, and that of the circle into 400 degrees. We ought to be able to do better by it.

THE POWER OF GREAT BRITAIN

THE extract which you publish in this day's *Scotsman* from the *Hamburger Nachrichten* on 'The Power of Great Britain' is of vital interest to all our fellow-countrymen. Like every statement emanating from, or inspired by, Prince Bismarck, it is plain and intelligible; and its truth is borne witness to by the uncomfortable feeling produced in the reader's mind. It is a "fact" which everyone who has lived with foreigners in foreign countries is perfectly familiar with, "that England has made herself hated all over the world," and this cannot be otherwise than inconvenient to our Ministers, whether they admit it or not. They have only to look across the narrow seas which wash our coast, and "they see the Continent bristling in arms, and when they compare England's comparative defencelessness with her wealth—gained to a large extent at the Continent's expense—and with her possessions all over the globe, most of them captured from other nations, grave anxieties arise in their minds. They say to themselves that the Continental Powers may one day grow tired of England's little game, throw aside their petty disputes, and turn against her with an united front." It is evident that, even if Great Britain were in the position of preparedness for self-defence, which all other nations, whether poor or rich, consider a necessity and a duty, the prospect thus held out is one which might well cause grave anxiety to her Ministers. But even without any great coalition such as is above suggested, the perilous position of England is clearly pointed out in a few sentences in which the combatant strength on which she has to fall back when attacked is figured up. "It is true that England has built many new ships of late, has even doubled her navy; but Mr Goschen himself had to admit lately that she has only 100,000 seamen in active service, and a reserve of 25,000 men and 10,000 pensioners, and that all the men needed, over and above, would have to be got from abroad. But foreign countries know very well that 25,000 men were wanting even before the doubling of the navy, and this deficit has, of course, proportionately increased in consequence of that measure." Many of us will be surprised at the extensive and precise information which foreign nations possess with regard to our supplies of men and material, and not a few will be shocked at their want of generosity in calling attention to the holes, and even rents, in our armour, instead of admiring the beautifully polished pieces which we love to contemplate. The Hamburg newspaper strikes the root of the whole matter when it continues—"The British navy suffers from the same evil as the British army. Without conscription a nation can no longer maintain its position in the world, and it is now too late for England to adopt that method, as it takes many years to train an adequate number of men. We have repeatedly shown, moreover, that the British army does not suffice for the protection of the mother country and the colonies when England has to deal with antagonists up to date both on land and at sea." Exception must here be taken to the word "conscription." It is, no doubt, a mistranslation, and should be "universal service," because conscription, which implies selection, and was tried and failed conspicuously in the Franco-German war, is as out of date as the voluntary system in practice in this country. Had we done what every other nation after 1870 felt it its interest and its duty to do, and adopted the principle of equality in the defence of the country, under which rich and poor alike contribute a man, what a position we should now occupy! With admittedly by far the most powerful navy, and with an adequate supply of men, it would be twice as powerful, and all our male population, which is at least as good as that of any other country, trained and in readiness, to put forth effectively its force

where and when wanted, we should then be in the enviable
position of a nation to' be let alone. We should have no
question with the United States of America and Venezuela,
with France and Newfoundland, Siam and Egypt, with
Russia and Afghanistan, or with the German Empire and
Paul Kruger. Our old Scottish motto, often quoted by
Prince Bismarck, would express a warning as well as a
fact—"Nemo me impune lacessit." What did Great Britain
do at that critical time in the history of Europe? It accepted
Lord Cardwell's Act, which doubled the quantity of paper
between the red covers of the army list. And now, our
activity is confined to every man encouraging his neighbour
to serve his country.

Exception must also be taken to the statement of the
Hamburg newspaper, that "it is now too late for England
to adopt this method," etc. The world is going to last for
many thousands of years yet; and it is not too late for
Great Britain to make provision for continuing to occupy
the ·paramount position, which has been hers in fact, and
which we believe is hers by right. But Bismarck long ago
taught us that "Might is Right." The truly advantageous
situation for a nation is to start with both the might and
the right, and to use the might to improve the right. It no
doubt takes many years to arm, train, and equip a nation
at all points so that absolutely the whole force of the nation
is ready to be exercised at once at the call of one man;
but a great deal can be done even in a year. For instance,
the personnel of the navy could be put on a perfectly satisfactory provisional footing, which would at once reduce our
war risks and give us time. What we stand to lose in the
event of our navy being over-matched is shortly but very
bluntly told by the Hamburg paper. Reminding us that
most of our foreign possessions have been captured from
other nations, and are, therefore, held under the provisions
of the *Lex talionis*, it points out "that in case of a great
war Russia and France would at once cut off England's
route to India by the Suez Canal, that she would soon
lose that dependency, Ireland, South Africa, Gibraltar, and

Canada, and probably be attacked on her own soil, the invasion of which is no longer deemed impossible." Let any one who feels not fully instructed as to what being attacked on our own soil means turn up a file of the *Scotsman* of the winter 1870–71 and read the experience of the French.

Only a few months ago Lord Dufferin used the first opportunity on which he was free to speak in public without official restraint to express the conviction produced by his observation and experience, that no nation is safe in its possessions except just in so far as it is able to defend them by its strong right arm. Success in the defence means that its right arm is stronger than the one that attacks it. It is, therefore, the duty of every nation to see that its right arm is as strong as possible, and always in best striking form.

It is often said "the country will never stand anything like conscription or universal service." But there it is assumed that the country forgets that it is subject, like other countries, to the action of natural laws. If it is attacked by another country which puts ten men into the field where it can only put one, then it may by chance escape with its life at first, but with such odds eventual destruction is certain. But it is also said our country is very rich, and we will vote ten millions for the navy, or if that is not enough, a hundred millions. But, in case of emergency, a hundred million of sovereigns is not as good as a regiment of soldiers. So that, in fact, unless we have strength to set against strength in the hour of trial, we shall be robbed of our possessions, and have to endure many, and worse, evils, as well. By doing nothing more than is done by the inhabitants of every other European nation in the exercise of self-denial, and the performance of public duty, we can put ourselves in a position to assure ourselves absolutely against these risks. The advantage of having large and well-organised defensive forces is that, if they are really effective, they are not likely to be attacked. If they are known to be weak, and especially if rich plunder is behind them, attack is

invited, and, with a favouring opportunity, certain. One of
the greatest advantages of the German military system is
seldom noticed. It is very little understood in this country,
because it is independent of the war value of the army and
navy of the country. If it were possible to be absolutely
assured that European peace would not be broken for the
next hundred years, it would pay the Germans to continue
their present system of universal military service. Look at
it apart from the uniforms, and guns, and swords, what takes
place in the country? Every year all the young men who
have arrived at a certain age (which for simplicity we may
call twenty years) are called to appear. Those that are
physically unfit are rejected. All the useful male population of that early and teachable age are taken charge of
by the country, in what is neither more nor less than a
great and well-appointed public school, and, for one or
two years, as the case may be, are compelled to live lives
both physically and morally wholesome. Their bodies are
developed by exercises, their ordinary education, where
deficient, is supplemented; they are taught cleanliness,
obedience, and order, and every German, even although he
may have been taken from the lowest slums of Berlin, starts
fair with two years of healthy and wholesome life, which
cannot fail to influence his whole life.

1916.—During their two years' service, the young Germans
have been taught, and have been forced to learn, how to do
one thing thoroughly and well, namely, the business of a
soldier. This fits them for learning how to do other things
well, and it is this acquired faculty which makes their services
in request out of their country. Our young men have made
themselves the equals of Germans as soldiers by earnest,
thorough work, and, in a time shorter than that demanded
of the selected Einjähriger. If they hold fast to this earnest-
ness and thoroughness after the war is over, they will be able
to fill all the posts and do all the jobs themselves, and there
will be no room for the foreigner. The civil re-occupation of
the country will be automatically prevented[1].

[1] See Contents, p. xxxv.

AND THE HOUSE OF COMMONS?

In expectation of the campaign against the House of Lords to be opened by the Prime Minister, it may not be out of place to look a little closer at the House of Commons and its methods, so that, when the worst has been said about the House of Lords, we may be able to form an independent judgment as to which Chamber is really the more useful to the people.

At all times it has been the custom of autocrats, when the abuses of personal rule at home have become too evident to pass much longer unnoticed, to involve their country in a foreign war. The generous instinct of a patriotic people is then appealed to to sink minor matters, and to let no consideration divert them from their primary duty to overcome the country's enemy.

Our autocrats are of a different type. When a Parliamentary dissolution is impending, it has been the practice, from time to time during the last forty or fifty years, for the Government of the moment to start some burning question and so to belabour the people with speeches on this one subject that many of them are deluded into the belief that they themselves have originated it, and that the matter is really of greater importance to them than the calm revision and deliberate appraisement of the value to themselves as a community of the legislative and administrative work of the expiring House of Commons.

It is commonly observed that these burning questions furnished for use at general elections rarely have anything to do with the welfare of the British people; and this has hitherto been an essential item in their preparation. If the question touched a real wrong suffered by the people, it might tempt them to give serious heed to their own affairs, and this is the last result which is desired.

When the burning question concerns only the people of a foreign country, remote from the British shores, this danger is negligible, and the device usually succeeds. The devisers get the people's votes, and the people have been diverted from considering their own affairs.

This geographical limitation has now disappeared. The question which is to be set alight to illuminate the apparently imminent general election is the abolition of the House of Lords, or its mutilation to such an extent that it will cease to be of use to the people as an independent Chamber.

The need to the nation of a Chamber such as the House of Lords increases every year, as the confidence of the people in the wisdom of the House of Commons diminishes. The House of Commons no longer fulfils its purpose in the British Constitution. It does not represent the third estate of the realm.

Leaving out of sight for the moment the question of female suffrage, the male population is divided into two classes—namely, those who possess the franchise and those who do not. The line which separates these classes has never been drawn with reference to any natural principle. On each occasion it has been drawn to meet the electioneering needs of the party in power at the moment.

The unfranchised males in a constituency have, owing to their necessity, a nearer knowledge of the needs of the people, and especially of the poor, than those who enjoy the franchise. Yet at an election we do not observe that they receive the attention which they deserve at the hands of either Parliamentary candidate; in fact, they are completely neglected by both. The candidate successful at the polls becomes member for the constituency—that is, for all its inhabitants, no matter whether enfranchised or not, or whether they have voted for or against him. Unfortunately, in practice the member acts usually as if he represented only those who actually voted for him, so that at the best his vote in Parliament represents nothing but the will of a small proportion of a selected minority of the population of the place, and can in no sense be taken as the will of the people of the constituency. The interests and the will of the voters who did not vote for him, and of those who were not permitted to vote at all, are not represented. What is true of one member and his constituency is true of the House of Commons and the country.

But why does the country contain so many men without the franchise? Because in Britain poverty disfranchises the citizen. For the House of Commons those that are too poor to have a vote are as though they were not. They have no representation except in the House of Lords. The House of Lords is for all; the House of Commons for the few.

Again, the House of Lords consists of made men, with the scientific advantage of having been bred for the work or introduced as new blood. The members of the House of Commons are taken at haphazard, and are for the most part in the making. We think that too liberal use is made of us in the construction.

Further, we observe that the business which arrives at the House of Lords is despatched in a business-like way, which we look for in vain in the transactions of the House of Commons.

The absolute power of the House of Commons to put its hand in our pocket and to take as much as it pleases or can find out of it, and then to spend it in any way that it likes, was acquired at a time when nobody recognised the fundamental irresponsibility of a popular Government. The reckless way in which this power is now exercised fills us with alarm, and we hope, though we do not expect, that the Prime Minister will reveal to us some large measure of devolution in this respect in favour of the House of Lords.

But it would be absurd to maintain that the many able men who are at the top of their professions, or who are almost by-words for success in their own businesses, are not able, if there were no impediment in the House of Commons itself, to put through the business of the country as well as the members of the House of Lords.

When a man of this class, especially the creator or head of a famous mercantile concern, is introduced to us as a fit and proper person to represent us in Parliament, stress is always laid on the advantage which the country will gain by its business being always under his practised eye. He enters Parliament, and, no doubt, looks into the colossal business of the country, and perhaps calls attention to methods which, if applied in his own business, would, he thinks, lead to bankruptcy. We wait and look anxiously for some of the effects promised at his election; but we wait in vain. When he comes down to address us in the recess he tells us to wait a little longer.

The determining advantage which the House of Lords has over the House of Commons in matters relating to the despatch of business is that it is exempt from the disturbance produced by general elections. This permits its members to do their work, whether legislative or administrative, according to their ability and their conscience.

If Parliaments were elected for seven years *certain*, there would be no excuse for a member of the House of Commons not attending solely to his public duties. It is the feeling that at any moment the caprice of a Prime Minister may spring an election on the House and people, which unsettles the members, and deprives the people of the advantage of at least nine-tenths of the useful energy of every House of Commons.

If the disease has been thus correctly diagnosed, there are several treatments which would promise beneficial results. The most radical would be to abolish elections altogether. But it is doubtful if the patient would be able to support anything quite so radical.

A milder and more conservative treatment would be to let the member who has sat in two consecutive Parliaments be ineligible for the third. In the first Parliament he would, as at present, work mainly for his re-election. In the second, as he would have no immediate re-election to look forward to, he could afford to disregard the crack of the whip, and to give effect to the results of his own observation and the dictates of his own common sense in matters affecting the public welfare.

As in every Parliament the number of members serving in the first of their two Parliaments would be about equal to that of those serving in the second, the people could always depend on getting the undisturbed services of one-half of the House of Commons.

Perhaps this is not much, but it is something. In any case, it is our duty at this juncture to raise high our voices; and if a cry is to go over the land it must be for *Reform of the House of Commons.* We must permit no tinkering at the House of Lords, even by its own members. It is our property.

1916.—Later events showed that I misinterpreted the value which the House of Lords put on its own independence.

LORD MILNER AND IMPERIAL SCHOLARSHIPS

I HAVE not had the opportunity of reading Mr Vaile's article in the *Fortnightly*, but I have read with interest Lord Milner's letter in reference to it in the *Morning Post* of October 4. In it Lord Milner deals only with the Rhodes bequest in so far as it applies to British subjects in the Oversea Dominions, and affords to a certain number of them opportunity of studying at the University of Oxford. But the bequest was cosmopolitan, and offered the same opportunity to the subjects of foreign Governments. With this qualification I agree with Lord Milner's words: "It would be a true analogy if, as he (Rhodes) gave money to enable Canadians, Australians, etc., to know the Homeland, wealthy men in these countries were to enable men born on this side to become better acquainted with the Oversea Dominions."

Having had the good fortune many years ago to be selected as chemist and physicist of the "Challenger" expedition, I was able in the years 1873 and 1874 to become acquainted with the important Oversea Dominions, South Africa and Australia, and I well remember stating the impression which they made on me in conversation with some prominent public men in New Zealand during our short stay there. The gist of my remarks was that in both South Africa and Australia I had seen lands of the greatest fertility and mineral wealth, but with so small a population that, if all the inhabitants had been distributed uniformly over the whole area, the curvature of the earth alone would have been sufficient to hide every individual from his neighbour;

while in the country that I had left there were at least three times more inhabitants than it could find food for. I then asked if it would not be a grateful task for the then Chancellor of the Exchequer, who, I think, was Mr Robert Lowe, to obtain from Parliament a grant to enable the poor of Great Britain to visit the countries which public-spirited statesmen of other days had acquired for them, and to assist them in settling there. When I had finished the same answer came from all: "Why, that is just what Sir George Grey always advocated and tried his best to bring about." Sir George Grey's name occupies a very high place on the roll of Colonial Governors, yet his efforts on this occasion met with no success.

It is advantageous to the Dominions to send a certain number of their young men to the English Universities, and it is an advantage to these Universities to receive them. But the suitable return for this is not the sending of young men from the English schools to study at Australian or Canadian Universities, or of Englishmen who have taken their degrees at their own University to the Dominions for some undefined purpose. The Dominions do not require University men, nor would University men as a rule benefit much by Colonial experience. What the Dominions do want is young men and young women, strong in body and sound in mind, and with their lives before them. If they have received the average education which the children in all countries receive nowadays, and if their nature is such that whatever they take in hand they do it with their might, then nothing can stand in the way of their success.

According to my recollection of the spirit, if not of the terms, of Mr Rhodes's bequests in connection with the University of Oxford, they were intended, in due time, to have an educative effect on that University, rather than on the young men for whom he provided the means to attend it. In imposing on the members of the teaching staff the duty of carrying further the education of young men who had been brought forward in the schools of their own country to the point when they had acquired the qualifications for proceeding to their home University, he furnished them with a kind of education which had previously been denied to them; and I cannot doubt that broadening the outlook of his own University was uppermost in his mind when he made his will. No corresponding benefit would be conferred on Colonial or foreign Universities by sending young Englishmen to study there. The reactive effect of the pupil on the teacher is possible only in Oxford and Cambridge, where the tutorial relation between the teacher and the taught exists in its highest development.

The true complement of Rhodes's scheme is to furnish the teachers at Oxford and Cambridge at the very beginning of their career with outside experience of what is to be the business of their lives. At present I suppose that at least nine-tenths of the teaching staff at these Universities date their start from the day when they were elected to a college Fellowship. In Oxford and Cambridge the Junior Fellow is the key to the situation. Yesterday his position was indefinite, he existed on approbation; to-day he is a Don. The change is very sudden, and he must have time. As he is now one of the managing body of the college it is of importance that he should know by personal experience how similar work is done in other countries, and there is no time for acquiring this knowledge better than the first year of his Fellowship. Absence from his college and University for a year is then an advantage. At the end of it he returns with greater authority than if he had continued his former life without a break. It is therefore of the greatest importance that he should use this critical year to the best advantage.

The only way to learn how things are done in other countries is to go there and do them; therefore the newly-made Fellow should lose no time in entering himself at a foreign University, not in the first instance to study the subjects of his own *fach*, but with a view to study whatever subject brings him most in contact with the students and the teaching staff. The subjects which do this best are those which require work in a laboratory. It does not matter whether it is a chemical or physical or physiological laboratory or even a dissecting-room. The primary object is not to learn the subject taught but the language in which it is taught. Hearing lectures is not sufficient for this. During them the student keeps silence, and a foreign language is not learned by listening but by trying to speak. When the Junior Fellow has heard his lecture, let us say, on chemistry, he goes to his work in the laboratory, where he finds himself in the company of twenty or thirty other young men of his own age and engaged for the time being on the same pursuit. From the very first day it is impossible for him not to listen to what the others are saying around him, and before the first week has passed he will be doing his best to let them hear what he has to say. This discipline should be continued for at least one year, during which he should not return to his own country. At the end of this year he may return to his own University to take up the duties of his Fellowship, and he will not fail to feel that his activity abroad has made him a much more efficient member of his college and University than if he had remained at home and only shifted his place at meat from a lower to a higher table.

But it will be said that the young man, before he gets his Fellowship, has been spending considerable sums of money on his education, and his opportunity of getting it back and earning a living dates from his election. I admit this, but the election to a Fellowship does not require residence, and I understand that this was done in order to give the newly-elected Fellow the opportunity of doing what I recommend, and although not common it is still far from rare. All objections to the proposal would disappear if a fund were instituted for supplementing the stipend attached to the Fellowship by such amount that, while it would afford no temptation to extravagance, it would constitute such an augmentation of the stipend that the extra expenses of living abroad and the University fees and charges would be rather more than covered. In these circumstances the temptation to take advantage of it would be very great, especially after the return of one such Fellow who had made conscientious use of the opportunities thus afforded him.

In a college which is endowed with such an augmentation fund, when a new ordinary Fellow is elected, he would be offered the augmentation for one year on his engaging to spend the first year of his Fellowship at a foreign University to study some subject which will bring him as much as possible in contact with the students and teachers and to make himself thoroughly acquainted with the University administration and educational system in the country where his selected University is situated. With regard to ways and means, it is not necessary to provide a perpetual annuity for the purpose; it is enough to provide a fund which shall be sufficient for furnishing the annuity for such a number of years as will enable a judgment to be formed as to whether the scheme is going to be a success or not. Provisionally, I take it that fifteen annuities of £100 each would meet the case. An annuity of this amount for fifteen years is bought and is every year paid into the college chest. When a new Fellow is elected he is offered one of these annual sums as an augmentation of his first year's stipend on his agreeing to comply with the conditions indicated above. Next year perhaps another new Fellow is elected and the augmentation is offered as before, but it is declined; it therefore remains in the college chest. Hearing of this the first Fellow who received the augmentation applies for it to be continued on the same terms for the next year. Should his request be granted? I think not. The object of the fund is to produce as large a body of Fellows with sound foreign experience as possible, and it is more likely to be profitably expended in giving a new Fellow the experience than in increasing that of the old Fellow. In due course a third new Fellow would fall to be elected, and I assume that he and those following him will take the augmentation. We may take it that fifteen Fellows would be added to the list in twenty years.

The advantage accruing to the college would be that after twenty years its management would be in the hands of men who had made themselves personally familiar with foreign methods and had been kept constantly in touch with them and their variations by younger colleagues as they went and came from abroad. If all the colleges adopted the system, then the University would be as up to date as the colleges.

1916.—Although changes can be made at once by an Act of Parliament, no real reform of a popular institution, like a university, can be effected in less time than the period of a generation.

No. 23. [*From The Morning Post, October* 2, 1912.]

HISTORY IN HANDY VOLUMES

THE different tastes in the matter of reading revealed by your correspondents are very interesting. May I add my contribution? I never was much either of a reader or writer. I always preferred original observation and experiment. But there is much to be learned by the observer and experimenter from history, and history must be read. History is the more authentic the more nearly it is contemporaneous and the more closely connected is the author with the events which he chronicles. This reflection was borne in upon me in a not unpleasant way many years ago on a tour in Norway. I think it was in Odde that I had to pass a day of such continuous and heavy rain that excursions were out of the question. In the reading-room of the hotel I found a number of the early volumes of the *Illustrated London News*, and they covered the duration of the Crimean War. At that date the telegraph was hardly available for correspondents, and as the mail was weekly, a weekly print with a good correspondent was as good as a daily. I never spent a more agreeable or instructive day in my life. Each week I learned what had passed in the previous week, and, what was even more interesting, what was expected to happen in the next, and the following week was full of surprises and revelations. I advise everyone who has access to these volumes to repeat my experiment.

Such volumes, however, cannot be carried in the railway carriage, and the Crimean War is a thing belonging entirely to the past. But there is history which is fortunately written in more handy volumes, which goes even much further back than the Crimean War and is yet of profound influence on

what is taking place to-day. It is the history of the rise
and progress of the Prussian State and the creation of
Germany under its hegemony.

In order to understand it, it is necessary, but it is also
sufficient, to read three books. They are: *L'Histoire de
Mon Temps*, by Frederick II, called Frederick the Great;
Ma Mission en Prusse, by Benedetti, published in 1871;
and *Gedanken und Erinnerungen*, by Prince Bismarck. The
Tauchnitz translation of this work is very good. As a com-
mentary in English on the later history nothing is more
instructive or better reading than *Friendship's Garland*, by
Matthew Arnold, published in 1870.

In the first of these works the King of Prussia explains
with perfect frankness the methods by which he succeeded
in extending Prussian territory. He began by perfecting
his own military forces, then he awaited the opportunity
which the Sovereign of the coveted territory was sure to
give him sooner or later, and when the moment arrived he
had the courage to use his whole might in taking advantage
of it. Bismarck's *Gedanken und Erinnerungen* tells exactly
the same tale, and the light thrown on the same events
by Benedetti during his mission in Prussia, which covered
the Austrian War of 1866 and terminated with the outbreak
of the Franco-German War in 1870, is most illuminating.
After studying these three books there is no difficulty in
perceiving how it is that Europe presents the aspect that
it does.

1916.—Although the present war is still being waged with
all its fury it has been going on for over two years, and its
history is minutely recorded, *de die in diem*, by the Newspapers of the world. Of these daily reports, the historian of
the war will naturally go to the records in the newspapers of
neutral nations, and, among them, to those having the greatest
resources at their back. Among these I can speak from
experience only of one; namely, the *New York Times*.

The daily reports cabled by its correspondents on both
sides, and from all the different seats of war, in a week would
equal in bulk all those furnished to the *Illustrated London*

News during the whole of the Crimean War which lasted through two winters. These correspondences are subject to censorship and have to be interpreted. This has been done with great discretion and success in the daily summary of the *Times'* military expert and in occasional leaders.

When one compares the correspondence by letter filling about a column of the *Illustrated London News* every week in 1854 with the columns upon columns cabled from the seat of war to the *New York Times* every day since August 1914, one obtains a conception, which is almost stupefying, of the advance made in practical journalism in these sixty years.

It is a satisfaction, perhaps a melancholy one, to know that it has all happened during my own life, for I read the accounts of the Crimean War with regularity, and with such attention, that I believe I could pass an examination now on the details, from week to week, of the bombardment of Sebastopol. But who would be the examiner?

SUMMARY OF CONTENTS

The Table of Contents,.pp. vii to xl, gives much detail and contains all the notes and comments which occurred to the author when collecting the papers for publication.

PAGE

No. 1. RECENT ANTARCTIC EXPLORATION. (From the *Quarterly Review*, October, 1906, p. 1.)   I

This article contains principally a critical discussion of the discoveries made by Captain Scott in the "Discovery" in the year 1902, with those made by Sir James Ross in the "Erebus" and "Terror" in the year 1842, in the region of the great Antarctic Ice Barrier.

No. 2. CHEMICAL AND PHYSICAL NOTES. (Contributed to the *Antarctic Manual*, 1901.)   25

These Notes were prepared for the use of the Chemist and Physicist of an Antarctic Expedition. They were suggested by the author's own experience, and they are confined to matters of observation and experiment, all hypothetical matter being excluded.

No. 3. ON ICE AND BRINES. (From the *Proceedings of the Royal Society of Edinburgh*, 1887, Vol. XIV, p. 129.)   . 130

The result of this experimental investigation was to prove that the ice formed during the freezing of sea-water and similar saline solutions is pure ice, and that its saltness is due to residual brine which is enclosed in the crystals ; further, that snow or other pure ice melts in a saline solution at the same temperature as that at which the solution freezes.

No. 4. ON STEAM AND BRINES. (From the *Transactions of the Royal Society of Edinburgh*, 1899, Vol. XXXIX, Pt III, No. 18.)   151

The results recorded in this paper run parallel with those obtained for Ice and Brines, and establish the fact that steam, water and salt can be used to produce Boiling Mixtures of constant temperature, higher than that of the saturated steam itself ; that steam produced by a boiling saline solution leaves that solution with the same temperature as that of the boiling solution itself ; and that pure steam passed into a saline solution raises it to a maximum temperature, which is its own temperature of ebullition. Blagden's Law is shown to hold in the case of boiling saline solutions in the same measure as it holds in the case of freezing saline solutions. It is demonstrated that Blagden's Law, in both cases, is identical with the thermal law of mixture.

PAGE

No. **5.** THE SIZE OF THE ICE-GRAIN IN GLACIERS. (From *Nature*, August 22, 1901, Vol. LXIV, p. 399.) . . . 226

No. **6.** ICE AND ITS NATURAL HISTORY. (From the *Proceedings of the Royal Institution of Great Britain*, 1909, Vol. XIX, p. 243.) 233

No. **7.** BEOBACHTUNGEN ÜBER DIE EINWIRKUNG DER STRAHL- UNG AUF DAS GLETSCHEREIS 280

No. **8.** IN AND AROUND THE MORTERATSCH GLACIER : A STUDY IN THE NATURAL HISTORY OF ICE. (From the *Scottish Geographical Magazine*, 1912, Vol. XXVIII, p. 169.) . . 283

These four papers are a portion of the author's work on Ice out of doors, principally on the Morteratsch Glacier. The phenomena observed in the open are found to be regulated by the laws established in the laboratory. Papers Nos. 6 and 8 are illustrated.

No. **9.** THE USE OF THE GLOBE IN THE STUDY OF CRYSTALLO- GRAPHY. (From the *Philosophical Magazine*, 1895, S. 5, Vol. XL, p. 153.) 313

The blank globe, whether black or white, with the divided circles belonging to it, is a calculating machine adapted to the solution of all the problems to which the analytical methods of spherical trigonometry are usually applied.

No. **10.** ON A SOLAR CALORIMETER USED IN EGYPT AT THE TOTAL SOLAR ECLIPSE IN 1882. (From the *Proceedings of the Cambridge Philosophical Society*, 1901, Vol. XI, Pt I, p. 37.) 337

The instrument is a Steam Calorimeter, which depends for its indications on change of state and not on change of temperature. The final result of experiments made with it on the uneclipsed sun is that the calorific value of a sheaf of the sun's rays, having a section of one square metre, is at least 9698 gram-degrees centigrade per minute, at the ter- restrial sea-level ; and it is certain that heat can be obtained from them there at this rate, in a useful form, by mechanical appliances of simple construction. Immediately after the total phase of the eclipse was passed, the instrument was pointed to the sun. Steam was not raised until twenty minutes after totality, and from that time onwards the output of steam increased and was recorded until the eclipse was ended.

No. **11.** SOLAR RADIATION. (From *Nature*, September 5, 1901, with Postscript in 1911.) 383

This is a short article on the subject, and attention is directed to the diminishing value ascribed to the Solar Constant by different observers, since the year 1900.

PAGE

No. **12.** THE TOTAL SOLAR ECLIPSE OF AUGUST 30, 1905. (From *Nature*, December 21, 1905, Vol. LXXIII, p. 173.) . 399

This eclipse was viewed from a point where the central line of eclipse cuts the east coast of Spain. Its duration was over four minutes. The display of Protuberances at second contact was very brilliant, but when the time of mid-totality arrived not a trace of them was visible to the naked eye. Therefore these Protuberances had an apparent height of less than 45 seconds of arc.

No. **13.** ECLIPSE PREDICTIONS. (From *Nature*, October 19, 1905, Vol. LXXII, p. 603.) 402

The predictions respecting the solar eclipse of August 30, 1905, as issued by the British *Nautical Almanac* and by the French *Connaissance des Temps* are compared and their want of agreement is illustrated by a Table.

No. **14.** THE SOLAR ECLIPSE OF APRIL 17, 1912. (From *Nature*, May 9, 1912, Vol. LXXXIX, p. 241.) 404

This remarkable eclipse was observed from a northern suburb of Paris. It was impossible to say whether it was completely total or not. At the moments of second and third contact the phenomenon called Baily's beads was well seen. It was found impossible to offer a satisfactory explanation of their nature, but they were clearly not due to the interruption of the solar rays by the mountains of the Moon.

No. **15.** THE PUBLICATION OF SCIENTIFIC PAPERS. (From *Nature*, August 10, 1893, Vol. XLVIII, p. 340.) . . . 410

A plan is sketched in this paper, whereby what is at present inefficiently and extravagantly done by a multitude of amateur publishers might at much less cost be efficiently done by a central publishing office as a matter of business. One great advantage of this would be that it would eliminate the censorship of the Councils of the Societies.

No. **16.** THE ROYAL SOCIETY. (From *Nature*, January 28, 1904, Vol. LXIX, p. 293.) 413

This paper is a summary of what the author said at a special meeting of the Fellows, and deals particularly with the practice of *referring* papers received from Fellows of the Society, the effect of which is to render doubtful the authenticity of the authorship of papers, as declared in the title.

No. **17.** NOMENCLATURE AND NOTATION IN CALORIMETRY. (From *Nature*, May 12, 1898, Vol. LVIII, p. 30.) . . 416

The general principle advocated is that compound units should be expressed by compound names ; and these should be self-explanatory : thus, instead of *calorie*, we should use gram-degree.

PAGE

No. **18.** THERMOMETRIC SCALES FOR METEOROLOGICAL USE. (From *Nature*, August 17, 1899, Vol. LX, p. 364.) . . 420

This paper was published in anticipation of the Meteorological Congress of 1899, which met in Berlin. Its object was to direct attention beforehand to the advantage, in securing accuracy and in relieving labour, which Fahrenheit's scale offers over that of Celsius, when used for meteorological purposes.

No. **19.** THE METRICAL SYSTEM. (From the *Times*, February, 1903.) 424

In this paper the history of the origin of the Metric System is given, and attention is directed to its advantages and disadvantages.

No. **20.** THE POWER OF GREAT BRITAIN. (From the *Scotsman*, March 26, 1897.) 430

This paper was an answer to an article which appeared in the *Hamburger Nachrichten* in 1897, in which the supply of men and material for our Navy and Army was detailed, and the conclusion was arrived at that, without compulsory service, England could not hope to withstand attack by any first class European Power, and that it was then (1897) too late for her to make the change. It is pointed out in the paper that this is inexact, because a great deal can be done even in one year : and this has been abundantly proved in the present war.

No. **21.** AND THE HOUSE OF COMMONS? (From the *Scotsman*, October 5, 1907.) 435

This letter appeared in the *Scotsman* in the morning of the day when Mr Asquith was to hold the first meeting in his campaign against the House of Lords. It is pointed out that the House of Lords may be as useful to the country as the House of Commons, so long as it maintains its independence.

No. **22.** LORD MILNER AND IMPERIAL SCHOLARSHIPS. (From the *Morning Post*, October 15, 1909.) 440

In this letter it is pointed out that the action of the English University on the colonial or foreign student must necessarily be accompanied by the reaction of the scholars on the University and that it must be productive of benefit to both.

No. **23.** HISTORY IN HANDY VOLUMES. (From the *Morning Post*, October 2, 1912.) 446

In this letter the practice of going always to the fountain head for information about all facts is recommended. As an example, the weekly letters in the *Illustrated London News* of 1854–55 are recommended to the student of the Crimean War; a comparison is made between the extent of war-reporting in 1854 and in 1916; and, as a neutral journal in the present war, the *New York Times*, with its many columns of cabled matter daily, is cited.

Printed in the United States
By Bookmasters